The English Rural Landscape

The English
Rural Landscape

EDITED BY *Joan Thirsk*

Oxford New York
OXFORD UNIVERSITY PRESS

OXFORD
UNIVERSITY PRESS

Great Clarendon Street, Oxford OX2 6DP

Oxford University Press is a department of the University of Oxford.
It furthers the University's objective of excellence in research, scholarship,
and education by publishing worldwide in

Oxford New York

Athens Auckland Bangkok Bogotá Buenos Aires Calcutta
Cape Town Chennai Dar es Salaam Delhi Florence Hong Kong Istanbul
Karachi Kuala Lumpur Madrid Melbourne Mexico City Mumbai
Nairobi Paris São Paulo Singapore Taipei Tokyo Toronto Warsaw

with associated companies in Berlin Ibadan

Oxford is a registered trade mark of Oxford University Press
in the UK and in certain other countries

Published in the United States
by Oxford University Press Inc., New York

British Library Cataloguing in Publication Data

Data available

Library of Congress Cataloging in Publication Data

Data available

ISBN 0-19-866219-X

1 3 5 7 9 10 8 6 4 2

Printed in Great Britain by
T.J. International Ltd
Padstow, Cornwall

Contents

Notes on Contributors   6

Introduction   9
    JOAN THIRSK

Part 1: Panoramas of Landscape

1. Downlands   27
    JOSEPH BETTEY

2. Wolds
    The Wolds Before *c.*1500   50
    HAROLD FOX
    A Longer View of the Wolds   62
    BERNARD JENNINGS

3. Lowland Vales   78
    CHRISTOPHER DYER, DAVID HEY,
    AND JOAN THIRSK

4. Woodlands and Wood-Pasture in
    Western England   97
    CHRISTOPHER DYER

5. Forests and Wood-Pasture in
    Lowland England   122
    BRIAN SHORT

6. Marshes   150
    ANNE REEVES AND TOM WILLIAMSON

7. Fenlands   167
    CHRISTOPHER TAYLOR

8. Moorlands   188
    DAVID HEY

9. Common Land   210
    ALAN EVERITT

10. Frontier Valleys   236
    CHARLES PHYTHIAN-ADAMS

Part II: Cameos of Landscape

11. Stonor: A Chilterns Landscape   265
    LESLIE W. HEPPLE AND
    ALISON M. DOGGETT

12. Hook Norton, Oxfordshire:
    An Open Village   277
    KATE TILLER

13. Eccleshall, Staffordshire:
    A Bishop's Estate of Dairymen,
    Dairy Wives, and the Poor   290
    MARGARET SPUFFORD

14. Staintondale, North Yorkshire:
    A Moorland Estate of the Knights
    Hospitaller   307
    BARRY HARRISON

15. Fen Drayton, Cambridgeshire:
    An Estate of the Land Settlement
    Association   323
    PAMELA DEARLOVE

Further Reading   336

Picture Credits   342

Index   343

Notes on Contributors

JOSEPH BETTEY was formerly Reader in Local History at Bristol University. He is the author of numerous books and articles on landscape, farming, rural society, and church life, particularly in Wiltshire, Dorset, and Somerset.

PAMELA DEARLOVE is a lifelong local historian who combines her family commitments with extensive involvement in the community. She completed in 1992 a dissertation on 'The Effect of the Policies of the Land Settlement Association on Fen Drayton', and works currently on the manorial estate of Hemingford Abbots (where she lives) and Hemingford Grey and on the social history of Fen Drayton.

ALISON M. DOGGETT is a landscape photographer, and was joint author with Leslie Hepple of *The Chilterns* (1992), where she has lived since the age of 10. She graduated in Geography at Bristol University, and teaches at Berkhamsted Collegiate School.

CHRISTOPHER DYER is Professor of Medieval Social History in the University of Birmingham and Fellow of the British Academy. He has published articles, chapters, and books on many aspects of medieval social and economic history, including *Standards of Living in the Later Middle Ages* (1989). His *Hanbury Settlement and Society in a Woodland Landscape* (1991) is a study of a West Midlands landscape.

ALAN EVERITT is a Fellow of the British Academy and was formerly Hatton Professor and Head of the Department of English Local History at the University of Leicester. His publications include *Landscape and Community in England* (1985), *Continuity and Colonization: The Evolution of Kentish Settlement* (1986), and chapters in *The Agrarian History of England and Wales*, vol. iv: *1500–1640* (1967). He is presently working on the history and use of common land in England.

HAROLD FOX is Senior Lecturer in English Topography in the Department of English Local History at the University of Leicester. He has written extensively on English landscape history, agrarian history, and social history, largely on the Midlands and on Devon, Somerset, and Cornwall.

BARRY HARRISON is Senior Lecturer in History in the School of Continuing Education at Leeds University. He has written and edited books on Yorkshire vernacular architecture, landscape history, and local history, together with articles on medieval field-systems and settlement patterns.

LESLIE W. HEPPLE is Senior Lecturer in Geography and Director of the M.Sc. programme on 'Society and Space' at Bristol University. He is author of *A History of Northumberland and Newcastle upon Tyne* and (with Alison Doggett) *The Chilterns*.

DAVID HEY was formerly Professor of Local and Family History at the University of Sheffield. His most recent publications include *The Oxford Guide to Family History* (1993), *The Oxford Companion to Local and Family History* (editor, 1996), and *A History of Sheffield* (1998).

BERNARD JENNINGS was Professor of Adult Education from 1974 to 1990 and of Regional and Local History from 1990 to 1993 at the University of Hull. He has published in two main areas: on adult education policy and history; and on regional history, on which he has edited several substantial books that were researched and written by WEA students, including one on Nidderdale. He has been National President of the WEA.

CHARLES PHYTHIAN-ADAMS was Professor of English Local History at the University of Leicester, where he was a member of the Department of English Local History for three decades, and for fifteen years its Head. His publications include works on early Cumbria, late medieval Coventry, popular culture, and the content of local history. He is currently writing a history of provincial England.

ANNE REEVES is a landscape historian with a special interest in marshlands. She studied first at the University of Kent, then at the University of East Anglia, and now works from home in the Romney Marsh area of Kent, where she has carried out extensive fieldwork and research.

BRIAN SHORT is Reader in Human Geography and currently Dean of the School of Cultural and Community Studies in the University of Sussex. He is author of many works on the historical environment and rural society of south-east England and in particular of the Weald of Kent, Surrey, and Sussex. He has an equal interest in evaluating historical sources on rural life in the nineteenth and twentieth centuries. His latest books are *The Ashdown Forest Dispute, 1876–1882: Environmental Politics and Custom* (1998), and *An Historical Atlas of Sussex* (co-edited) which appeared in 1999.

MARGARET SPUFFORD is Professor of Social and Local History at Roehampton Institute, London, and a Fellow of the British Academy. She began to know Eccleshall when she taught at the University of Keele, and ran an extra-mural class there with her husband. Her books include *Contrasting Communities* (1974); *The World of Rural Dissenters* (1995); and *Poverty Portrayed* (1995).

CHRISTOPHER TAYLOR in 1993 retired early from the Royal Commission on the Historical Monuments of England, having served it for thirty-three years, latterly as Head of Archaeological Survey. He is author of a number of books and papers on archaeology and landscape history, editor of a revised edition of Hoskins's *Making of the English Landscape* (1988), and is a Fellow of the British Academy.

JOAN THIRSK, CBE, is a Fellow of the British Academy and was until retirement Reader in Economic History at the University of Oxford. She specializes in agrarian history, and is General Editor, and editor and contributor to two volumes of *The Agrarian History of England and Wales*, 8 volumes (Cambridge, 1967–2000). Her most recent book is entitled *Alternative Agriculture: A History from the Black Death to the Present Day* (Oxford, 1997).

KATE TILLER is University Reader in Local History in the Department of Continuing Education at the University of Oxford, and a Fellow of Kellogg College. She established the Undergraduate Certificate, and the postgraduate Master's and Doctoral programmes in Local History at Oxford. Her published work includes studies of enclosure, religious dissent, open and closed villages, and a book on *English Local History: An Introduction* (1992). She has a particular interest in landscape history and has worked on Blenheim Park and on Wychwood Forest in Oxfordshire.

TOM WILLIAMSON is Lecturer in Landscape Archaeology at the Centre of East Anglian Studies, University of East Anglia. He has written widely on landscape and garden history. His publications include *Polite Landscapes: Gardens and Society in Eighteenth-Century England*; *The Origins of Norfolk*; and *The Norfolk Broads: A Landscape History*.

Acknowledgements

All the authors in this book have benefited from the many insights, comments, and criticisms of their fellow-authors. But they also wish to thank others for their help. Joseph Bettey wishes to give special thanks to Michael Aston and Jim Hancock; Christopher Dyer to David Crowley, David Cox, Simon Esmonde-Cleary, Paul Stamper, Nigel Tringham, Jane Laughton; Leslie Hepple and Alison Doggett to Madeleine Hammond; and Margaret Spufford to Professor Andor Gomme, who retraced her footsteps in Eccleshall and produced finished professional photographs to match her roughs, and James Went who did the analyses of the Eccleshall probate inventories. The editor and authors also wish to give special thanks to their editor at the Oxford University Press, Alysoun Owen, for her constant support.

Introduction

Joan Thirsk

We all as individuals look at landscapes differently. Old and young, men and women, country folk and town dwellers, we all register some things sharply, and fail to notice others. The landscapes of childhood plainly affect us deeply. But then other experiences and places enlarge our sympathies and sensitize us to more variety, while leaving us still relatively undiscerning when looking at the rest. Perceiving landscapes historically, however, poses another challenge, for then a fresh dimension has to be accommodated. Through the centuries, every corner of landscape in the kingdom has been altered many times over, and layer upon layer is hidden from sight by modern developments. Yet innumerable traces remain of the past, if only we train ourselves to see. We shall greatly enrich our enjoyment of landscape if we have at least some of the past at the back of our minds as we survey every new scene.

Some of the transformations that have occurred in our countryside have become clearer through historical investigations that have dug ever deeper in the last fifty years. As more layers of history have been peeled away, and families travel nowadays much more freely than ever in the past, they have shown their pleasure in getting behind the present-day scene and knowing more of the people who changed it. Here, in this study of *The English Rural Landscape*, we try to advance another stage, taking two different directions. First we describe some of the major types of English landscape, in order to assemble the generalities that unite them, and confer a similarity on, say, all downlands, all fenlands, and all moorlands, which we instinctively recognize wherever they occur. Second, in cameos of landscape, we explore single places, to demonstrate the pleasure of knowing some of them more intimately and to celebrate their individuality. The cameos, of course, illustrate some of the generalities about landscape already rehearsed, but they also exemplify the singularities of every place. Sometimes they even introduce us to people with recognizable names and faces who lived and shaped the present scene. We hope that the two alternative approaches will spur another generation to look more discerningly at the landscape in which they live, ask more questions, and go in search of new explanations.

The different viewpoints taken by our authors here are as intriguing as the different histories they record. We rejoice in that diversity, hoping that we shall uncover new seams of information and stimulate others to follow fresh paths of discovery. Anne Reeves is addicted to marshland scenery, and has spent months, nay years, walking over the ground, finding pottery, and detecting innumerable other traces of human activity, in Romney Marsh. Bernard Jennings lives on the Yorkshire wolds, and wakes every morning to the scene that he describes. Harold Fox looks out on other wolds in Leicestershire, but being a medievalist his imagination is stirred by such different elements in the landscape from those engaging Bernard Jennings that we have not attempted to weld them into one whole. They vividly demonstrate our different perceptions of the same scene. David Hey was born and brought up in moorland country. Alan Everitt has a special sympathy and admiration for the modest, resourceful people who have made a living over the centuries by using the natural resources of their commons. Pamela Dearlove has talked with many a smallholder who lived through the first and the last days of a truly pioneering land settlement at Fen Drayton, in Cambridgeshire, set up during the great depression of the 1930s, which thereby transformed an old village. We do not give biographies of all our authors, but they all have scenic preferences and distinctive insights. Notice, too, that we have women writing about landscape as well as men. Women are not often asked to write on the subject, and, from choice, they seem to prefer to make it a background, often memorable, to fiction. Perhaps it is because people are as important to them as the natural features, and essays on landscape tend to leave them out of sight. On the other hand, menfolk excel when writing about large landscapes, assembling the features that bring them together and create a unified impression. Those writing here stand in a long line of illustrious predecessors going back to Tudor times. But they break new ground, for their historical investigations all reveal fresh ways of reading our maps, and interpreting the significance of roads, dykes, hedges, and buildings.

From different vantage points, we are all building up a better understanding of the antiquity of our countryside and its cultural diversity. Thus we move on from the simple view of the eighteenth-century poet William Cowper (1731–1800) that 'God made the country and man made the town'. Assuredly people built the towns, but they made the countryside too. In our more densely populated world nearly 200 years after Cowper's death, and better served now by many knowledgeable historians, few would make such a confident distinction between the making of the rural and the urban landscape. People built towns, but none of us would assert with confidence that any piece of countryside stands untouched in its primeval state. Sometimes fragments of woodland are labelled as ancient, but knowing how quickly once-cleared woodland can regenerate, we are wiser to say only that some landscapes have been less disturbed than others.

Over the centuries from prehistory to the present, a distinction can fairly be made between phases of more and less dramatic disturbance, usually driven by the rise and fall of populations needing to be fed, housed, and employed. In the better-documented periods we fasten on the thirteenth, the sixteenth, and the nineteenth centuries when landscapes were profoundly altered, by more ploughing, by the felling of forest, and the drainage of fen and marsh; at the same time, we also recognize less profound alterations, scarrings would be a better term, when roads, canals, and railways were built. Then at other times, the urge to clear and cleanse has weakened, and people have even withdrawn from the inroads earlier made, as they did when the Black Death halved the population in the mid-fourteenth century, and the land enjoyed a brief respite.

At present we live through a phase that stands between those two extremes, when industrial and commercial needs, and the desire for more housing, still make pressing demands for changes in the use of land. But we also show more positive appreciation of our varied landscape and strive to conserve the wild, the wooded, and the wetland. We may be standing at a watershed in time, facing two ways at once, seeking a better balance between the need to accept change and the need to conserve some memory of the natural vegetation and wildlife that existed before human hands strove to tame it.

Robert Nisbet has suggested that the mood to start anew, in order to achieve a balanced ecological community, as he expresses it, has recurred several times in our history, and he cites as earlier spokesmen for that objective St Benedict in the seventh century, Sir Thomas More in the sixteenth, and Prince Kropotkin in the nineteenth century. Past parallels with our present concern for the environment certainly deserve reflection, for, in fact, the Heritage Commission has recently set out the same objectives all over again, without citing those who have said it all before.

In consequence of current debate, our eyes are steadily being sharpened to see the distinctive features of our countryside that should not be destroyed. We notice with more discernment than ever before clues that lie all around us to significant episodes in our past, and know that more are mirrored in the landscape, which we yet lack the eyes to see. So we need in our baggage as much information as we can carry if we are to penetrate the layers and perceive the detail. The writers of this book offer a companion on that journey.

The most influential author in the present century, who opened our eyes to historical features in the landscape, was W. G. Hoskins. In tribute to his achievement, he was described in a *Sunday Times* survey in 1991 as one of 'a thousand makers of twentieth-century opinion'. His most memorable book, *The Making of the English Landscape*, published in 1955, made a lasting impression both on those who like to survey the large tracts of countryside which make our distinctive regions, and on others who prefer to study the more intimate local scenes which they traverse daily. On the one hand, he drew attention to strong regional contrasts, as between Devon and Leicestershire, for example, and explained them historically; on the other hand, he fixed his gaze on humble features of the landscape, old hedges, green roads, deeply sunken lanes, and verges wide or narrow, asking questions about their form and direction, and showing where explanations for their origins might be found. As a result of his path-breaking books and television programmes, Britain may be said to lead the way in Europe in telling the history of its landscape, and it is not surprising, in consequence, to see the strength of public opinion in favour of conserving that diversity.

Major features that distinguish the large regions in our landscape are singled out in the first ten chapters, giving us the headlines, as it were, before engaging with the small print in the five chapters that follow. Identifying the main regional types calls for more than historical knowledge, and other skills have been developed and applied to this task. At the base of the large regions, whether moorland, vale, fen, or wold lies a structure of rock which we can never greatly alter. These determine soil type and soil depth, and limit the vegetation that each supports. As we pass swiftly from one region to another by road, or better still in a railway train, we have a splendid view of the contrasts in scenery imposed by those constraints. I once heard a geologist, so sensitive to changes of rock and vegetation that he claimed to feel it physically: a cold shudder ran through him as he moved onto the coal measures from the limestone. Most of us have less strong sensibilities, but we cannot fail to see some of the most striking changes as we drive, say, into the fenlands of East Anglia from the clays of Kesteven (Lincolnshire), leaving a dissected country of hills and lowland, to stumble suddenly into a flat land that belongs to another world, criss-crossed with dykes, lined with willow; or, moving in a northerly direction, we drive along the sharp-topped limestone ridge towards Lincoln, once dominated by sheep, and look down on a clay vale, comprising the Trent valley to the west, once dominated by cattle; yet another thoroughly familiar scene to the southerner is the abrupt difference between the heathland of Ashdown Forest in Sussex and the bright, upstanding chalk of the South Downs as we near Brighton.

At the beginning of this century the geological skeleton of the country attracted remarkable investigative attention, when the geologists were eager to share their excitement with everyone else by describing the significant landmarks. The essence of that phase was distilled in *Stanford's Geological Atlas of Great Britain and Ireland*, in which Horace B. Woodward compiled an atlas, and went on to help his readers by pointing out the geological features that could be seen from a railway carriage. That book, first issued in 1904, and then enlarged, was still being reprinted in 1927; it was a most instructive companion on rail journeys, and, indeed, as interesting now as then; the idea could deservedly be revived for the train traveller's pleasure nowadays. The account of the journey from Charing Cross to Hastings (Sussex), for example, signalled the chalk of the North Downs, first visible at Orpington (Kent), and described as 'smooth hills with thin, scanty hedgerows here and there, . . . quite bare at the surface', while at other moments one might see the surface 'modified by coverings of clay-with-flints, brick earth and gravel'. The descriptions for southern England were markedly helpful in picking out the narrow beds of greensand which can so easily escape the traveller's eye, but which harbour such an interesting history of flexible land use. On the line of the London, Midland, and Scottish railway, the journey to Lancaster and Carlisle was made memorable by its crossing many different geological strata, including the bunter pebble beds on the borders of Cannock Chase in Staffordshire, the keuper marls with sandstones that led to the rock salt and brine springs near Crewe, and, much further north beyond Carlisle, the further bands of keuper marl, limestone, and peat—yes, the line runs over peat—before the train reached the Solway Firth.

In 1913, with the fourth edition of the geological atlas, readers wanting to match the text with the actual landscape were still better served, when Hilda D. Sharpe, Science Mistress at the Edgbaston High School for Girls, joined Woodward in compiling a Photographic Supplement, in which she and her women friends supplied the bulk of the photographs, showing and explaining significant cross-sections of rocks, folds, and faults, sometimes capturing their shots opportunely when new railway cuttings left clean rock faces, or when mining and quarrying were actually in progress. Two years before this work appeared, Miss Sharpe had demonstrated her zeal for practical field observation by publishing a *Field Notebook of Geological Illustrations*, designed to help her pupils in school who found difficulty in matching the lessons in class with the scenes they encountered out of doors. Her photographs were a halfway house on the road to the fieldwork that we nowadays take for granted.

Another opening into the history of the landscape was made from a very different direction when scholars began to look more critically at place names, and saw the possibility of uncovering the sequence in which different areas of England had been settled. It seemed that this might be done by studying the names given by natives or invaders to the places that they colonized or transformed. Swedish scholars were notably active in this branch of study in its early days, and work started seriously in England when the philologists joined the historians and set up the English Place Name Society in 1923. The work proceeds by the study of counties, one by one, and depends on gathering all the oldest forms of names. Advancing knowledge since the 1960s and 1970s has changed many of the original assumptions about place-naming, and thus has revised many inferences that used to be drawn about early settlement history. But it has become an exciting and richly illuminating field of study, in which Margaret Gelling acts as our best interpreter, keeping the rest of us in touch with the larger picture. It is not a study for amateurs, and we can only look on and listen. But it has proved to be an essential tool in uncovering the action of early peoples in making the countryside serve their needs.

The next steps in opening our eyes to landscape were taken at the same period, in the 1920s and 1930s, when the more readily visible surfaces of the land were described by walkers with literary gifts, and their distinctive features were identified. More people could now grasp the contrasts as they travelled quickly by bus and car across larger distances, while the writers who set these features in a frame became almost household names. H. J. Massingham, C. Henry Warren, Adrian Bell, and H. E. Bates are among the best remembered from this period, encapsulating the spirit of a region in their books. H. J. Massingham achieved this alone with *The English Downland* (1936), and then, in a volume of essays, he joined with eleven others to tackle the whole country in *The English Countryside: A Survey of its Chief Features* (1939). It was he, as editor, who perceived two significantly different attitudes to landscape among his writers, some seeing it as a work of art, and being content to describe and enjoy its aesthetic beauty, others seeing it as the work of people, and wanting to see the painters as well as the pictures. Massingham confessed himself to have been once a disciple of the first school, looking only for scenic distinction, but was then converted to joining the second, wanting to see the people who had made the scene. We have already noted those separate viewpoints, and the watchful reader will discern them again in the pages that follow.

The timing of Massingham's publication in 1939 was unlucky, for war soon put such calm appraisals of landscape out of sight and mind. But it is worth noticing that the essayists were writing then about a countryside that was deeply depressed agriculturally. One earlier publication in 1931 was actually entitled *Why the Land Dies*; Massingham saw it crumbling into decay; and Charles Fry in a historical survey thought the current depression so severe that only some form of land nationalization would rescue it. Yet before the year was out, the countryside was lifted out of gloom, for the land now became the indispensable source of our food. Every possible acre had to be turned to use, green pastures were ploughed, and allotments were assiduously cultivated. That book on landscape in 1939 deserves to be reread now as a thought-provoking reminder of how the wheel can suddenly turn; in that single year war led to dramatic changes in the landscape.

Finally, a move towards our better understanding of landscape history was made in the 1920s and 1930s by the archaeologists who began to relate archaeological remains, such as forts, settlements, and burial mounds, to the landscapes in which they stood. Among the most important of these archaeologists were Heywood Sumner, whose books about the earthworks of Cranborne Chase and the New Forest had appeared as early as 1913 and 1917, O. G. S. Crawford, and C. E. Curwen, while Jacquetta Hawkes played a memorable part in bringing to the bare bones of archaeology in the landscape philosophical reflections and a popular appeal. Their successors have made still more remarkable strides since 1945, showing how much of our landscape, not just on the chalky downlands but everywhere else, was fundamentally shaped in prehistoric times, so that some features from that time even now persist and shape our present-day fields. New scientific techniques have, of course, immeasurably advanced those earlier insights: aerial photography, geophysical prospecting, and ground-penetrating radar illuminate much that could never otherwise have been visible to us. But simple field-walking remains another highly valued aid to discovery, and, indeed, is regarded by some who have plenty of experience of more highly technical methods as 'the single most important method of data collection'. Meanwhile television programmes that show archaeological sites to us all have further deepened our knowledge about the making of landscape, and open endless further possibilities to those with a consuming curiosity and sharpened powers of observation. As recently as 1998 Bronze Age paddocks dating back to 1500–1400 BC were recognized on Salisbury Plain by a discerning eye, the army's occupation of the land having preserved them from ploughing.

As for the historians, after 1945 they turned their attention away from surveying large regional prospects to explaining smaller features recurring in scattered places everywhere. M. W. Beresford, W. G. Hoskins, and J. G. Hurst focused on deserted village sites, running into thousands all over England; others traced drovers' roads, A. R. B. Haldane first of all in Scotland (1952), K. J. Bonser in England (1970), and Shirley Toulson in Wales (1977). The plants composing our hedges fired the imagination of those who did not drive in cars but walked and cycled round the lanes. It was realized, for example, that the more species of shrubs and trees present in a hedge, the older it was. Where stone walls rather than hedges defined boundaries, David Allen, as a botanist, saw (in 1971) the significance of distinguishing the 387 recognized species of brambles, for when they tangled themselves over old ruins and along old paths, the number of species again gave a clue to the age of the tracks and the sites.

This move to examine the landscape at closer quarters has been greatly stimulated in the last fifty years by Hoskins's writings, for they have set every intimate local inquiry in a larger frame, and given it place and significance in our national history. To look more discerningly at small features on the ground was to place the local people who put them there in a much larger historical context of long-term change and sometimes even fix those people firmly in a particular period, possibly a precise year. Thus, we are much more alert nowadays to question the reason for wide roadways, the size and sometimes the rough tell-tale surface of fields, sprawling banks and ditches, or rectilinear lines in the grass that show up only in fading light. On investigation they may turn out to be boundaries of ancient estates or of deer parks, burial mounds, mounts that once looked out over old gardens, even, in the right place, salterns, all contributing something to the precise history of one place and then illuminating the general history of our nation. Some memorable discoveries made in recent times have resulted from the sheer chance of observing small details, leading on to the discovery of large features in past landscapes. A woodhenge, for example, was identified at Stanton Drew in Somerset in 1997, accidentally found during a routine survey. It was a wooden temple of about 3000 BC, which showed itself unexpectedly through magnetic measurements of the soil, and proves to be the precursor of the more visible ring of stones still standing there. Elsewhere in Somerset a farmer, Roy Sweet, clearing his ditches in 1970, found the fragment of a track of wooden planks laid out over the fens and dating from about 3806 BC. It lay under some 5 feet of peat; a replica, called the Sweet track, has now been built, and is explained at the Peat Moor Visitor Centre at Westhay.

In recovering features of past landscapes that have now disappeared, we are not usually able to identify the individual people who put them there. But it is sometimes possible to get close when looking at landscapes that still survive. Shirley Toulson, writing on the drove roads of Mendip, put a name to one drover, John Butt, on his way from Cheddar to Wells (Somerset). He must have traversed the ancient way on the southern fringe of Mendip, now the A371, so it is difficult to recapture there any flavour of the older scene. But another driftway preserves the lonely environment in which such routine travellers moved, with their animals and bags of oatmeal, a bakestone, and brandy, to sustain them on a journey of two or three weeks. The Ridgeway across the chalk downlands of Berkshire, passing through Blewbury, Aston Tirrold, and Ilsley, still conveys the atmosphere, while the wide verges signal a droveway, and documents exist to show that its width was once measured at 132 feet, allowing generously for the wayward rovings of large herds of cattle and sheep from Wales.

Everyone walking our countryside today has the chance to enter in imagination into the local history of a place they know well. They may then go further, and link otherwise disparate observations with others, so as to establish some general propositions about our landscape history, for as we understand things better, we ought, without being too dogmatic about such things, to be able to construct a few more general theories, inserting another cornerstone into the general landscape history of the whole kingdom. How many sites, for example, can we identify, possessing still an old marker tree or two, representing almost certainly the meeting place of intercommoning parishes, or even the meeting place of the Hundred? Are they chequer trees (wild service trees) by any chance, or whitebeams, and what is their age? Can we associate the choice of tree with any one period rather than another? In how many places can we still find significant *rows of trees* marking parish boundaries, and again which trees were they? The parish boundary between West Peckham and Hadlow in Kent, for example, has ancient hornbeams lining its course. Why were hornbeams chosen, and are they a clue to the date of the boundary? A puzzle lurks here, in any case, for the parish of West Peckham is a small triangle of land that has obviously pushed itself into place at the top of a hill in the Hurst Woods, and upset a former symmetry. Three large parishes originally converged at one high point on the wooded Gover Hill. But then the parish of Hadlow was pushed out of place and so deprived by West Peckham of access to the beech and chestnut woods at the summit. Here is a mystery in people's making of the landscape that may only be solved when we assemble more such examples of shifting but ancient boundaries, and compare the trees that were chosen.

A multitude of questions in the Hoskins tradition will in future lead us to explore more detailed points in the landscape, and anchor them in time and place. We may yet go further still for the wealth of documents gathered in our archive offices nowadays give us the chance to elucidate more of the role of people in moving our landscape around, whether as members of a particular social class, or as individuals. Such an inquiry could prove to be an unexpectedly satisfying experience in a world that is being made more and more impersonal by technology. H. E. Bates saw the countryside in these human terms. Apethorpe in Northamptonshire struck him as an excellent example of a village, planned, governed, and preserved according to the standards of a great lord dwelling in the great house; he might equally well have cited Rockingham in the same breath. In contrast, he viewed the clay lowlands of the Weald of Kent as the creation of yeomen farmers. How right he was; the traces strongly persist. Wordsworth in the Lake District saw that countryside as 'a perfect republic of shepherds and agriculturists, proprietors for the most part of the lands which they occupied and cultivated'. Cotswold villages differ yet again. They have been described as the mirror of a relatively harmonious collaboration between gentry, farmers, and labourers, harmonious because all classes accepted their place, whether low or high, in a hierarchy that allowed and acknowledged both duties and rights. In many such parishes, of course, things were made easier because one manorial estate defined the whole parish and embraced the whole community. So Joyce Cary in his *Selected Essays* (1976) saw the resulting Cotswold village as 'a balance in society, a tempered ambition, a concord of work, of religion, of public office, and private concern'. For him, it was this region that lay at the heart of England.

The chapters in this book add grist to the mill of those who wish to refine their perceptions of the mix of social classes making different kinds of countryside. David Hey shows the contribution of that class of farmers standing below the gentry, the sturdy independent yeomen who so clearly made the moorlands that we see today. Yet, plainly, they were not always firmly in charge. Rather, he shows how monastic houses in the thirteenth century created big grazing farms, until changing times after the Black Death caused the monasteries to lose interest in direct farming on a large scale, and allowed former tenant farmers to take over, and divide the land into smaller units. They made butter and cheese, and their womenfolk knitted stockings from the local wool; thus they earned a livelihood in their own way making the most of their local resources. Their descendants preserve some of the traditions that struck root then. More than a trace survives in the villages that still make cheese, while in Muker in Swaledale (Yorkshire) some of the womenfolk in the 1970s revived the ancient tradition of knitting with the local wool, not for stockings as in the past, but for cardigans and waistcoats.

Another landscape, having a quite different social ancestry, is creatively examined here by Alan Everitt. Some extensive wastes and commons were, in fact, not pressed into cultivation before the early nineteenth century, and so for centuries they offered a precious freedom to individuals needing to squat, build shacks, and resourcefully wring a living from scraps and scrapes. Despite their poverty, such people nourished their independence and individuality in untamed, unkempt places, building up a distinctive landscape of which the remnants can still sometimes be detected today. A colourful story takes us to Padworth in Berkshire, where one-fifth of the parish was common in the eighteenth century, and over its boundary lay another stretch, Mortimer Common; both were pieces of a still larger stretch of common land, forming one of the chains described by Alan Everitt. Religious life at the end of the eighteenth century was said to be at a low ebb when a young turf-cutter inhabited one of the Padworth cottages on the common in 1778, and was converted to fervent Christian belief by a minister at neighbouring Tadley. He started preaching, and gathered a congregation too large for his cottage. So a chapel was built for him by a rich local man of gypsy descent, who claimed for his coat of arms 'three moles and a molehill.' 'My great ancestor', he said, 'was molecatcher to William the Conqueror.' The turf-cutter's death in 1803 stands as the first burial entry in the chapel register, and in 1826 this chapel became the property of the Countess of Huntingdon Connexion. It is one of many dissenting chapels across the country that stand as memorials to the poor but strong-minded, stoutly independent, squatters of the past. In the Padworth case, we may note that the next-door parish is Aldermaston.

The social history and landscape of common and waste lands is poles apart from villages which have an elegant manor house and ancient church at their core. Alan Everitt's essay will cause many of us to scrutinize our maps with more discerning eyes henceforward, differentiating the two zones of influence, and recognizing signs of different village origins in the landscape. But in reconstructing a more complete social map of the past, we could, in fact, construct several different images of England, separating or matching regions according to their particular social mix. Some of our forebears must have made such maps for themselves for thoroughly serviceable, day-to-day purposes. The gentry, plainly, had a mental map of the length and breadth of England, showing exactly where their many kinsmen and friends lived who would accommodate them, seemingly at a moment's notice. They lived in villages rather than hamlets, and their households had large food stocks, stables, and servants always ready. George Sitwell, at Renishaw in Derbyshire, was not the only gentleman enjoying this environment, yet claiming to live without anyone else in sight: 'no one between us and the Locker Lampsons,' he told Evelyn Waugh, blithely overlooking the fact that 'in the valley at our feet,' wrote Waugh, 'still half hidden in mist, lay farms, cottages, villas, the railway, the colliery and the densely teeming streets of the men who worked there'.

In contrast with that map of England was another made by people whose living and lifestyle meant that they rarely slept at home, or had no roots at all; they were always itinerants, casting their eyes across the kingdom in search of a countryside having generous commons and wasteland, where wanderers could be certain of finding a place to linger undisturbed, for short or long spells, not always being moved on by authority. The gypsies were the largest cohesive group in search of such places, and the map of England that survives in their folklore must be unique. Some of its familiar place names were Erith and Belvedere along the Kent side of the Thames, Chatteris in Cambridgeshire, an enormous parish of 15,000 acres in the fenland, whose independent spirit shines out in its highly varied religious loyalties (how many other places rival, or perhaps surpass it, in possessing seven nonconformist chapels, for Independents, Methodists, Baptists, and Quakers in the nineteenth century?), Pocklington in the East Riding of Yorkshire where some local folk still recall a large gathering of gypsies at a funeral a generation ago, and Appleby in Westmorland, notable for its horse fair. Moving into the West Midlands, comparable places that were being sought out still in the 1920s and 1930s were Merrington Green, north of Shrewsbury, Prees Heath near

Whitchurch, and Sound Common near Nantwich in Cheshire. By then, however, the gypsies were finding Ireland more congenial: there 'we could camp almost anywhere', they said.

The large commons were gradually nibbled away, to give more permanent homes to longer-term squatters, living on the plants, fruits, wood, and minerals that lay freely around, and valiantly succeeding in turning their shacks into more substantial dwellings. Thus the broad areas of commons were broken down into smaller fragments, contained within parishes. In Kent Gillian Rickard has identified in the single parish of Chilham in the 1820s and 1830s three such places, Godmersham Hill, White Hill, and Old Wives Lees. What an evocative name is that last, though it is, in fact, a corruption of Oldwood Lees! It is a small heath in Chilham parish, fairly close, we are not surprised to learn, to the Forest of Blean. Bequests to the poor by local people in Chilham clustered in the years 1599–1638 when poverty was a central national and local concern; but evidently the parish preserved space for the vagrant poor for at least another two centuries.

Cattle drovers were another group carrying in their mind's eye another map of England, not entirely different from that of the gypsies. Again it signalled ample commons that might give grazing en route during the day on their long journeys from Wales and Scotland to the south-east. But they also needed more confined and sheltered grazing at nightfall, and so their maps had to signal grass for a night and, if possible, a halt for themselves at an inn or wayside house, where a friendly woman in the kitchen might boil a pudding to sustain them the following day. A former drovers' inn at Stockbridge in Hampshire still bears its large inscription in Welsh offering good hay, sweet pasture, fine ale, and comfortable beds (*gwair tymherus perfa flasus cwrw da a gwal cysurus*). Is there any support for the notion that a group of three Scots pines signalled to them the presence nearby of shelter for men and animals? We could test it if we looked more closely at the drovers' routes. From Wales one of their mapped journeys named Banbury, Buckingham, Leighton Buzzard, Markyate Street, Redbourn, and Smithfield. Coming from a more northerly direction, they headed for Daventry, Northampton, Wellingborough, and Hertford, and if they intended to sell their animals to a farmer for fattening in Essex, then they moved east to Ongar and Chelmsford. Cottesbrooke in Northamptonshire has a house with a tradition of feeding the drovers, and there, indeed, three Scots pines stand alongside. At Beckhampton in Wiltshire, on one of the wildest stretches of the road from Bath, it is taken for granted locally that their Scots pines have the same significance. But we need to interpolate a word of caution here before bestowing too much antiquity on this tradition, for some tree historians tell us that Scots pines were not planted in England until about 1660.

Another glimpse of social variety in the landscape also raises questions about the chronology of change. Different areas of the country have been gentrified at different periods. The arrival of gentlemen to live in places where they had not chosen to live before, and the shaping of the environment to suit their tastes, could have a profound effect on local landscapes. Sometimes the effect endured and settlements became what we call 'close' villages under the authority of one lord, in contrast with the 'open' villages in the hands of several gentry or of freeholders, who could not command single authority. But the arrival of a gentleman did not necessarily alter the village style for good. In some places and periods, the structure of the economy and society resisted the change. The gentry, for example, moved into the Northamptonshire forests in the later Middle Ages, and Arthur Throckmorton, brother-in-law of Sir Walter Raleigh, was one of those who toiled enthusiastically on house and garden in the later sixteenth century at Paulerspury. But his handiwork did not survive the test of time, and the house has gone. In a similar case, my own village in the Weald of Kent had a manorial estate next to the church, but it had no deep tradition of living under the sway of a powerful resident gentleman, for the owner of the land held court instead at Tonbridge Castle. So when an aspiring gentleman did arrive to take up that role and built an impressive house in the later eighteenth century, he did not succeed in striking deep roots in this countryside of yeomen and small farmers. In less than two centuries the gentry had faded from the scene, and people of the 'middle sort' moved in to share and reshape his house and garden between them.

If we strive to see more features in the landscape in these social terms, we may well perceive fresh subtleties by matching buildings and landscape with more precise documents and dates. I ponder this possibility whenever I pass the entrance of Mereworth Castle in the Weald of Kent. Lying back from the road, and formerly surrounded by a moat, the castle is a celebrated Palladian mansion, built by John Fane, Earl of Westmorland in the early 1720s, when the family, after living in the district for some 200 years, had risen high over the surrounding circle of modest local gentry. The central dome of Mereworth Castle connects it with Palladio's house at Vicenza, to which two free-standing buildings were added, designed originally as service pavilions, but emerging as handsomely adorned as the domed centrepiece itself. But another cluster of associations draws my eye to two entrance lodges. They were built in the same style some twenty years later, with columns and portico facing the road, and because of their

Lodge leading to Mereworth Castle

One of two lodges that stand on either side of a drive that led to Mereworth Castle, in the Weald of Kent. It was built in the 1740s when the owner, Lord Westmorland expected to house a senior official in it. C. Henry Warren lived in Mereworth and in A Boy in Kent *(1937) remarked that 'the lodge by the side of the gates had had more attention paid to its pillars and porticoes than the two tiny rooms where the lodgekeeper lived with his ailing wife'.*

scale, room heights, and elegance, were presumably intended to house higher-ranking officials of the estate, perhaps even some occasional guests, for not many could be accommodated in the castle. This was at a time when gatehouses were first finding their purpose at the entrance to parkland, protecting the hidden house, yet also setting the tone, making 'a statement of minor magnificence'. Very similar to the elegant façades of the two gatehouses at Mereworth, adorned with columns and portico, was a model gatehouse inserted by T. D. W. Dearn in 1807 in his book entitled *Sketches in Architecture*. Dearn was architect to the Duke of Clarence, and was also at home in the Weald, living in Cranbrook, and writing in 1814 another book, *An Historical, Topographical and Descriptive Account of the Weald of Kent*. For Dearn, as for the Fanes, gatehouses had the double purpose of 'embellishment and use', but Dearn offered his design not for the dwelling of a steward or estate official but for 'two cottages for labourers as an entrance to a park or grounds'. The frame of mind in which his design was commended to landowners at the beginning of the nineteenth century was very different from that in which the Fanes had built theirs in the 1740s. Dearn firmly believed that labourers' living quarters should sit on the street, at the park gates, in order to be conspicuously visible, because 'experience teaches that there are none more liberally endowed with low cunning, and, I will add, none possessing a more predatory disposition than farmers' servants'. So living quarters for them were 'of all others best calculated to meet the end in view . . . where their conduct will be most liable to observation'. The significance of entrance lodges plainly deserves finer social analysis in time and place when once we move from generalities to examine a particular landscape in more detail.

Another view of people shaping the landscape holds sway when walking in the grounds of Everingham Hall and Londesborough House in the East Riding of Yorkshire. The landscape of Londesborough is inextricably bound up with the name of Thomas Knowlton, who was engaged as gardener there and came from southern England (he was born in Chislehurst, Kent). He lived on the estate between 1728 and his death in 1781, and planned its layout for the Earl of Burlington. In his spare time, when the Earl of Burlington was away, he went over

Lodge

This lodge was commended by T. D. W. Dearn, a resident of the Weald, in his Sketches in Architecture, *1807. It was designed to house and keep an eye on labourers, for none, he said, were 'more liberally endued with low cunning and . . . a predatory disposition'.*

to Everingham, and gave advice on another park to Sir Marmaduke Constable. In 1731 Constable wanted a line of elms 'to be planted down Jacksons Garth from the church stile to the back way to my house'. In February 1732, possibly in the coldest of weather, Knowlton spent three whole days planting the elms himself; that was his habit to make sure the job was well done. The layout of parkland between the house and the public road has undergone considerable change since then, but, in the mind's eye, Knowlton is still in sight at Everingham, tending the many fine trees that remain. At Londesborough itself, even more surely, the beautiful parkland of the early eighteenth century is a monument to Knowlton. Judging by his correspondence, he himself was most excited by his success in growing pineapples, coffee trees, and other exotic rare plants in the hothouses. But it is the parkland that has preserved his handiwork longest. And the surviving walnut trees that once made a whole avenue, and the turkey oaks that were not introduced until about 1728 and made two avenues, proclaim to us still another fact: they were two of the tree fashions of the age.

A final, very different social glimpse of individuals making a landscape takes us to Rixton Moss in Lancashire, a 5-mile stretch of land between Warrington and Irlam. In 1879, when deep depression struck the farming world, some people turned to the life-saving occupation of growing vegetables intensively. Rixton Moss was then 'a crust of peat on which heather and stunted birch struggled for existence above many feet of slimy ooze'. People heard that the landowner, Lord Winmarleigh, was willing to let land at a nominal rent to stout-hearted young men who would cultivate it. So a group of takers, of whom the oldest was 25 and the youngest was 20, took 20–25 acres apiece, working at first together to make roads and dig ditches to drain the land. They worked 'for long, back-breaking hours, but it seemed to be a healthy life'. A young lad, Marshall lived in a house made of turf sods, sleeping with matches under his pillow to have light in the morning. Others lived in old railway carriages or wooden railways sheds. They spread tons of marl with hand-shovels and forks to sweeten the acid soils, and Lord Winmarleigh built a siding to bring nightsoil from Manchester. A muck wharf on the river also brought cow dung from the live cattle that were shipped from America. The land dried out and was enriched with these nutrients, until it began to pay after about twelve years, growing potatoes and peas, then lettuce and celery, radishes, and spring onions. Nowadays people marvel at its soils, growing, it is said, the best lettuce in the world. Where better than here to be persuaded that all our soil is man-made?

Snatching the chance to see people in their social context, and making the landscape with their own hands, or at least making it by their decisions, yields imaginative insights into the lives of many colonists of land in the past, whether Anglo-Saxon in the seventh century, English in the thirteenth, or French and Flemish in the seventeenth century. In a small corner of Cumberland we can even see it happening today. Ian Johnston inherited Thortergill near Alston after lead mining had ravaged the valley, and mine shafts, railway culverts, underground workings, and thousands of tons of rock and spoil lay all around. His family, he maintains, were cattle farmers and one-time raiders over the Scottish border, selling the valley to a lead mining company in the 1780s when mining gave a chance of a different kind of income. Lead was taken out until the 1920s, when the workings were abandoned as they lay. Ian Johnston's uncle bought the land back, and Ian with Jennifer, his wife, and their two sons, Andrew and David, resolved to pay their debt to Nature. They had the help of machines that our ancestors did not have. But

**The lead mine at
Whitesyke, Garrigill,
Alston, Cumbria**

*This mine exploited among
others the Thortergill Syke
Vein, and was worked
for over 100 years by The
London Lead Company.
The photograph, taken
c.1890, depicts a scene
repeating the one which
Mr Johnston inherited on
his land at Thortergill.*

still on such steep slopes they did plenty of hand barrowing as well. Their heroic toil is hardly imaginable now that it is at an end and Nature is clothing every inch of bare soil in its own way. The old mine workshop is their family home, their sons are blacksmiths and forge iron, and in 1996 the journal *Country Life* named Mr Johnston Countryman of the Year. He stands at the end of a long line of valiant makers of landscape, winners from the waste, improvers, driven by many different impulses, but combining in different proportions the same mixture of desires, to make the most or the best of the land's resources and potential.

Historians retrieve this story through documents and archaeology, and some-times artists assist them. But artists do not always depict a landscape faithfully. The reputation of taking no liberties with the truth belongs to Samuel Palmer in his Golden Valley near Sevenoaks at Underriver, after he had returned from two years in Italy. Before that, in the 1820s and 1830s, he had seen the landscape around Shoreham as a visionary, and that is the style that Geoffrey Grigson applauded; he

The disused lead mine at Thortergill, Garrigill, Alston, Cumbria

The mine exploited the same veins as its sister mine at Whitesyke and was worked by the same company. The photograph, taken c.1990, shows a collapsing culvert, which conveyed Garrigill Burn (Thortergill Syke) underground for the entire length of the mineyard. It was overfilled with thousands of tons of mine waste.

lamented the realism that came afterwards. Historians of landscape will take a different view, and rank the later Samuel Palmer, 'reporting what he had actually seen', among their best aiders and abetters. Similarly, the poets separate into the visionaries and the realists, none more successful in the realistic school perhaps, than Vita Sackville-West writing about the Weald of Kent in her poem *The Land*. But the many other writers who have depicted landscape in their reminiscences and in fictional and semi-fictional writing in the nineteenth and twentieth centuries lie waiting for the historian to tap their work fully for its insights. The fact that we all see landscape differently holds out a bright and expanding prospect for landscape history, promising to take it in yet new directions.

Lead mine at Thortergill

The same scene as the photograph on p. 23, after the culvert was excavated, the hole was backfilled, and a new reinforced concrete bed was being made for Garrigill Burn. The entire area is now returned as near as practicable, to its appearance c.1820 when the mine first operated.

Cottage on Kettles Hill

Samuel Palmer is considered one of the great landscape painters of the nineteenth century. His early paintings were visionary, his later ones more faithful portrayals of the landscape. The latter were often painted in his Golden Valley in Kent, at Underriver. Shown is the Primitive Cottage on Kettles Hill, which is still recognizable and unspoilt.

Part I

Panoramas of Landscape

1 Downlands

Joseph Bettey

The chalk downlands of Southern England present one of the most distinctive and easily recognizable landscapes in the country. At the heart of this region of smoothly contoured hills, freely drained soils, dry combes, and deeply cut river valleys is Salisbury Plain, the largest expanse of unbroken downland in England, mostly between 450 and 550 feet in elevation. From this central core long fingers of chalkland stretch out north-eastwards through the Marlborough and Berkshire Downs and on into the Chilterns, while south-westwards the chalk downs extend across Cranborne Chase into Dorset, reaching the English channel coast between St Aldhelm's Head and Abbotsbury. To the south and east from Salisbury Plain the chalk continues across the Hampshire Downs and beyond through Sussex, Surrey, and Kent as the North and South Downs, culminating in the chalk cliffs of Dover and Beachy Head.

The underlying chalk gives a unity to the topography, farming, and settlement pattern of the region, but there are considerable variations in its consistency and composition. In some parts there are deposits of flints, gravel, loam, or clay-with-flints; elsewhere occasional outcrops of hard, grey sandstone occur known as sarsens or 'grey wethers' from their distant similarity to a flock of sheep. These sarsens were much used for prehistoric monuments and ritual sites such as Avebury and Stonehenge. Below the chalk and appearing around the edges of the escarpments and in the valleys are fertile bands of greensand from which many of the chalkland springs or streams emerge and where most of the settlements are to be found. Although much of the downland today presents an aspect of bare hills, with occasional plantations or shelter belts of beech and yew, nonetheless in some parts the overlying soil provides support for extensive tree cover, and areas such as the north downland of Kent and Surrey, the Chilterns, the West Sussex downs, Savernake Forest, the copses or 'hangers' along the sides of the steep combes of north Hampshire, and Cranborne Chase are heavily wooded with beech, oak, hazel, yew, hornbeam, or holly. This verdant woodland gives an unwarranted appearance of fertility to the landscape, although the tree cover generally survives only on the poorest or most difficult ground. Characteristic of the chalk downlands are the clearly defined escarpments, making the climb up on to the downlands very obvious, while the settlements are strung out along the lower slopes. For example, the northern escarpment of Salisbury Plain between Westbury and Pewsey rises up steeply from the clay vales of west Wiltshire, the conspicuous chalkland escarpment overlooking the vale of the White Horse in Berkshire, with far-ranging views across the Thames valley, has a string of settlements along its base, while similar clearly marked and deeply indented escarpments show the boundaries of the downland overlooking the Vale of Wardour (Wiltshire) and the Vale of Blackmore (Dorset). In Surrey the chalk of the North Downs between Farnham and Guildford is only 400 yards wide and forms the Hog's Back, with its remarkable distant views northward over the sandy heaths of Aldershot and Bagshot and across the Thames valley to the Chiltern Hills. In Kent a long string of settlements marks the scarp-foot of the North Downs close to the Pilgrim's Way which runs along the downland edge, and the chalk escarpment drops steeply to the Vale of Holmesdale. In Sussex the steep northern escarpment of the South Downs provides a dramatic contrast with the totally different landscape of the Weald beyond.

Apart from flint, the downland provides little good quality building stone. In some parts a hard band of chalk known as 'clunch' has been used for building, while all over the region great ingenuity has been shown in using the irregular flint, either whole or 'knapped', to reveal a flat, attractive surface. Elsewhere buildings have been constructed of timber, cob, brick or have used stone imported from adjacent regions such as Portland, Purbeck, and Ham stone in Dorset, or Chilmark, Bath, and Doulting stone in Wiltshire. The white chalk lying only a few inches beneath the downland turf has proved irresistible to the creators of hill figures and other emblems, ranging from the Bronze Age White Horse of Uffington (Berkshire), the enigmatic Cerne Giant (Dorset), and the Long Man of Wilmington (Sussex) to the numerous White Horses of Wiltshire, the figure of George III above Weymouth (Dorset), the Kiwi at Bulford, the regimental badges on the escarpment at Fovant (Wiltshire), and the Royal Crown at Wye (Kent) which commemorates the coronation of Edward VII in 1902.

Many authors have been impressed by the beauty, solitude, and wide expanses of the chalk downlands and by the remarkable variety of plants, birds, and animals which they support, and notable descriptions of the downlands occur in the novels of Thomas Hardy and in the writings of W. H. Hudson, H. J. Massingham, and Richard Jefferies, while Hilaire Belloc and Rudyard Kipling celebrated in verse the landscape and vistas of the Sussex Downs. Earlier observers were well aware of the distinctiveness of the chalk downlands, of their characteristic farming and society, and of their marked contrast with the surrounding regions. Writing of the Dorset downland in c.1620 a local gentleman, Thomas Gerard, described the clay vales as 'verie subject to Durt and foule Wayes' whereas the well-drained downlands were free of such inconveniences 'for that consisteth altogether of Hills … all overspread with innumerable flockes of Sheepe, from which the Countrie hath reaped an unknowen Gaine …'. The marked contrast in Wiltshire between the downs and the vale or between the 'chalk' and the 'cheese' was mentioned by both William Camden and John Speed in the sixteenth century, and the seventeenth-century antiquarian John Aubrey contrasted the

The Cerne Giant

One of the best-known of the numerous chalk-cut figures on the downland, the Giant dominates the site of the former Benedictine monastery. On stylistic grounds he has been dated to the Romano-British period, but there is no documentary reference to him until 1694, and recent research suggests that he may have been cut in the seventeenth century.

society of the 'cheese' and the 'chalk' districts of Wiltshire, suggesting that people from the claylands or 'cheese' country were 'melancholy, contemplative and malicious', given to nonconformity and lawsuits, while of those living on the 'chalk' he wrote that

> On the downes, the south part, where 'tis all upon tillage, and where the shepherds labour hard, their flesh is hard, their bodies strong: being weary after hard labour, they have not leisure to read and contemplate of religion, but goe to bed to their rest, to rise betime the next morning to their labour.

In a letter of 1773 published in his *Natural History of Selborne*, the Revd Gilbert White described the South Downs in Sussex as 'that chain of majestic mountains' and his pleasure at the 'noble view of the weald, on one hand, and the broad downs and sea on the other'. He went on to speculate on the forces which had shaped these chalk hills and had caused them 'to swell and heave their broad backs into the sky so much above the less-animated clay of the wild below'.

Over the centuries the distinctive landscape of the chalk downlands has given rise to several characteristic features in its farming, land tenure, field systems, and settlement patterns. Among these are the long predominance of agriculture in the economy and the absence of large-scale industry, the close control exercised by great estates and a few major landowners, the continuing importance of manors and strong manorial authority over the tenant farmers and numerous landless labourers, the concentration of settlements along the streams in the chalkland valleys, and the comparative absence of habitation on the open downland. Other features include the long survival of common-field farming, and a farming system which for centuries was based on 'sheep and corn husbandry'; that is a system of grain production made possible by the use of large sheep flocks which fed all day on the downland and by night were folded on the arable lands enriching the thin chalkland soils with their dung which until the nineteenth century remained the only easily available source of fertilizer. It was the large flocks of sheep which so impressed travellers, and it was the effect of their grazing which created and maintained the characteristic short turf and plant-rich sward of the downland, with its distinctive flora such as bird's-foot trefoil, cowslips, harebells, orchids, and the ubiquitous wild clematis or 'old man's beard'.

The Prehistoric Landscape

Another major characteristic of the downland landscape is its wealth of archaeo-logical features and the evidence which they provide for the long period of human habitation and land use. The imprint of prehistoric farmers upon the landscape can more easily be appreciated in the downlands than anywhere else in England, although recent archaeological work is increasingly demonstrating that this is only because so much of the chalk downs has remained under grass, while the evidence of early land use and settlement in more favoured sites on the sheltered and fertile lower land and valley bottoms has been obliterated by later occupation. No longer credible is the statement of H. J. Massingham in *English Downland*, published in 1936, that 'the chalk was for twenty centuries the princi-pal home of prehistoric man'. Nonetheless, the fact that so much evidence of human activity survives on the highest land shows the intense and prolonged impact which prehistoric settlers had made upon the development of the land-scape long before the arrival of the Romans in the first century AD.

Recent analysis of surviving pollen grains and mollusc shells indicates that during the long post-glacial period much of the chalk downlands was covered by deciduous forests, and that it was the slow clearance of this woodland cover by the earliest farmers, followed by erosion, which has created the shallow-soiled open grassland. The effect upon the downland landscape of the labours of Neolithic farmers from *c.*4500 BC, using stone axes and fire to make their slow and laborious inroads into the forest in order to create areas for pasture or culti-vation and to provide timber for fuel, shelter, and ceremonial structures, can hardly be exaggerated. Over many centuries, their toil was to produce a more dramatic effect than any subsequent human activity. The introduction of agri-culture and pastoral farming by the Neolithic settlers during the fourth millen-nium BC was followed by the first large-scale works in the landscape. These include the causewayed enclosures, which are roughly circular earthworks with multiple lines of ditches broken by frequent gaps and are generally regarded as communal ritual centres. The earliest phase of the Neolithic period is named after the causewayed enclosure at Windmill Hill near Avebury (Wiltshire), while the numerous other examples include Whitesheet and Knap Hill (Wiltshire), the Trundle (West Sussex), Whitehawk and Offham (East Sussex), and those under-lying the Iron Age hill forts at Maiden Castle and Hambledon Hill (Dorset).

The Neolithic need for good-quality flints to make cutting tools of all sorts led to the development of the flint mines on the Sussex downs, at Blackpatch, Church Hill, Findon, and most notably at Cissbury near Worthing, where the 50-foot deep shafts of more than 200 mines with their associated galleries, provide dramatic evidence of Neolithic energy and enterprise. Also from the Neolithic period are the distinctive long barrows such as West Kennet near Avebury, Pimperne near Blandford (Dorset), Badshot (Surrey), Chilham (Kent), Wayland's Smithy (Berkshire), those at Danebury and Toyd Down (Hampshire), and between the Adur and Beachy Head (Sussex). The Neolithic people were also responsible for the strange 'cursus' earthworks such as the Dorset Cursus on Cranborne Chase which consists of two parallel banks about 70 yards apart and running for more than 6 miles from Thickthorn Down almost to Bokerley Dyke. This is one of the major monuments of prehistoric Britain although there is little agreement as to its purpose.

The massive 'henge' monuments which remain such prominent features of the landscape are also the work of the later Neolithic period, including Avebury, Marden, Woodhenge, Durrington Walls (Wiltshire), and Mount Pleasant and Maumbury Rings (Dorset), as well as the first phase of the complex stone circle at Stonehenge, the most celebrated of all prehistoric monuments. Whatever the precise meaning and purpose of these impressive sites, it is certain that the people who could undertake their construction had an extraordinary degree of organization as well as motivation or compulsion to complete the huge labour involved.

From c.2000 BC the Bronze Age people continued and extended the farming activities of their Neolithic predecessors, and intensified the onslaught upon the woodland cover of the landscape. Their most prominent memorials right across the downland are the many thousands of surviving round barrows in which they buried their dead, often sited on high downland overlooking the valleys, from which they can be seen silhouetted against the skyline. Throughout the Bronze Age the monuments at Stonehenge and Avebury continued to be developed, while at Silbury Hill the largest prehistoric mound in Europe was constructed, providing further evidence of the sophisticated level of organization and authority which the society had achieved.

Less dramatic but hardly less significant is the evidence which survives on the downland for the farming activities of the later Bronze Age from c.1500 BC. Numerous settlement sites and farmsteads, as well as the traces of large field systems, and long linear banks and ditches or 'ranch' boundaries enclosing very large areas of downland, possibly marking areas of arable and pasture land, survive and reveal an intensive use of the whole landscape. The detailed excavation and examination of the surrounding area of sites such as Black Patch near Lewes on the East Sussex downland, Shearplace Hill on the downland above Sydling St Nicholas (Dorset), and Rams Hill within an Iron Age hill fort on the Berkshire Downs, have shown that large areas of downland were being farmed, and that sheep were already very important in the economy.

The Iron Age, extending from c.700 BC to the Roman invasion of AD 43, saw the continuation and intensification all over the downland of trends which were already apparent in the Bronze Age. The most characteristic and impressive monuments of the period are the Iron Age hill forts, the product of fierce local rivalries and growing instability, with prodigious defences demonstrating an increasingly complex system of construction. They were created by the back-breaking toil of successive generations of prehistoric men and women, using the most rudimentary tools. Many of the larger hill forts remain as well-known and dramatic features of the landscape, such as Maiden Castle and Hambledon Hill in Dorset, Yarnbury, Barbury, and Bratton Castle in Wiltshire, Blewburton in Berkshire, Danebury in Hampshire, and the Trundle, Chanctonbury Ring, and Cissbury in Sussex. But in addition to these spectacular examples with their huge fortifications, scores of smaller hill forts are scattered all over the downland, many of them having much less impressive defences, and possibly intended only as stock enclosures.

The number, complexity, distribution, and defences of the hill forts provide abundant evidence for the sophisticated political and military organization of the period, as well as for its warlike nature. Less obvious are the many farmsteads of the Iron Age which consisted of round wooden houses within roughly circular enclosures, with storage pits for grain, and surrounded by small arable fields and extensive areas of downland pasture. These provide evidence of rising population and increasing pressure upon available land. Examples of early field systems are found all over the chalklands, and include Little Woodbury near Salisbury, Farley Mount, and Chilcomb Down (Hampshire), Gussage All Saints (Dorset), Park Brow (Sussex), and a great many others. All the evidence from

Maiden Castle, near Dorchester

The numerous Iron Age hill forts across the downlands reveal the military rivalries and inter-tribal disputes of the pre-Roman period, and obviously involved immense labour and organization. Maiden Castle is the largest, occupying 47 acres, with formidable defences, although they failed to withstand the disciplined onslaughts of a Roman army.

modern archaeological research points to the fact that the downland was thickly settled and intensively used before the Roman Conquest in AD 43 and that the landscape had already been substantially altered, extensively divided, and organized into units.

The Roman Occupation

The well-known effects of the Roman conquest and four centuries of rule in Britain have left numerous features which are still evident in the landscape of the downlands. In addition to the roads, towns, villas, amphitheatres, and aqueducts resulting from Roman occupation came also the suppression of the internecine strife among the Iron Age tribes, the introduction of markets, efficient distribution systems, and improved farming methods and a greatly increased demand for foodstuffs to feed the army of occupation and to export to other parts of the Empire. As a result the downland landscape was more extensively used for arable farming than at any subsequent period until the twentieth century.

The downlands also preserve evidence of the speed and efficiency of the Roman conquest. The disciplined skill of the Roman army in overcoming the formidable defences of the hill forts was demonstrated by the rapidity of their advance across southern England; the fall of the hill forts through the downlands of Hampshire, the battle cemetery at Spetisbury hill fort above the Stour in Dorset, and the overrun defences at nearby Hod Hill, all bear impressive witness to the slaughter inflicted by the Roman army. Most dramatic of all is the evidence from Maiden Castle where the hastily contrived cemetery at the east gate contained the graves of twenty-eight native warriors, many with terrible injuries, and one with the head of a Roman ballista bolt still fixed in his backbone.

In few other places is the observer made more aware of the ruthless efficiency of the Roman army than at Maiden Castle, where in spite of the immense fortifications which are still so impressive after nearly 2,000 years, the disciplined power of the Roman soldiers rapidly overwhelmed the defenders. The principal effect of the Roman occupation upon the downland was to accelerate trends which were already evident during the later Iron Age. Cultivation continued to be extended, and the so-called 'Celtic' fields reached their fullest extent. These are the small square or rectangular fields used for prehistoric arable farming, the relics of which survive over many thousands of acres of downland in spite of enormous destruction by nineteenth- and twentieth-century agriculture. Remaining examples can still be seen on Overton and Fyfield Downs near Marlborough (Wiltshire), at Nutwood Down and Streatley Warren (Berkshire), Pertwood Down (Wiltshire), above Sydling St Nicholas (Dorset), and on many parts of the South Downs in Sussex.

No doubt many of these farms were subordinate to the fine villa residences situated in the valleys. The villas generally avoided the bleaker parts of the downland and were in any case most tightly grouped around the Roman towns. They are absent from the central chalklands of Salisbury Plain and the Hampshire and Sussex Downs, or the still thickly wooded downland of Kent, although many villa owners favoured the fertile lower slopes or fertile areas of the Upper Greensand around the North and South Downs, the Kent foothills, or along the valleys of the chalkland streams in Dorset. Notable Sussex examples are the villas at Bignor,

Southwick, West Blatchington, Angmering, and Wiggonholt, others include the Dorset villas at Hinton St Mary, Tarrant Hinton, Frampton, and Maiden Newton, or at Downton (Wiltshire), Bramdean, and Rockbourne (Hampshire).

The farming associated with the large estates dependent upon many of the Roman villas was in advance of the traditional native practices, using improved ploughs, better facilities for corn drying and storage, more efficient mills, and a complex system of marketing and distribution. But many native farmsteads continued to exist on the downland, often on or near the site of Iron Age farms. An example is the settlement on Berwick Down near Tollard Royal (Hampshire) which continued throughout the Roman period. Another site is Chalton on the expanse of Hampshire downland between Petersfield and Portsmouth where a late Iron Age settlement developed and prospered during the Roman occupation, growing to the size of a small village. Here the area of cultivated land was extended greatly during the Roman period, covering much of the surrounding downland.

A similar small village developed on the Wiltshire downland at Chisenbury near Great Bedwyn and eventually consisted of some eighty huts laid out along a village street and surrounded by a carefully laid out field system covering more than 200 acres. Other native farmsteads have been excavated on the downland of East Sussex at Park Brow and Bullock Down, and at Gussage All Saints, Woodcutts, and Rotherley in east Dorset. By the end of the Roman occupation such farmsteads were thickly spread across the downland, often less than a mile apart, providing evidence of an intensively farmed landscape, and displaying a few sophisticated Roman features such as masonry, tiles, painted plasterwork, and window glass.

The Anglo-Saxon Invasions

The five centuries following the departure of the Roman legions, the so-called 'Dark Ages' were to have a profound effect on settlements, boundaries, defensive earthworks, administrative units, religious structures, ecclesiastical divisions, field-systems, and all other aspects of the downland landscape. The Jutish invaders of Kent and the Saxons who settled further to the west arrived sporadically and penetrated inland only slowly, generally settling on the richer lands in the river valleys or lower slopes, avoiding the high downland where Romano-British farmsteads continued to exist. One effect of the invasions on the downland landscape was the construction of major defensive barriers or territorial boundaries in an ultimately unsuccessful attempt to repel the newcomers. The most remarkable and visible of these long linear frontiers is the Wansdyke, the eastern part of which runs for 12 miles across the Wiltshire downs and dates from the fifth century. The massive earthworks face north and were no doubt intended to deter invasion by Saxons who had penetrated up the Thames valley and into Berkshire. Also dating from the fifth century are the massive earthworks of Grim's Bank at Aldermaston (Berkshire) which attempted to protect Silchester and its environs. More successful in delaying the Saxon advance into Dorset were Bokerley Dyke on Cranborne Chase and Combs Ditch on the Dorset downs, which between them checked the advance of the invaders until the mid-seventh century.

By far the most important contribution of the Saxon invaders to the development of the downland landscape came not in the form of the farms and hamlets which they created, nor in their numerous barrow cemeteries, but in the precise boundaries and landscape divisions which they established. The conversion of the Saxons to Christianity during the sixth century led in the seventh and eighth centuries to an outburst of Christian piety, the building of parish churches, and the founding of numerous monasteries with large endowments of land. The charters confirming these endowments and setting out the boundaries of the lands are among the earliest and most informative of all documentary sources and provide a mass of detail about the Saxon landscape. In many places the boundary marks can still be recognized today, and it is evident that the Saxons had taken over the existing boundaries of villa estates and Romano-British farmsteads, and that they had also developed further the complex patterns whereby territories were arranged in long narrow strips stretching from a river valley up to the high downland, thus ensuring that each community had a share of meadow land in the valley bottom, arable land on the fertile greensand or gravel terraces of the lower slopes and access to downland grazing for their sheep flocks. Such long narrow units are common throughout the chalkland. The twenty parishes along the dip slope of the North Downs between Croydon and Guildford consist of long strips of land no more than 2 miles wide but as much as 5 miles long incorporating the whole series of geological formations from the London clay to the chalk downs. Some parishes such as Godstone and Tandridge on the southern side of the downland east of Redhill are less than a mile wide and more than 7 miles long. Likewise along the northern escarpment of Salisbury Plain in Wiltshire from Westbury through Bratton and Edington to West Lavington the carefully planned layout of each parish is evident from their long narrow shape running from the clay vale up through the fertile greensand to the high chalk downland. East Coulston, for example, is half a mile wide and nearly 5 miles long, rising from 150 feet above sea level up to 700 feet on the downs. Many of these downland parishes were very large, often stretching for several miles across the open countryside. In contrast, the parishes of the Kentish downlands remained small, many of less than 1,500 acres. These originated as subsidiary settlements or isolated farms probably founded in clearings within the downland forest by the younger sons of families from the richer lowlands, or when permanent habitations were established on summer pastures. Many have the place-name element *stead* meaning 'stock-farm', and tiny settlements, scattered farmsteads, and small isolated churches remain a distinctive feature of the Kentish downland landscape. Colonization and the creation of subsidiary settlements with dependent chapelries were fostered by the continuance of partible inheritance, and by the difficult, infertile terrain of the high downland which Alan Everitt has described as 'mantled with that terrible clay-with-flints which has been the despair of so many generations of Kentish farmers'. Much of the Kentish downland remains heavily wooded and quite unlike the wide vistas and smoothly contoured landscapes of the classic chalk downland.

Parish and estate boundaries have profoundly influenced the development of the downland landscape and settlements and are a lasting legacy of the Saxon period. But the careful division and allocation of resources did not end there. The evidence of Saxon charters and later of the Domesday Survey and of medieval records shows that care was also taken to arrange that the numerous downland manors which lacked supplies of timber had access to woodland elsewhere, or that necessary land for grazing and swine forage was available even though it was a considerable distance away. Such an arrangement of linked estates, providing access to essential natural resources, is to be found throughout the chalklands from Kent to Dorset and offers further proof of the complex administrative systems and carefully planned boundaries which remained as the legacy of the Saxon period. In Kent, downland estates with well-established rights in the distant swine pastures or 'dens' of the Weald can be identified in charters from the eighth century. In Sussex a charter of *c*.675 relating to Stanmer on the downland near Brighton shows that the farmers there also possessed pasture rights some 20 miles away in the wooded region around Lindfield. The men of Kingston Buci beneath the Iron Age fort of Thundersbarrow in West Sussex had rights in the Wealden parish of Shermansbury more than 7 miles away. Evidence from charters and the Domesday Survey likewise show that the tenants of the royal estate of Wilton in south Wiltshire enjoyed rights to timber and pasture in the forest of Melchet 15 miles distant; while the tenants of Overton in the valley of the Hampshire Test had rights in the woodland at Tadley several miles away in the forest of Pamber. Throughout the Middle Ages the tenants of the royal manor of Fordington near Dorchester (Dorset) enjoyed rights of timber and pasture at Hermitage a dozen miles off in the forest of Blackmoor, and similar rights of access to distant resources can be found throughout the downland region.

The Middle Ages

The Domesday Survey of 1086 provides an invaluable view of the settlement pattern on the downlands. Most settlements were strung out along the valley bottoms where the chalkland streams provided a source of water, or were situated along the escarpments beneath the downs where the springs emerged and where the upper greensand and lower chalk formations provided a fertile arable soil. In Surrey the Domesday Survey records a line of prosperous villages along the northern edge of the North Downs with each settlement extending northward on to the London clay and southward up on to the chalk downs. Similarly in Berkshire and Wiltshire a line of settlement occupied the northward-facing escarpment of the Berkshire Downs and Salisbury Plain. In Berkshire, Hampshire, Wiltshire, and Dorset other settlements were also strung out along the river valleys, for example, the Lambourn and Pang in Berkshire, the Test, Itchen, and Meon in Hampshire, the Avon, Nadder, Wylye, and Ebble converging on Salisbury in Wiltshire and the Cerne, Piddle, Tarrant, and Gussage in Dorset. In contrast, most of the high downland was empty of settlement, and was used for feeding the large sheep flocks which were now settled as a notable feature of downland farming, providing the dung of the sheepfold for the arable land as well as mutton and the wool which was the essential raw material for the most important industrial activity of the region.

The Domesday Survey also shows the continuing domination of the downlands by large landowners and great estates, a feature which was to remain a major characteristic. The lands of the Crown were widespread as were those of the Church: the ecclesiastical estates included those of the wealthy monastic houses, Bishop Odo of Bayeux's enormous estate centred on Minster (Kent), the

The South Downs

The unspoilt beauty of the downlands with their dramatic escarpments and wide views can be appreciated in many places through Sussex and Hampshire. The long vistas and contrasting scenery were features which impressed travellers such as the naturalist, Gilbert White of Selborne, during the eighteenth century.

Archbishop's estate of South Malling stretching from the South Downs across the Weald, and the bishop of Winchester's estates north and south of Farnham. The crown lands included the vast areas designated as royal forests, of which several extended on to parts of the chalk downlands. Numerous deer parks were created by the nobility, clergy, and gentry; and since formidable fences were required to prevent the escape of the agile deer, the deer parks have left a considerable imprint upon the landscape. Harbin's Park in Tarrant Gunville parish near Blandford Forum (Dorset) covers 115 acres and is still completely enclosed by a massive earthen bank with an internal ditch; the huge park surrounding the former royal palace at Clarendon near Salisbury was more than 3 miles across; and the bank along the south side of John of Gaunt's deer park at Kings Somborne (Hampshire) is still 12 feet high. The bank and ditch which surrounded the 145-acre park of Milton Abbey (Dorset) is still well preserved in the thick woodland of an isolated valley east of the former monastery, while the boundary bank of the deer park at Hampstead Marshall near Newbury can be traced for several miles. The creation and maintenance of these fences imposed a heavy burden upon the tenants of many downland manors.

During the later Middle Ages many of the deer parks also included rabbit warrens as rabbits were increasingly valued both for their flesh and their fur, and were carefully nurtured in artificially created burrows. These burrows or 'pillow-mounds' can still be seen on many parts of the downland, for example at Ashey Down and Headon Warren on the Isle of Wight, at Longwood Warren near Winchester, and at Aldbourne and Liddington on the Marlborough Downs. The survival of so many 'warren' place names, throughout the downlands, especially in Hampshire and Berkshire, or of 'lodge' names as a reminder of the former warren-keepers' dwellings, provides further evidence of the importance of the rabbit in the downland economy. The medieval bishops of Winchester possessed several warrens on the Hampshire downland as well as at Swainston, Bowcombe, Thorley, and Arreton on the Isle of Wight, and a study of the lands of Quarr Abbey on the Isle of Wight has concluded that 'the very frequent documentary references to rabbits almost give one the impression that the Isle of Wight was one large warren'. At Aldbourne during the fifteenth century nearly half the revenue came from the sale of rabbits.

By the twelfth century much of the arable land on the lower slopes of the downs was being farmed in open fields within which the tenants had their holdings in scattered strips, a complex system of communal farming which was to survive in many parts of the downland until the parliamentary enclosures of the eighteenth and nineteenth centuries. Many of these field-systems were operated on a two-course rotation whereby half the land was left fallow each year when its fertility could be restored by the dung of the sheepfold. The large sheep flocks were to remain an essential characteristic feature of the downlands, together with the continuing dominance of the great estates, the survival of strong manorial control, large demesne farms, extensive labour services, and husbandry based on the twin pillars of sheep and corn.

The surviving records of the great ecclesiastical estates show that throughout all the chalklands the sheep flocks were carefully regulated with highly organized systems of breeding, grazing, folding, and transfer between different manors. The sheep/corn husbandry of the downlands was dominated by such large estates, while a huge social divide separated the wealthy landowners or institutions from their villein tenants owing heavy labour services. The system also relied on a large number of landless labourers.

All across the downlands there is abundant evidence of the population growth, agricultural expansion, and pressures upon the land during the thirteenth century. Among the commonest of all man-made features in the chalkland hillsides are the *strip lynchets* or cultivation terraces. These steep hillside terraces were fashioned by the immense labour of countless farmers, and they provide dramatic evidence of population pressure and increased demand for food. Evidence of thirteenth-century land hunger and the expansion can also be seen in the *assarts* or lands newly reclaimed for cultivation which were created around the borders of the chalklands, and of which details are provided in the exceptionally full documentation of the great ecclesiastical estates. At Whiteparish (Wiltshire) on the edge of the chalklands during the thirteenth century successive generations of peasant farmers made constant inroads upon woodland and waste ground to bring more land under the plough; at Downton (Wiltshire) the Bishop of Winchester created two new farms at Timberhill and Loosehanger, while on his manor of Wargrave (Berkshire) nearly 1,000 acres were brought into cultivation during the first half of the thirteenth century, and a further 680 acres between 1256 and 1306. Numerous new farms were created on the Bishop of Winchester's manors of Merdon, Burghclere, Highclere, and Woodhay in north Hampshire, and at Woodhay the increased number of tenants raised the bishop's rent revenue from £4 in 1209 to £29 in 1255. Waverley Abbey in Surrey paid fines of £667 during the period 1171–81 for making encroachments on the royal forest, and assarts totalling 184 acres were made into the royal forest on the downland edge at Hermitage (Dorset) in 1314. All around the downlands in the thirteenth century the same process can be discerned in the surviving documents. The great barns built by the monastic houses also provide impressive evidence of wealth and productivity, as the barn at Great Coxwell on the edge of the Berkshire downland which belonged to the Cistercian abbey of Beaulieu (Hampshire) demonstrates; it is 152 feet long and 51 feet high, with a noble roof covered with stone tiles. The barn at Alciston (East Sussex) which belonged to Battle Abbey is 170 feet long; Shaftesbury Abbey's barn at Tisbury (Wiltshire) is nearly 200 feet long; the abbey barn at Abbotsbury (Dorset) was originally 272 feet long, and the barn at Boxley (Kent) which belonged to the Cistercian abbey is 186 feet long. The need for such huge barns to store their produce shows the efficiency of monastic farming on the downlands.

The contraction and desertion of settlements and the retreat from marginal lands which occurred as a result of the decline in population during the fourteenth century following a series of poor harvests, famines, and plagues, is also very evident in the downland landscape. The most dramatic manifestation is the deserted village of which many examples can be seen on the chalklands, in places where exposed positions and thin soils with marginal viability forced farmers to abandon their land. Families whose enterprising predecessors had established settlements on the marginal soils of the high downlands during the thirteenth century either succumbed to famine and plague or gradually withdrew to occupy vacant farms in more fertile situations. For example, the thirty-five households on the Bishop of Winchester's manor of Northington on the high downland near Overton (Hampshire) in the mid-fourteenth century had shrunk to four tenants by 1485, and in that year a lease was granted to one husbandman, William Aycliffe, who occupied the whole of the now deserted village and its lands. Even in the more favourable situations and richer soils of the chalkland valleys there is evidence of fourteenth- and fifteenth-century population decline. Much recent research has demonstrated that the process of contraction and desertion was frequently long drawn out and complex, and that many settlements continued to exist into the fifteenth and sixteenth centuries although greatly depleted or reduced to a few cottages or a single farmstead.

As well as the earthworks of abandoned houses and farms, further evidence of late-medieval population decline is provided by the number of downland churches which are either ruined or greatly reduced in size or which now stand alone amid the relict features of the houses they once served. A study of the Kent downlands concludes that churches 'were at one time substantially more numerous than they are now; for many failed to survive the demographic and economic decline of the later Middle Ages'. Many of those that have survived stand isolated or are close to a manor house but remote from other dwellings. In Sussex the church at Lullington (near Beachy Head) consists of no more than the chancel of the original church, nearby at Alciston the chancel has been shortened, at Botolphs, south of Steyning, the former thirteenth-century aisle has been demolished and similar reductions occurred in Wiltshire, at Upavon, Sutton Veny, and Chitterne. The parish of Shrewton on Salisbury Plain consisted of several medieval settlements, Shrewton, Elston, Maddington, and Rollestone, each with its own church. Today Elston church has gone and the others have all contracted or are no longer used. Throughout the later Middle Ages the great landed estates continued to dominate the landscape of the downlands. As well as the widespread royal and ecclesiastical landholdings some secular estates were also vast. By the fifteenth century the lands of the Fitzalans, Earls of Arundel, com-

Ashdown House, near Lambourn, Berkshire

Built during the 1660s by the Earl of Craven, and intended as a refuge for Elizabeth, Queen of Bohemia. Unusually for a major house, it is built of chalk blocks with ashlar dressings. The copper-covered cupola gives views across the Berkshire Downs and the Vale of the White Horse.

prised sixty-four manors, as well as their castles at Arundel, Lewes, and Reigate; the lands of the Hungerfords stretched right across Salisbury Plain from their castle at Farleigh Hungerford on the borders of Wiltshire and Somerset, and are still recognizable from the parish churches they rebuilt and the chantries they founded, each decorated with the Hungerford badge of a sickle.

The Post-medieval Landscape

The abundant documentation surviving from the sixteenth century reveals more detail of changes in the landscape, the pressures and personalities involved, and also frequently records the final disappearance, or major contraction, of settlements whose slow decline had begun with the catastrophes of the fourteenth century. Several examples show the way in which landlords increased their income by converting their estates to large-scale sheep farming, leading to the evictions of tenants from their holdings and the consequent desertion of downland settlements. In 1521 the tenants of Sir William Fyllol in the settlements along the Winterborne valley south of Dorchester complained to the Court of Star Chamber that their lands were overrun by his sheep and that they would not be able to pay their rent 'nor be able to Abide in their countrey by cause of the saide great oppressions'. It is clear that their pleas were in vain, for all along the valley there is a string of deserted village sites—Winterborne Ashton, Herringston, Farringdon, Germayne, Came, and Whitcombe. It was of these places, and particularly of Winterborne Farringdon, that the Dorset topographer, Thomas Gerard, wrote in c.1630, describing 'a lone Church for there is hardlie any house left in the Parish, such of late hath beene the Covetousness of some private Men, that to increase their Demesnes have depopulated whole parishes'. At Iwerne Courtnay, a chalkland manor near Blandford Forum which belonged to the Earl of Devon, a manorial survey of 1553 records that the arable lands had been enclosed in 1548 and that as a result half the tenants had left, thus greatly increasing the size of the remaining holdings. 'The custumarye tenants were so small and so lyttle lande longinge to them that the tenants were not able to pay the lordes rent, but the one half of them departed the towne, … and then every tenant inclosed his owne landes and … occupyed his grounde severall to hymself …' Similar piecemeal enclosures of common arable fields on the lower slopes of the chalk downs occurred in Wiltshire, Hampshire, and Sussex, and, for example, by 1640 twelve out of the forty-two Sussex downland parishes were fully enclosed.

Often the fate of a settlement is apparent from the history of the parish church. Sutton, near Seaford (Sussex), was evidently very badly affected by the adversities of the fourteenth century, and its population declined dramatically; by the early fifteenth century the church was in a bad state and in 1509 it was reported that 'the church is utterly destroyed and [there are] no parishioners, save for a few neat-herds'. At West Blatchington, near Brighton only sixteen taxpayers were left in the parish in 1377–8, by 1428 no one remained and in 1596 the church was said to be unused. At Frome Whitfield (Dorset) the parish was almost completely deserted by the later sixteenth century, there was no resident parson and 'the church is filled up with hay and corne and is so far in decay that it is like to fall downe'. In 1608 the parish was united with Dorchester and the church was demolished. Elsewhere the system of common field agriculture remained unchanged and on much of Salisbury Plain and the Hampshire Downs the ancient pattern of open fields with their strips and furlongs continued. On the Surrey downland open fields and communal agriculture continued to operate around the spring-line villages of the downs escarpment from Leatherhead to Guildford including Ashtead, Headly, Fetcham, Great and Little Bookham, and East and West Clandon. The vast areas of open sheep grazing on the higher downland also continued to exist with little change, but as the pressure for arable crops increased so too grew the numbers of folding sheep, and the larger flocks meant ever more competition for downland grazing.

Friction between commoners competing for more grazing sometimes erupted into violence, or led to attempts to alter ancient customs, especially on the open downland where there were few obvious boundaries. At Charminster (Dorset) in 1616 the inhabitants made a cunning attempt to lay claim to some 50 acres of downland grazing attached to the neighbouring parish of Stratton. When a new vicar arrived at Charminster he led his first Rogationtide procession round the parish bounds, and the parishioners 'did forsake their old and wonted way and course … and did appoynt the said Curate to goe farther in upon Stratton by some fifty acres or thereabouts than they did before', giving rise to a long dispute before the Exchequer Court. Such quarrels illustrate the increasing value which farmers attached to their downland pastures and the pressure created by increased numbers of sheep.

It was essentially for the sheep that a major change in the downland landscape of the seventeenth century was made. This was the introduction and rapid spread of water meadows along the chalkland valleys. Water meadows were carefully constructed with a network of channels and drains, and with hatches built across the fast-flowing chalkland streams so as to bring the water over the surface of the meadow, covering the grass with a shallow, rapidly moving blanket of water. Operated during the winter months, this protected the grass from frost, deposited valuable nutrients around the roots, and stimulated a much earlier growth than would have occurred naturally, thus providing feed for ewes and lambs during the early spring, the most difficult period of the year for livestock. Later in the year, the water meadows also produced much more abundant hay crops than would otherwise have been obtained. The streams of the chalkland valleys were ideal for the creation of water meadows, and from their first appearance early in the seventeenth century the idea spread very rapidly, so that in spite of the large capital cost of laying out a meadow and constructing the necessary hatches and channels, within a few decades water meadows were to be found along all the chalk streams. In 1669 the agricultural observer, John Worlidge of Petersfield (Hampshire) could describe the watering of the meadows as 'one of the most universal and advantageous improvements in England within these few years'. The water meadows continued in use throughout the eighteenth and nineteenth centuries, and were of vital importance in the agriculture of the chalkland districts, enabling farmers

Farm buildings at Hurstbourne Tarrant, near Andover

These timber-framed and thatched barns are typical of seventeenth- and eighteenth-century farmsteads of the chalklands where good-quality building stone was not available. The walls of the house and barns are a mixture of timber, flint, and brick. Typical also is the timber-framed granary built on staddle-stones to protect the stored corn from rats and mice.

to keep larger sheep flocks and so enrich more of their cornland by use of the sheepfold than would otherwise have been possible. Writing in 1794 Thomas Davis, the steward of the Longleat estate in Wiltshire, described the value of water-meadows as 'almost incalculable', and concluded that 'the water-meadows of Wiltshire and the neighbouring counties, are a branch of husbandry that can never be too much recommended'. Most of the water-meadows went out of use from the late nineteenth century because of the long agricultural depression, the high labour costs of maintaining and operating the meadows, and the availability of alternative sources of early spring feed for livestock. But throughout the chalk-land valleys the relict features of the water-meadows survive over many thousands of acres and the remains of the hatches, channels, and drains can be easily recognized, although very few are still in operation.

The downlands continued to be dominated by large estates, employing numerous landless labourers, with much larger farms than were to be found in the adjacent clay vales. It was from cereals that most profit was derived, and these could be most efficiently produced on large farms. This situation was already clear by the sixteenth century. For example William Poore of Longstock on the Hampshire downs had over 200 acres of arable land and more than 1,000 sheep in 1590; Robert Loder of Harwell on the Berkshire downs, whose account book survives for 1610–20, concentrated on the production of wheat and barley, sending large quantities of malting barley down the Thames to Reading and London; Robert Wansborough who farmed at Shrewton on Salisbury Plain also kept an account book during the 1630s, which shows that it was from corn-growing that his main profit came and that his large sheep flock was an essential and highly valued element in his farming practice. This emphasis on the large-scale cultivation of wheat and barley, made possible by the use of the sheepfold, was to remain a major characteristic of downland farming. Writing of Sussex in 1808 the Revd Arthur Young commented on the fact that several downland parishes were almost completely occupied by single farms.

The Transformation of the Downland Landscape

Although many of the arable fields on the lower slopes had already been enclosed, much of the high downland in Berkshire, Wiltshire, Dorset, Hampshire, and Sussex remained entirely open and unenclosed at the end of the eighteenth century. The vast uninterrupted vistas of the downland, the absence of houses and of clearly marked roads, the archaeological remains, and the great flocks of grazing sheep were the features which most impressed travellers. The diarist Samuel Pepys recorded how he and his party got lost on Salisbury Plain in 1668, even with a guide, and how they also experienced great difficulty in finding their way across the Berkshire Downs. Early in the eighteenth century Daniel Defoe described the 'fine carpet ground' of the South Downs, the 'delicious' Surrey Downs around Epsom with their widespread views, the open downland of Hampshire, and the vast expanses of Salisbury Plain with 'neither house nor town in view all the way'. Defoe was obliged to rely on the directions given by the numerous shepherds he encountered for finding his way across the Plain, and was greatly struck by the contrast between the 'wild and uninhabited' downs and the chalkland valleys, 'the most pleasant and fertile country in England'. Defoe also commented upon the extension of arable land on the downs of Hampshire and Wiltshire and on the excellent crops of wheat which were grown by intensive folding of the large sheep flocks:

> the vast flocks of sheep, which one every where sees upon these downs, and the great number of those flocks, is a sight truly worth observation; 'tis ordinary for those flocks to contain from 3 to 5000 in a flock; and several private farmers hereabouts have two or three such flocks.

In 1768 Arthur Young crossed Salisbury Plain and wrote that 'in twenty miles I met with only one habitation, which was a hut'. John Claridge in his *Survey of Dorset Agriculture* of 1793 wrote that 'The most striking feature of the County is the open and uninclosed parts, covered by numerous flocks of sheep, scattered over the Downs'. But within the next few years all over the downland the scene was to be transformed and much of the 'open and uninclosed parts' were divided and separated by hedges, and had been converted to arable.

The impetus for this dramatic change of land use was rapid population growth and the high prices which could be obtained for arable crops during the Napoleonic Wars. The previously empty downland was enclosed, then ploughed, and new farmsteads were built. These well-built farmhouses, barns, and other buildings often bear names characteristic of the period such as Botany Bay, Quebec, Canada, Trafalgar, Waterloo, New Zealand, and California, or are called Old Down, New Down, or New Barn. On the Hampshire Downs, for example, more than half of the forty-eight enclosures by Acts of Parliament between 1700 and 1815 were passed during the Napoleonic War period, and farmers and labourers undertook the massive task of creating new large rectangular fields enclosed by hawthorn hedges, replacing the huge expanses of open downland. The newly cultivated fields were used to grow crops of wheat and barley or turnips, sainfoin, clover, and mangolds as feed for the sheep flocks, especially for the new breeds such as the Southdowns and Hampshire Downs which now became popular. In Dorset, of the fifty downland parishes with Enclosure Acts for the period after 1750, thirty included huge areas of downland while ten others were for downland alone. Some farmsteads established on the thin soils of the high downland under the impetus of high corn prices bear satirical or ironic names. Thus farms on the Kentish downland include Starvecrow, Starveacre, Owl's Castle, Rat's Castle, and Heart's Delight; while the Berkshire downland has several Starvealls, a Rook's Nest, and a Lonesome Farm, all indicating a similar origin.

Other parts of the downland landscape were also transformed during the eighteenth and early nineteenth centuries by the planting of very large numbers of trees, many of which survive as features of the modern landscape. A great deal of planting was carried out as part of the newly planned landscapes around the mansions of the gentry; elsewhere belts of woodland, predominantly of beech, were planted to provide shelter for livestock or for the new arable fields, while small, often circular coppices were planted on the downland in many districts to provide shelter for pheasants or for foxes as a result of the increasing popularity of large-scale shooting and fox-hunting. Other woods of coppiced hazel provided the spars which were essential for sheep hurdles and for use in thatching.

This was the period when the great estates were at the height of their power and influence, their lands extending over much of the downlands. Early in the eighteenth century Defoe had commented on the estates in Surrey, 'The ten miles

Foxhounds at The Vyne, Hampshire c.1930

The number and variety of the participants show the way in which fox-hunting remained a focus of interest for all classes in the rural community. It also had an effect on the landscape through the planting of small woodland coverts which remain a feature of many parts of the downlands.

from Guildford to Leatherhead make one continued line of gentlemens houses … their parks or gardens almost touching one another.' In Wiltshire the two vast estates of the Herberts of Wilton and the Thynnes of Longleat between them controlled much of Salisbury Plain, while the South Downs in Sussex were dominated by the widespread lands of the Duke of Richmond at Goodwood, the Earl of Egremont at Petworth, and the Duke of Norfolk at Arundel. It was landowners such as these who could afford to have their grounds landscaped by Lancelot Brown and Humphry Repton, who could hold sway over all aspects of life on their properties, and who could build the hospitals, almshouses, schools, and picturesque estate cottages which are still such a distinctive feature of many downland towns and villages.

The landscape, farming and society of the downlands during the 1820s were graphically described by William Cobbett during his *Rural Rides* which took him across much of the chalk region. Throughout his journeys he commented upon the numerous great estates and large mansions, and upon the excellence of much of the farming which he saw with fine cattle, immense flocks of sheep and well-grown crops of wheat and barley. In the Pewsey vale of Wiltshire he exclaimed, 'This is certainly the most delightful farming in the world. No ditches, no water-furrows, no drains, hardly any hedges, no dirt and mire, even in the wettest seasons of the year: and though the downs are naked and cold, the valleys are snugness itself.' Riding between Heytesbury and Warminster in Wiltshire, Cobbett declared that the downland provided 'everything that I delight in':

> Smooth and verdant downs in hills and valleys of endless variety as to height and depth and shape; rich corn-land, unencumbered by fences; meadows in due proportion, and those watered at pleasure; and, lastly, the homesteads, and villages, sheltered in winter and shaded in summer by lofty and beautiful trees; to which may be added roads never dirty and a stream never dry.

Like other commentators, Cobbett emphasized the crucial importance of the sheep flocks in the chalkland farming: 'The farms are all large … because *sheep* is one of the great things here; and sheep, in a country like this, must be kept in flocks, to be of any profit. The sheep principally manure the land. This is only to be done by folding …' In spite of his admiration for the farming of the downlands, however, Cobbett was also well aware of the polarization of society, for a few extremely wealthy estate owners, and tenant farmers with large farms, depended upon the toil of numerous landless labourers; and he was horrified by their long hours of work, poor housing, and very low wages. In the valley of the Salisbury Avon he wrote of his 'deep shame … at beholding the general *extreme poverty* of

Sheep Fair at Poundbury, Dorchester c.1900

Although the number of sheep was declining by the late nineteenth century, they remained important in the farming economy and sheep fairs continued to attract large numbers of buyers and sellers. Poundbury fair, like several other sheep and cattle fairs in the region, was held within the ramparts of an Iron Age hill fort. The sheep were confined within hurdles produced in the widespread coppiced woodlands.

those who cause this vale to produce such quantities of food and raiment. This is, I verily believe it, *the worst used labouring people upon the face of the earth.*'

Cobbett's indignation was echoed by many others, and the sober reports of parliamentary commissioners during the mid-nineteenth century testify to the fact that conditions were very slow to improve. Reporting on Dorset in 1867 a commissioner wrote that 'the cottages of this county are more ruinous and contain worse accommodation than those of any other county I have visited except Shropshire … [the labourers' cottages] are a disgrace to the owners of the land'.

Some landowners did make a valiant effort to improve the housing of their labourers, and nineteenth-century estate cottages remain a feature of the downland landscape, as do the fine farm buildings constructed by many landowners. Notable examples include Lord Wantage's farms and cottages on his Berkshire estates at Lockinge and Ardington, the cottages built by the Marquess of Bath and the Earl of Pembroke on their Wiltshire estates, the well-built and distinctive cottages of Lord Portman on his estates around Blandford Forum, and more than 600 cottages of the Earl of Shaftesbury on his estate around Wimborne St Giles (Dorset).

These changes in the landscape of the downlands after 1790 transformed its appearance more dramatically than at any time since the Neolithic destruction of the woodland cover. The characteristic short springy turf of the downs which for 2,000 years had been kept short and dense by the constant grazing of large sheep flocks was destroyed within a few decades in order to create large rectangular fields surrounded by quickset thorn hedges, linked by new straight roads. In much of the Kent and Surrey downland farms remained small; but elsewhere the corn-producing farms had traditionally been large, and enclosure was to lead to amalgamation and a further increase in farm size. On the Wiltshire chalklands between 1781 and 1831 the number of farms declined by 12.5 per cent, while farms of up to 5,000 acres were created. On the North Downs of Kent and Surrey, cultivation was extended in places to the summit of the chalkland, the multitude of flints thereby exposed being used by farmers to build new farmhouses and barns amid the newly enclosed arable fields. The increase in farm size, and the decline in the number of farms with all the social consequences which followed, were deplored by the Dorset poet, William Barnes:

> Then ten good dairies were a-ved,
> Along that water's winden bed,
> An' in the lewth [shelter] o' hills and wood,
> A half a score farm-housen stood
> But now—count all o'm how you would,
> So many less do hold the land,
> You'd vind but vive that still do stand,
> A-comen down vrom gramfer's.

William Cobbett also deplored the social effects of farm amalgamations, and gave as an example the parish of Burghclere in north Hampshire where 'one single farmer holds by lease, under Lord Carnarvon, as one farm, the lands that men now living can remember to have formed fourteen farms, bringing up in a respectable way fourteen families'.

The depopulation of downland parishes was accelerated during the later nineteenth century as labourers and their families escaped the poor conditions and low wages of farm work and either moved to the industrial towns or took advantage of the opportunity for emigration. The result was that many downland parishes had considerably fewer inhabitants in 1901 than they had in 1801.

The onslaught on the downlands, its enclosure, and the conversion of the grassland to arable, continued throughout much of the nineteenth century, and the process was hastened with the introduction of the steam plough during the 1860s. Writing of Dorset in the *Journal* of the Bath and West Society in 1861 a local farmer, Joseph Darby, emphasized the enormous changes which had occurred on the downs, 'the chalk hills formerly presented to view one vast sheet of downs … but what a change has been effected in their appearance!' He went on to describe the thousands of acres which had been broken up and enclosed for arable, and declared that 'the ancient landmarks are obliterated'.

The profitable sheep/corn farming system of the chalklands was severely affected by the agricultural depression which began during the late 1870s and which was to last with a few brief intermissions, until 1939. A run of cold, wet summers coincided with a sharp fall in the price of corn and meat as imports flooded into England. As a result, more and more land was laid down to grass, many farmers were unable to pay their rent and the income of estate owners fell dramatically. The retreat from arable farming, together with the availability of artificial fertilizers, brought to an end the long supremacy of the downland sheep flocks which were no longer required for folding on the cornlands. Only the production of milk remained profitable, since this could be transported by the new railways, and many downland farms became dependent upon their dairy herds. In Dorset, for example, the number of sheep fell from about 500,000 in 1895 to 200,000 in 1914 and to 46,000 in 1946; many hundreds of acres of water-meadow fell into disuse and became overgrown, farmers' incomes slumped, there were many bankruptcies, and between 1872 and 1914 the average rent of farm land fell by 29 per cent with a devastating effect upon the income of estate owners. A parliamentary commissioner reporting on Dorset farming in 1895 described the falling rents, the unlet farms, and the chalk downland abandoned to scrub and rabbits; he concluded that 'the ownership of land is rapidly becoming a luxury which only men possessing other sources of income can enjoy'. One alternative use for the downs was as a military training area, and military establishments have had a profound impact on the landscape of the downs during the twentieth century. By 1890 the army had already begun to use parts of Salisbury Plain as a training ground; the camp at Bulford, for example, grew from 341 persons in 1891 to over 4,000 in 1931 and a large village was created to house the soldiers. Didcot became a military barracks and Ordnance depot; Royal Air Force bases were built at Figheldean, Rollestone, Upavon, Chiseldon, and Tidworth, while in Hampshire aerodromes were constructed for the RAF at Abbots Ann and Worthy Down.

During the Second World War military installations and airfields stretched across the downs from Aldershot to Portland, and along much of the South Downs; the villages of Imber on Salisbury Plain and Tyneham on the Dorset coast were commandeered for military training and remain in military hands, as do large areas of Salisbury Plain.

The demand for home-grown food during the Second World War brought renewed profitability to downland farming, and led once more to the cultivation of large areas which had reverted to grass. Today much of the chalk downs which once supported the great flocks of sheep now yields huge annual crops of wheat and barley through the copious use of artificial fertilizers, pesticides, and fungicides and by the employment of elaborate powered machinery needing only a tiny proportion of the workforce once employed in agriculture, but demanding the sort of capital investment that is only possible for large firms and agricultural combines.

Continuing military use, coupled with all the changes in farming and the decline in sheep numbers in face of a vast extension of arable land on the downs during the later twentieth century mean that only a few areas of traditional grass-covered chalk downs remain unaltered. In a few places, however, it is still possible to find tracts of down with the characteristic short turf and rich catalogue of plant species still surviving. Among the most interesting examples are Box Hill Country Park on the edge of the North Downs in Surrey with unspoiled downland and magnificent views to the South Downs. Large areas of downland also remain on the Isle of Wight. Various stretches of the South Downs Way across Sussex provide excellent views and examples of undisturbed downland, particularly in East Sussex around Ditchling Beacon and the Cuckmere valley; Pewsey Down (Wiltshire) and the Queen Elizabeth Country Park near Petersfield in Hampshire both contain extensive open downland, the latter including a reconstructed Iron Age site. On parts of Salisbury Plain and around Tyneham on the Dorset coast military occupation has preserved the downland which remains rich in wild life and plants. At the Weald and Downland Museum situated on the South Downs at Singleton, near Chichester, houses and farm buildings from Kent, Sussex, Surrey, and Hampshire have been painstakingly re-erected, providing an excellent introduction to the building styles, farming, crafts, and history of the region.

Shepherd and sheep on Salisbury Plain 1901

One of the last survivors of the once-numerous shepherds whom travellers like Defoe found such an invaluable help in pointing out the route across the Plain. Note the unmetalled chalk road, the indispensable dog, and the shepherd's cloak which provided the traditional protection on the bleak downland.

2 Wolds
The Wolds Before *c.*1500

Harold Fox

In a thirteenth-century saga, written in Iceland, the saint Olaf is said to have landed at a Yorkshire harbour 'off the wold', showing that the Yorkshire Wolds were not unknown then among educated people living in a distant land. The twelfth-century Welshman Giraldus, topographer and historian, referred in one of his writings to 'the hill of Cotswold' assuming that his readers would be able to place the region in their mental map of the country. The concept of a wold as a distinctive type of region must also have been in the minds of less lettered men and women who dubbed places such as Wold Newton and Welton le Wold in Lincolnshire, Wold (now Dry) Drayton in Cambridgeshire, and Wold Dalby on the Nottinghamshire–Leicestershire border. The outlines, however imprecise, of other regions of wold may be reconstructed not from 'Wold' as an addition to a name, as in the examples above, but where it is fully incorporated into a place name, sometimes compounded with an element of some antiquity, for example Horninghold (once *Horningewald*, the wold of the people called the *Horningas*) in High Leicestershire, a rolling pastoral region stretching, so place names tell us, eastwards from the fringe of the Soar Valley towards Rutland. The same approach gives another wold region, High Northamptonshire, a great arc of countryside centred on the lowlands around Northampton and in the west spilling over into the hilly lands of the Warwickshire border which were, as Dugdale noted in the seventeenth century, 'called *Wouldes*, as many other of that kind are … in other counties'. Finally, straddling the western boundaries of Huntingdonshire and Bedfordshire is Bromswold, a region with several ancient names incorporating *wald* and others of the 'Leighton Bromswold' type; this was probably the remote region where the outlaw Hereward the Wake took refuge in the late eleventh century. A few other, very small, regions of wold are not much discussed in the first part of the chapter, which tries always to take examples from places in the cores of more extensive wolds (rather than from parishes which take in both wold and vale and which, thereby, have their own special characteristics).

The pastoral heritage

Ultimately our term 'wold' comes from an Old English word *wald* meaning 'wood'. Old English has a rich vocabulary to describe subtle differences in natural resources and there are hints, in the evidence of literary sources, place names, and Domesday Book, that the term was applied to regions of relatively lightly spread woodland. A stranger coming into a wold one evening in the seventh century or the eighth would have entered a wood-pasture (a term used in Domesday Book to describe the resources of some wolds manors), a landscape dominated by those two types of land use rather than by ploughland. He would have seen clumps of wood casting long shadows over the great open spaces and, everywhere on the pastures, domestic animals of all kinds. Now people come into view, the keepers of those animals returning to their summer dwellings as night closed in. Everywhere traces of older landscapes, all the stranger in the evening light, showed that people had been there before in times long distant even in the seventh century, had buried their dead there, and divided up the land. These marks the stock keepers did not understand, giving them names such as Grim's Hill, an Iron Age camp in Sevenhampton (Gloucestershire) or the 'Rood Stone' (modern Rudston on the Yorkshire Wolds), a Neolithic standing stone.

High Leicestershire, a lonely landscape

Slight remains of ancient woodland in valley bottoms; ridge and furrow possibly dating from fuller occupation before 1350; deserted village site of Lowesby (middle distance: hall in park) and few other villages in sight.

It is sometimes difficult to appreciate that wolds in the seventh and eighth centuries were relatively well wooded because the meaning of the term changed later on as their landscapes were modified by more intensive exploitation leading to loss of woodland. In the Cotswolds, studies by Della Hooke and Terry Slater, using the evidence of landscape features mentioned in records of perambulations of Anglo-Saxon estates, have shown that some woodland was once present in many (but not all) parts of the region. Again, Alan Harris, who made the landscape of the Yorkshire Wolds his favourite field for research, was convinced that as late as the seventeenth century some small patches of ancient woodland survived in that region. It must not be forgotten that those who gave the wolds their names were outsiders looking in, as from the Vale of Trent, already treeless in the eighth century, towards the steep rise of the wolds of the Nottinghamshire–Leicestershire border where ancient woodland survived simply because the land was, in places, too steep to cultivate easily; the same could be said of the scarp slopes of the Cotswolds and the Yorkshire Wolds. In the interiors of wold regions even quite small stands created wooded vistas, 'the eye forever being on the verge of a forest', to use William Marshall's words much later.

'With its fresh leaves, its young shoots, the grasses of the glades, the acorns and beech-nuts, woodland served above all as land for grazing.' So, evocatively, wrote the great rural and landscape historian Marc Bloch and, following him, a very good case may be made for a largely pastoral use of wolds in the early Anglo-Saxon centuries. Those who sent their animals into the wood-pastures were people from adjacent vales who had not yet evolved classic 'Midland' field systems with their great fallow fields for the safe-keeping of animals; all the more reason, therefore, for them to dispatch livestock into the fresh pastures of the wolds in springtime so that there was no risk of their trampling the growing crops at home. Next to Newton Bromswold (Northamptonshire) is a place, called Yelden today but once *Giveldene*, a name which relates it to the *Gifla*, a people mentioned in the eighth-century Tribal Hidage who lived in the Vale of the Ivel south of Biggleswade 20 miles away. In the Cotswolds part of the parish of Condicote (Gloucestershire) was a detached portion of the manor of Oddington, in the valley of the Evenlode 6 miles away, and Eycot (in Rendcomb) was a detached part of Bibury in the Coln valley 7 miles away; in the Lincolnshire Wolds the little vill of Enderby near Somersby had strong connections with Horncastle, in the valley of the Bain 6 miles to the west. These linkages strongly suggest that at one time the wolds were exploited by valley people at a distance. Trackways, determined in their direction and etched strongly into the landscape, linking vale and wold—such as those which reach eastwards from the Soar valley into High Leicestershire or the parallel tracks leading from the valley of the Ise into High Northamptonshire—were once droveways along which livestock were driven to the summer pastures. In the Cotswolds Herbert Finberg identified an ancient droveway, still called Greenway today, linking Badgeworth in the Vale of Gloucester to Coberley on the wolds, both properties of Gloucester Abbey, which was developing its sheep flocks in the eighth century.

Place names (some of them rather late, but no doubt recalling earlier practices) allow us to say a little more about how the wood-pastures of the wolds were used. In the Yorkshire Wolds there are several containing the Scandinavian element *aergi* which means a habitation occupied on a temporary basis, such as Argam, now a deserted village in the rolling countryside behind Flamborough

Head. *Skali* is another Scandinavian term meaning a place occupied on a seasonal basis (giving modern shieling) and is found in the Yorkshire Wolds, in Lincolnshire, and as far south as the wolds of Leicestershire and High Northamptonshire. These names all tell of the use of wold pastures on a seasonal basis and of seasonally occupied settlements. So too does the place name Somerby or the like ('place used in summer' according to one interpretation) in High Leicestershire and the Lincolnshire Wolds. The Cotswolds are too far west to have place names containing Scandinavian elements denoting seasonal use and one wonders if the Old English element *cot* was not sometimes used in that sense there. For example, Elmcote, once *Hulomonecote*, is 'the hillmens' cottage', Coldicote is 'the cold cottage', perhaps so named because of lack of a hearth. As for the people involved in summer migrations (transhumance) to the wolds there are suggestions in place names of a youthful and, in some cases, a female personnel. For example, Maidenwell is a high, bleak parish in the Lincolnshire Wolds, the Cotswolds have several names in Maidenhill, we find Childerley ('wood of the young men') in the wolds of Cambridgeshire and Chilcote ('humble buildings of the young men') in High Northamptonshire. These names should come as no surprise, because where movements to summer pastures survived until recently as in Scotland and Ireland, we know that young people were much involved.

In our snapshot of a region of wold in the seventh or eighth century there were few permanent settlements. Place names confirm this for the wolds contain few ancient names of the type which were given to seminal estates with a long history of occupation. There are some exceptions. For example in the southern Lincolnshire Wolds there are some place names of the types which linguists think were characteristic of early Anglo-Saxon England (certainly before the eighth century) and this is perhaps not surprising because here the hills are cut into by enticing valleys, in Joan Thirsk's words 'as sheltered and inviting to the present-day traveller as … to the early settlers'. Moreover, these southern wold fringes had deep-rooted connections with the fens to their south, still undrained in the Anglo-Saxon period and a highly attractive resource for everything from peat fuel to wildfowl. It is the same in Yorkshire, where ancient folk names such as Leavening and Gembling refer to territories which took in valuable low-lying land as well as some wold. We would expect some early names on the fringes of the wolds of these two counties, because the Wash and the Humber have long been considered as entry points for Anglian immigrants, in whatever numbers they may have come.

But in general ancient place names are rare, especially in deep wold countrysides. Their rarity does not point to a total absence of early permanent settlement: here and there isolated farms and hamlets may have been established by the eighth century. Yet the evidence for a good deal of woodland and open pasture on the wolds, not to mention their use as seasonal grazing grounds, suggests that permanent settlements were relatively sparsely spread in the seventh and eighth centuries, except in some favoured spots. At that time these regions were at a low point in one of those cycles of occupation followed by depopulation which they may well have witnessed before in prehistory, and which they were to witness once again in later centuries. It is important to stress this point about cycles made up of intensification of settlement and exploitation, followed by partial abandonment and less intensive use of the land, because some areas of wold yield many prehistoric features of all kinds and ages. These certainly speak of periods of intensive use in prehistory but they do not prove continuity of occupation over many millennia; that is unlikely in most cases, given the known cycles of settlement and partial depopulation which the wolds underwent in historic times. Nevertheless, to this generalization exceptions will always be found in favoured locations such as, perhaps, the sheltered and fertile valleys of the eastern Wolds of Yorkshire; in Burton Fleming and Rudston, writes Ian Stead, 'every year the ripening crops reveal an amazing range of ancient settlements, burials and ritual monuments'.

Developing landscapes, developing societies

The rarity of place names of the types given to seminal estates is contrasted with the presence of many elements frequently used in naming processes between the eighth and eleventh centuries. Examining Cotswold place names, both of parishes and of smaller places within parishes, one is struck by the way in which certain of these relatively young elements occur again and again. Among elements which denote habitations of some kind, *wic*, with pastoral connotations, recurs and so does *throp*, meaning a minor settlement; *cot* is very frequent and implies a relatively late and humble beginning for a settlement, perhaps in some cases, as already suggested, on a site once used on a seasonal basis.

The place names of the Cotswolds suggest, then, an intensification of permanent settlement between the eighth century and the eleventh. The same could be said of the wolds of eastern England where Scandinavian place names (e.g. names ending in *by* and *thorp*) which must date from later than the last decades of the ninth century, are noticeably densely distributed. Tennyson, who was born at Somersby on the Lincolnshire Wolds, was perhaps influenced by local nomenclature when he wrote of his brook flowing past 'twenty thorps | A little town | And half a hundred bridges'. A few of these Scandinavian names may possibly represent the renaming of pre-existing places but the most recently stated opinion, of Henry Loyn, is that pure Scandinavian names of types found so thickly spread on the wolds 'represent for the most part … settlements established by the Danes in the first full flood of their migrations'. A folk migration of people of Scandinavian origin (not necessarily in huge numbers) is likely because their native culture was strong enough to influence field names in subsequent centuries: for example, detailed study of these in Lincolnshire parishes around Wold Newton reveals many examples of names ending in *gata* (path), *garthr* (enclosure), and *dalr* (valley), all of which are Scandinavian dialect words. Discovery of a small pre-Scandinavian sunken hut at a place with a Scandinavian name at Salmonby (Lincolnshire Wolds) does not run counter to this argument in view of earlier seasonal settlement on the wolds.

By the end of the eleventh century the process of permanent settlement replacing seasonal use was already largely complete, for Domesday Book shows that most manorial centres in the wolds were in place with their ploughlands and ploughteams. We are first allowed to glimpse their two-field systems in twelfth-century charters. Even the Newtons (Cold Newton in High Leicestershire, Wold Newton in both Yorkshire and Lincolnshire, Naunton, a corrupt form, in the Cotswolds) were present in the landscape by 1086. A few wolds villages and field systems may have evolved during the twelfth century, as at Dry Drayton (once

Some patterns of spoke-like parish boundaries from (a) the Leicestershire Wolds, (b) the Lincolnshire Wolds

In each an area of intercommoning on wold pasture (around the hub) has eventually been divided among the participating parishes.

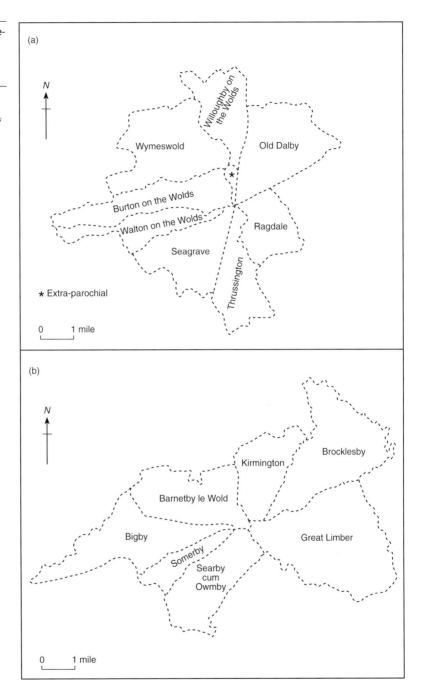

(a)

N

Willoughby on the Wolds

Wymeswold

Old Dalby

★

Burton on the Wolds

Walton on the Wolds

Ragdale

Seagrave

Thrussington

★ Extra-parochial

0   1 mile

(b)

N

Brocklesby

Kirmington

Barnetby le Wold

Bigby

Great Limber

Somerby

Searby cum Owmby

0   1 mile

Wold Drayton) in Cambridgeshire, for which there is a rare type of document from around 1150 describing the replanning of the fields. But most came into being earlier, as at Hawling (Gloucestershire) where detailed topographical and documentary research, by David Aldred and Christopher Dyer, shows that the nucleated village was probably in place by the tenth century. Most historians now agree that nucleated villages and field systems evolved together, a single cluster of farmsteads (replacing more dispersed settlement) being more appropriate for a township-wide common field system within which the strips of each peasant holding were widely scattered. Once they were formed, wolds villages and field systems were marked by two noteworthy characteristics. First, many wolds villages are on the small side. This could in part be a reflection of their evolution from relatively closely spaced sites once used as seasonal settlements in small, fragmented pastoral territories; on the wolds of eastern England small closely spaced sites may in part represent a Scandinavian cultural transmission, for in the ninth and tenth centuries hamlets and isolated farmsteads were the tradition of the homelands from which Scandinavian farmers migrated. Second, those who created the arable fields frequently left a significant acreage of rough pasture beyond the plough-land, way out near the township boundary, the gorse and whin covered grazings of the Yorkshire Wolds or the grassy 'slaits' of the Cotswolds (a local dialect term meaning sheep pasture). Many of these grazings are still named Wold (prefixed with the name of the villages they belonged to) but readers of the map must be aware that this is the term used in its more modern sense of 'open pasture'. In these cases the formation of field-systems with common grazing on the fallow cannot have resulted from absolute shortage of pasture, as I once believed; rather the two great prairie-like fields of a wolds village were created as safe compartments for livestock, keeping them at a suitable distance from the growing grain.

For the thirteenth century and early fourteenth centuries, we can provide a more focused snapshot of the landscapes of the wolds than was possible for earlier periods. A stranger, say a migrant harvest worker, entering a deep Cotswold parish for the first time, would have been struck by a certain monotony, by the bareness of it all. Absent were wide expanses of lush meadowland typical of the vales with the light greens of grass and willows and the broad shining waters of a wide stream beyond, absent the variegated colours of timber-framed houses. In the middle distance the stranger saw a small cluster of rather uniform peasant houses, the light browns of their walls (stone rubble at the base) and their thatch merging into the light brown of the bare fallow field beyond. And, beyond that, to the skyline, the remaining pasture of the village's wold.

By the thirteenth century we can begin to discern wolds as a *genus* of region. In all wolds most village communities based their husbandry on two-field rather than on three-field or more intensive systems. Indeed when Walter of Henley in the thirteenth century wrote of 'lands parted in two as they be in many countries' he may well have had wolds in mind (for 'countries' he used the Norman French word *pais*, modern French *pays*, i.e. a region within which the people share a common type of farm economy, culture of work and cultural landscape). Undemanding two-field systems reflected in part a wise reaction, by the farmers who formed them, to relatively unrewarding soils. But this had adverse effects on standards of living because, with half the land fallow each year, the acreage cropped for subsistence and income on, say, a 30-acre holding was less than that on a holding of similar size under three-field or more intensive systems. A chance reference from Bourton-on-the-Hill (Gloucestershire) shows that some peasant holdings there, vacant because tenants had died of plague, yielded miserably. A picture is beginning to emerge of a rural standard of living in wold regions which was below average in the thirteenth and early fourteenth centuries.

The two-field system of the wolds also reflected deep-rooted pastoral traditions for, with a larger area of fallow, it supported more livestock than a three-field system; in its adoption, by community after community, we can see the emergence of a common culture of work within each wold. Wolds were not in general dairying countries because valleys were usually short and narrow and meadow tended therefore to be in limited supply (there are Domesday references to 'small' or 'little' meadow from several Cotswold vills). Rather, the sheep was the principal type of livestock and the weights of fleeces from wold sheep were high, perhaps because of some localized expertise in flock management which we do not understand at this remove. The financial accounts of the de Lacy family's widely dispersed estate show high averages of 2.6 lb per fleece at Greetham in the Lincolnshire Wolds and of 2.3 lb at Thoresby in the same region, as compared with an estate average of 1.9 lb. The special qualities of wool from the wolds were widely appreciated by contemporaries. Already by the fourteenth century the word 'Cotswold' was used to describe wool from that region, and later its sheep were described as 'Cotswold lions' (presumably as a result of size and mane-like wool round their fronts and heads, not their ferocity). Again, Francesco Pega-lotti's early fourteenth-century treatise 'On the practice of trade', listing English religious houses and the price of the best wool to be had from each, gives 28 marks per sack for Stainfield Priory and 24 for Kirkstead Abbey, both with land on the Lincolnshire Wolds, as compared with an average of 18 marks.

Landlords, with their superior command over the pastures of the wolds, could make good profits from their large flocks, for example, the 420 sheep on the Archbishop of York's manor of Condicote (Cotswolds) in 1345, attended by two shepherds and other underlings, or the 500 sheep belonging to Meaux Abbey at Thwing on the Yorkshire Wolds. The position of peasant farmers was different. Stints restricting the size of peasant flocks, the ever-present tithe, the relatively small acreage of holdings, pressure in the thirteenth century to plough as much land as possible to feed an expanding English population: all of these factors tended to limit the size of an individual's flock. Studies of the taxation records of Wiltshire villages have revealed an average of about 15 sheep for each villager, and it is likely that the same was true of the wolds at this time. Sale of the fleeces of 15 sheep would yield about 9s. to a peasant farmer as compared with about 60s. in a good year which would be the value of the crops on a 30-acre holding (assuming a two-field system).

At this time sheep farming did not therefore yield much wealth to individual peasant farmers but rather to non-resident merchants who collected up large consignments of fleeces, like the members of seven Italian merchant companies which in 1275 exported about half of the wool leaving Hull to the south of the Yorkshire Wolds. Wool also sustained the lives of many, and enriched the lives of a few, in the great cloth-making towns of the thirteenth-century, many of which were close to wold countrysides. By this means wealth flowed from, and was not retained in, the wolds as is clear from taxation returns from Lincolnshire where the amount paid per acre by the people of the high wolds was one-third of that

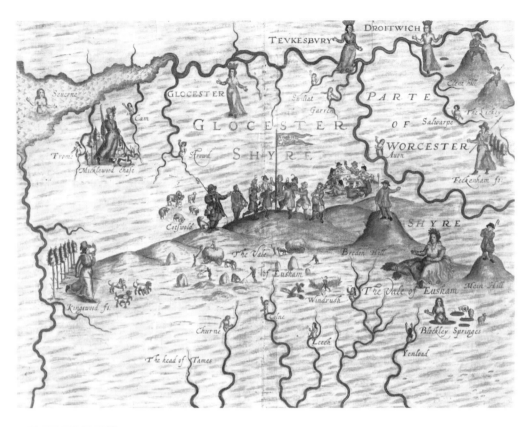

The West Midlands from Drayton's *Polyolbion* (1612)

Shown are the Vale of Evesham (grain harvest) and the Cotswolds (sheep). The high jinks may represent the Cotswold games, established (or re-established) by Robert Dover, a friend of Drayton.

paid by fenlanders to the south with their intensive farming systems, dairying enterprises and opportunities for earning in a diversified economy which included fishing, turf-cutting, and salt-making.

A lucrative, diversified economy was exactly what was not to be found on the wolds, with the exception of those rather few Cotswold parishes with a cloth industry; instead there was a certain monotony and austerity. Theirs were also rather isolated lives, because population densities tended to be relatively low by the standards of the thirteenth and early fourteenth centuries, and in the deep wolds towns must have seemed very far away (a sense still experienced by the traveller today). Population may have been maintained at a low level because of relative poverty. Another factor at work here, reflecting low standards of living, was emigration from the wolds, which has been studied in detail only for Leicestershire where David Postles has found that they were net losers of population in the early fourteenth century. Less comprehensive work on Lincolnshire at this time shows that people flocked from the wolds to the fenlands, to the great wool-exporting port of Boston and to nearby rural parishes with their diversified economies. Faced with an exceptional royal tax in 1342 inhabitants of several parishes in the wolds of Cambridgeshire complained of poverty, those of Longstowe adding that there were many empty houses, a sure sign of emigration; the same tax demand made some Cotswold villagers (at Naunton, Gloucestershire, for example) complain of a drain of people away from the countryside.

In a community prone to emigration there would have been little need for accommodation for sons waiting to take up a farm holding or for immigrants, so wold villages often lacked a large stock of landless cottages to serve those needs. Thirteenth-century evidence from the Cotswolds shows that at Aldsworth (Gloucestershire) the cottage ratio (landless cottage holdings to farm holdings) was 1 : 30 (excluding dwellings inhabited by widows), while at Duntisbourne Abbots (Gloucestershire) there were no landless cottagers at all to serve just over twenty peasant farmers, although a shepherd and one other tenant had only 4 acres between them. A relatively small number of labourers' cottages often indicates that young labouring men and maids were lodged as servants in their employers' farmhouses on contracts which compelled them to remain in the village at least for a year. At North Cerney (Gloucestershire) there were seven peasant farmers, five servants, and apparently no cottage labourers in the late fourteenth century; at the now deserted village of Marefield in High Leicestershire twenty peasant farmers were served by five living-in servants but only one cottage labourer. The presence of many living-in servants, notable in some wold regions well into modern times, may have been another reason for low medieval population densities, if servanthood was prolonged enough to delay marriage.

It seems therefore that a severe, rather austere, unvariegated social structure prevailed in many wold villages in the thirteenth and early fourteenth centuries, with few independent cottagers and with young male and female servants living under the rule of their masters. No *pays* has settlements completely uniform in type. The presence of a religious house would give a wolds village a more diverse social structure, although it should be said that these regions are not notable for large, ancient pre-Conquest foundations. Thus at Croxton Kerrial on the Leicestershire Wolds, seat of a small house of Praemonstratensian canons established in the 1160s, fourteenth-century evidence shows the presence of a sprinkling of craftsmen and retailers, some said specifically to have been in the service of the abbey, alongside the farming population. Although towns are noticeably absent from most wold regions (they exist on the borders, of course), the presence of a village with a periodic, weekly, market might lead to a diversified social structure. Such a place was Kilham, in its vale on the eastern edge of the Yorkshire Wolds. Formal markets and fairs were instituted there in the early thirteenth century and this act encouraged the permanent settlement of craftsmen; long after the Middle Ages Kilham had a significant non-farming element among its large population.

Farming type lay at the roots of the distinctiveness of wold regions in the thirteenth century and, as we have seen, this influenced cultural landscapes, standards of living, migration, and the social relations of labour, all of which developed in a special way within wold regions as a group. Later on artefacts belonging to cultures of work differed from one wolds region to another so that we find the eighteenth-century agriculturalist William Marshall in Yorkshire marvelling at the distinctiveness of methods of leading plough-horses and of grouping the horses used in harrowing, practices, he said, 'perhaps peculiar to the Yorkshire Wolds'. Such differences probably already existed in the Middle Ages and although many of the details are as yet unknown to us, we can point to longhouses on the Cotswolds as a possible regional house type (to judge from excavations at Upton and observation of earthworks at other deserted sites) and to the use of horses rather than oxen for drawing the plough from a very early date on the Yorkshire Wolds. Among other aspects of culture particular to

individual regions of wold was the unusual concentration of church dedications to St Margaret, the shepherdess, on the Lincolnshire Wolds and, perhaps, local dialects, all the more likely to retain archaic features in remote regions which received few incomers. When, some time before 1220, the inhabitants of Wardington (north of Banbury in Oxfordshire) gave the name Walder to an emigrant from the Cotswolds and when, some time before 1330, Thomas an emigrant to the Lincolnshire fenland vill of Market Deeping was given the name Oyewold (probably 'of the wold'), it is not unlikely that their unfamiliar dialects marked them as outsiders and, to use a later term, woldsmen.

Some of the difficulties facing the people of the wolds before the Black Death were greatly exacerbated by plague and population decline in England at large during the fourteenth and fifteenth centuries. Demand for grain grown on the arable fields of the wolds was reduced as the population level fell generally and the vale lands had the advantage as producers of crops. There were difficulties in retaining labour during a period when it was in short supply throughout most of England and, while landlords could afford to buy in migrant workers to help with the harvest (as did Catesby Priory in High Northamptonshire in the fifteenth century), poor peasants with declining incomes could not.

As farm incomes declined villages became unkempt and unfriendly, 'the howsinge falling into great ruyn and decay' according to a description from the 1470s of Burton Dassett on the edge of the wolds of Warwickshire. At nearby Chapel Ascote the last ale-maker ceased brewing in the 1450s, probably because of a shortage of customers, an indication of the declining provision of services, familiar in parts of the countryside today. The smallness of many wolds villages magnified this problem because their populations were likely all the sooner to decline to a threshold below which provision of a service was not viable. Now the fields begin to have a ragged appearance; crows, with no boy to scare them away, cry loudly over the meagre grain in a sown arable strip while next to it weeds grow on other strips belonging to farmers who have reduced their sown acreage, divisions between them becoming obliterated; this is probably what caused the court of Chapel Ascote (Warwickshire) to order tenants to make good their boundary marks in the 1450s. Studies of some villages in decline during the late fourteenth century have shown that women emigrated in greater numbers than men; these were young females for whom there were good work prospects in towns at that time. This resulted in highly skewed sex ratios, perhaps further encouraging the emigration of males, to produce that great scarcity of 'tenants to occupie their landes' reported from Burton Dassett (Warwickshire) in the 1490s.

Many villages in the wolds underwent these tribulations to some degree during the fourteenth and fifteenth centuries and in some, especially those which had always been small, the downward spiral eventually resulted in a deserted place. As a village began to expire a landlord might evict the few remaining tenants and turn the fields into a sheepwalk, as probably happened at Middle Ditchford (Gloucestershire) in the 1480s or at Leicester Abbey's Ingarsby, an exceptionally well-preserved site in the empty rolling green pastures of High Leicestershire close to me as I write this. Alternatively a few tenants might amass vacated tenures to become the occupiers of the new sheepwalk. Tenants with larger holdings could now own more livestock and tenant flocks of 300 sheep, such as those recorded for some places in the Cotswolds in the fifteenth century, could make some profit because demand for wool (unlike demand for grain) was high, the labour costs of a flock were generally low, and the product light in weight, thereby being able to bear the cost of transport away from the wolds. Deserted villages are especially concentrated in all wold regions and only in these do we find countrysides with constellations of adjacent shrunken and deserted sites—as in High Leicestershire, in the Cotswolds, and in the Yorkshire Wolds at Raisthorpe, Towthorpe, Thoralby, and Mowthorpe, all very close to Wharram Percy, the most famous of all 'lost' villages because of the excavations carried out there over many years by Maurice Beresford and John Hurst. In such countrysides remaining villages seemed to Laurie Lee, recalling the Cotswolds of his youth 'like ships in the empty landscape'.

We could end this account of the wolds in the closing years of the Middle Ages on a low note, evoking the landscape of a deserted parish as 'nothing but a wilderness for sheep, with a cote, a pastorall boy and his dog' or, as John Rous did at the end of the fifteenth century, by describing nestbeds for robbers and other undesirables in those tracts of countryside which became especially isolated because of the desertion of a significant proportion of villages. But some spirit still existed among remaining populations when the people of a wold region came together in common cause; many inhabitants of the central Lincolnshire Wolds took part in the Lincolnshire Rebellion of 1536 and one of their fears was that poorly endowed churches would be suppressed in this region where many parishes were small and poor, a reflection of settlement origins and economy. And there were spirited goings on at Burrough Hill near Somerby in High Leicestershire where John Leland, in the early sixteenth century, noted that people from the surroundings came in early May to matches in wrestling, running, dancing, and shooting. One wonders how old such a gathering was: could this May-time assembly of young men and women be a tradition as old as the phase of seasonal settlement, recalling the first reunion of the stock-keepers after a winter spent at home?

A Longer View of the Wolds

Bernard Jennings

This essay on the wolds traces the evolution of the landscapes of three major regions, in Yorkshire, Lincolnshire, and the Cotswolds, concentrating on the period since *c.*1500, but reaching back to medieval developments which have had a lasting influence.

Before the eighteenth century, the traveller approaching these three wold regions and mounting successively the escarpments which rise from the Vale of York, the central Lincolnshire Vale, and the Severn valley, would have noticed some important similarities. The scarp slopes were steep and well wooded, with beech predominating; the highest point of each region was found close to the escarpment; and extensive open pastures occupied the upper sections of the dip slopes. The topographical contrasts would also have been apparent. The Lincolnshire Wolds have the slightest relief, the highest point being about 550 feet above sea level. The Yorkshire Wolds rise to 800 feet, and the Cotswolds are topped by Cleeve Hill (Gloucestershire), at 1,083 feet.

Another important characteristic distinguishes the chalk wolds of Lincolnshire from those of Yorkshire. The northern end of the former has little surface drainage, but in the central and southern zones several river valleys have cut through the chalk to the underlying impermeable beds. In the extreme south the erosion of the chalk has exposed sandstone and clay in the hillsides. The Yorkshire Wolds are virtually devoid of permanent rivers, if streams emerging from spring-lines on the edge of the chalk are disregarded. The 'Great Wold Valley', running west–east towards the coast at Bridlington, contains the Gypsey Race—but a 'gypsey' is a river which keeps disappearing underground.

The Cotswolds are different again, being formed of bands of clays, sandstones, and limestones, the latter being composed of small round grains of calcium carbonate, often with a minute core of sand. From the appearance of the grains, the limestone is called oolite, egg-stone. The clays of the underlying Lias series form an impermeable layer, which provides the spring-lines on which many villages are located and supports the rivers and streams flowing down the dip slope. Several of the latter have a current sufficient to be harnessed for water power, which has in turn made an important impact on the landscape. To quote William Marshall (1789), 'Almost every dip has its rill, and every valley its brook. The sides of the hills abound with springs; and even on the highest swells, water generally lies within the reach of a pump. Benefits, these, with which few hill countries are blessed.'

Perhaps because the medieval churches of the Lincolnshire and Yorkshire Wolds are built of stone brought from outside the chalk zone, the importance of chalk as a building material has often been overlooked. The massive cliffs of Flamborough Head (Yorkshire) proclaim the resistant quality of some of the chalk, which is much in evidence in the older houses of Flamborough and some other villages near this stretch of the coast. The exterior walls of Watton Priory, near Driffield (East Riding, Yorkshire), were constructed of ashlar masonry, the interior ones of chalk. A tavern with butchers' shops above it in Kilham (Yorkshire) was built in 1420–1 of 'chalkstone', brought from Caythorpe, 4 miles away. Chalk can still be seen on the outside of farms and barns in both Yorkshire and Lincolnshire, and in both areas some houses built in the local vernacular style have a single skin of brick covering a thicker wall of chalk.

By contrast no building stone in England is better known than the Cotswold oolite, which weathers into grey, honey-coloured and brown shades. It is almost invariably allied with 'stone slates', thin limestone slabs used for roofing. The abundance and quality of the stone have encouraged the development of distinctive styles, with refinements of carving. One of the finest collections of such buildings is to be seen in Broadway (Gloucestershire), a honey-pot for

Chalk in the walls of a house in Wold Newton, Lincolnshire

tourists. As was the case with many villages bordering on the chalk and lime-stone hills, the pastures of Broadway extended well up the limestone escarp-ment, but the village itself is in the Vale of Evesham. Here we encounter a major problem of definition in a study of 'wolds' landscapes. In Lincolnshire and Yorkshire the wolds mean the chalk hills. To many writers the Cotswolds are defined in terms of domestic buildings, although not in a consistent way. Bishop's Cleeve, under Cleeve Hill, with its older houses built of Cotswold stone, appears in Denis Moriarty's *Buildings of the Cotswolds*, but is left out of Josceline Finberg's *The Cotswolds*.

The 'Cotswolds' of both these books include, as in this chapter, the strictly defined limestone uplands. These are the valley of the upper Evenlode, carved out of the clays of the Lower Lias series; a wide tongue of the land stretching from Moreton-in-Marsh almost to Milton-under-Wychwood, which does not qualify as 'wold' on either geological or topographical grounds; and the Redlands south-west of Banbury. The latter name derives from the reddish-brown colour of the marlstone of the Middle Lias. On the ground it is possible to define the limestone uplands with some precision by mapping the distribution of the drystone walls, built from local stones which have been carried for only short distances. On the clays and on the Redlands the stone walls are replaced by hedges.

Another problem for a book on the *rural* landscape is the degree of urbaniza-tion in the Cotswolds. There were no medieval boroughs in the chalk uplands of Lincolnshire and Yorkshire, and there are no settlements which would today be recognized as urban, with the possible exception, in both periods, of Caistor at the foot of the Lincolnshire escarpment. Six places in the Yorkshire Wolds had medieval charters for markets and fairs—eleven if townships extending on to the chalk uplands are included. Figures for Lincolnshire are five and nine.

The Cotswolds had between twenty-one and twenty-five medieval boroughs (according to different geographical definitions) and six other places with char-tered markets. However, the socio-economic structure of the boroughs varied considerably. The number of burgages *c*.1300 ranged from 177 at Chipping Sod-bury (Gloucestershire) to three each at Minchinhampton and Painswick. Some boroughs with few burgesses had rather more customary tenants owing labour services. None of them was taxed as a borough in the 1334 calculation of lay sub-sidy quotas. Cirencester and Winchcombe in Gloucestershire paid one-tenth of the assessment, instead of the normal one-fifteenth, but that was because they ranked as ancient demesne of the Crown. Cirencester had the highest assess-ment, £250, followed by Painswick, £186, and Burford (Oxfordshire), £155. However, Bloxham in the Redlands (Oxfordshire), not a borough, had an assess-ment of £173. Stow-on-the-Wold (Gloucestershire), £101, fell into the range of a group of townships in the Yorkshire Wolds, Kilham, £123, Sledmere, £110, and Rudston, £94. Minchinhampton and Broadway, on £70 and £64 respectively, ranked with a group of Lincolnshire townships, Binbrook, £81, Wold Newton, £66, and Nettleton, £61. One Cotswold borough, King's Stanley (Gloucester-shire), had an assessment as low as £32. Contrasts in the urban/rural balance continued into modern times. At the end of the nineteenth century farming was important at Painswick (1,059 acres of arable, 773 cattle, and 2,407 sheep and lambs in 1901), marginal at Burford (corresponding figures 307, 50, and 358), and non-existent at Stow-on-the-Wold, which reported to the collectors of agri-cultural statistics that it contained no farmland or stock.

William Marshall wanted to exclude industrialized, as well as urbanized, districts from the rural landscape. 'The Stroudwater hills … a lovely plot of country, but not a proper subject of rural study, as being a seat of manufacture.' However, the subtle interaction between industrial development and farming landscapes is a very proper subject for this study. All of the wolds areas produced the basic raw material of the wool textile industry. The differences in quality, with the Cotswold fleeces having a particularly high reputation, were not so great as to explain why the market production of woollen cloth developed in some areas but not in others.

One important factor was the availability of water power to drive fulling mills. The earliest record of such mills in England, the 1185 survey of the estates of the Knights Templar, lists two, one near Leeds (West Yorkshire), the other, built by the Templars themselves, at Barton on the upper Windrush in Gloucestershire. About 1200 Winchcombe Abbey had a fulling mill at Cleveley (east of Chipping Norton, Oxfordshire), and Evesham Abbey was operating another mill at Bourton-on-the-Water in 1206. By the early fourteenth century there was a cluster of mills in the Stroudwater/South Cotswolds district, with quite intensive development there by c.1450. Stroudwater had additional advantages: a deposit of fuller's earth between the Great and the Inferior oolite; woad growing in the Wotton–Painswick district; and soft water, emerging from the Milford Sands overlying the Upper Lias clay, which was ideal for washing and dyeing wool.

The efforts of the Bristol clothiers to prevent cloth being sent to rural fulling mills, prompted by fears that the industry itself would migrate to the streams, proved to be unavailing. The manufacture of fine-quality broadcloth came to be concentrated in an area bounded by Painswick and Cirencester (Gloucestershire), and Bradford-on-Avon (Wiltshire), which had easy access both to the port of Bristol and to the Thames Valley route to London. Sir John Fastolfe, the veteran soldier who was the model for Shakespeare's Falstaff, turned the village of Castle Combe (Wiltshire) into a major textile settlement, partly engaged in clothing his soldiers. In 1454 a survey by Fastolfe's steward listed seventy clothiers, some of them wealthy, who with their employees made up the majority of the population. Some fifty new houses had been built.

The decline of Bristol and Oxford as major textile centres was matched by that of Lincoln and Stamford, by the early fourteenth century, and of York and Beverley rather later, but their loss did not turn out to be the wolds' gain. Louth, on the edge of the Lincolnshire Wolds, emerged as the most important secondary centre in the county. It had thirteen 'walkers' (fullers) in the 1379 poll tax lists, whose customers may have included weavers in nearby wolds villages. There were, however, no fulling mills in the chalk hills of either Lincolnshire or Yorkshire. That was not the only missing ingredient. Sixteenth- and seventeenth-century inventories in both wold areas show that many farming households processed both wool and locally grown flax and hemp (especially the latter, better suited to light soils, and widely used for sheets, shifts, and shirts) to meet domestic needs. The majority without looms put out the weaving stage to a customer weaver, a figure represented in literature by Silas Marner. To develop a market-oriented textile industry from such a base required more than fulling capacity in the woollen branch.

In the Stroudwater area, flexible or slack manorial administration facilitated the multiplication of smallholder-weavers and cottager-outworkers. At Minchinhampton weavers had the customary right to enclose a portion of the common for a house and garden plot. Tenants in the manor of Bisley had no such privilege, but attempts to prevent encroachments in the sixteenth century by imposing fines were unsuccessful. In 1733 a Bill to enclose the commons in Bisley was defeated by a petition from small occupiers, said to represent 800 dwellings of people engaged in the woollen manufacture. Defoe noted that the textile towns of the Stroudwater area were 'interspersed with a very great number of villages ... hamlets and scattered houses in which, generally speaking, the spinning work of all this manufacture is performed by poor people ...'.

In those areas of West Yorkshire which developed a fully-fledged textile industry out of the sidelines of the farming household, the practice of dividing farmholdings, not necessarily equally, between sons, and the possibility of increasing the holding by licensed intaking or illegal encroachment, operated as interactive processes to produce dual-economy households. At the same time squatting on the common created the population of carders, spinners, and journeymen weavers. Such processes were blocked on the chalk by the absence of surface water. The provision of water through springs, ponds, and shallow wells in the valley bottom could not easily be replicated by would-be settlers on the chalk hillsides. Furthermore, in Yorkshire at least, the lands immediately beyond the open arable fields, which in other locations would have been the main target of the intakers and encroachers, were frequently earmarked for collective use through the periodic cultivation of the outfield.

The impact on the landscape of the growth of a textile industry is not confined to settlement and field patterns, or even surviving industrial buildings. In the Cotswolds, a leading player on the national textile stage until its steep decline in the early nineteenth century, the wealth of the clothiers was invested in splendid houses; and dual-economy dwellings needed to be relatively large. The profits of textiles, and the associated wool trade, built churches 'of a spectacular magnificence' (Herbert Finberg). The churches of the Yorkshire and Lincolnshire Wolds, often marked by substantial towers erected in what seems to have been an age of competitive tower-building in the fifteenth century, are much more austere by comparison. Apart from the payments to the farmers with their generally modest flocks, none of the profits of the wool trade were retained within the latter wold areas. They were taken by the monasteries, who bought up peasant supplies to augment their own clips, and by merchants invariably based outside the wolds.

It is in relation to specifically agrarian history—land use, field systems, commons, enclosure—that the Yorkshire and Lincolnshire chalk uplands and the limestone hills of the Cotswolds have most in common. 'Sheep and corn husbandry' is a description that applies from the high Middle Ages into the twentieth century. The arable land lay in common fields, the greater part of which (taking the three areas together) remained open until the mid-eighteenth century. As already explained, there was very little nibbling at the commons, except in the south Cotswolds, and therefore great scope for the improver after the pastures were enclosed by Act of Parliament.

The picture of 'wold' areas as bleak and barren can always be traced back to descriptions of the higher stretches of sheep walk, such as Arthur Young provided in the first edition of his *Northern Tour*. In fact the fertile valleys of the eastern Yorkshire Wolds and the southern Lincolnshire Wolds had a long settlement history, and were heavily populated in the thirteenth century. In the first of these areas villages were packed close together along the Great Wold Valley between

Naunton, Cotswolds

The name means 'new settlement' and is recorded in Domesday Book *(1086), so this new village, which always remained small, probably dates from the tenth century or the eleventh.*

Kirby Grindalythe and Butterwick. In this valley there was plenty of upland pasture, but further south, e.g. in Kilham, the extension of arable cultivation had, by the fourteenth century, left little common grazing outside the open arable fields. The south Lincolnshire Wolds were dotted with settlements which formed numerous small parishes. Corn growing was important in both areas. The available evidence suggests that barley was the favoured crop on the chalk, but supplies bought to provision a parliament to be held in Lincoln in 1301, and for Edward I's campaigns in Scotland in 1298–1304, included considerable quantities of wheat. In 1298, for example, Kilham provided 195 quarters of wheat, and 93 quarters each of oats and peas.

In the Middle Ages two-field systems seem to have predominated in the Lincolnshire Wolds and Cotswolds. For the latter H. L. Gray lists ten two-field examples, but only two places with three cropping divisions, Minchinhampton and Castle Combe, both fourteenth century in date. In the same source the six townships wholly within the Yorkshire Wolds all had three fields, but only one of these examples, Boynton, 1327/8, is medieval. Nearby Burton Fleming had two cropping divisions in 1299. In this area more than half of the townships had three fields on the eve of parliamentary enclosure. There is some evidence of a retreat of cultivation in the southern wolds of Lincolnshire in the first half of the fourteenth century. A survey of the yield of a tax on the assessed value of corn, wool, and lamb in 1341, compared to the valuation for papal taxation fifty years earlier, showed a shortfall in the deaneries of Hill and Horncastle much greater than can be explained by technical factors. At the same time some Cotswold villages, e.g. Naunton, reported a loss of population.

The heavy mortality of the Black Death in 1348–9 and the subsequent outbreaks of 1361–2, 1369, and 1375 may have permanently depopulated some villages. More often the plagues weakened small villages so that they later fell prey to sheep depopulation. It is often uncertain, however, when (and therefore to some degree how and why) villages were reduced to being only one or two farms. Brookend near Chastleton, and Asterleigh north of Woodstock (Oxfordshire Cotswolds) were the product of post-Domesday colonization of apparently marginal land. They had been abandoned by c.1440. Pockthorpe, Swaythorpe, and Octon were settlements on the ridges adjacent to the Kilham valley in Yorkshire. All had chapels before the Black Death which are not recorded after 1400. Two holdings in Swaythorpe were described as waste in 1421. Pockthorpe, apparently overpopulated in 1297, had a low quota fixed in 1334 (already in difficulties?), but still received a 45 per cent tax rebate in 1354. These three settlements were later reduced to single farms, but whether this resulted from the gradual migration of occupiers to more favoured ground, or from evictions, is not known.

The high levels of relief from taxation granted in 1352–4 to townships in the southern wolds of Lincolnshire reflect the heavy mortality there. As already mentioned, this was an area of small parishes. In 1397 the parish of West Wykeham was merged with the adjoining parish of Ludford Magna because its two small villages had between them fewer than ten households. In 1428 the deaneries of Hill and Horncastle listed eleven parishes with fewer than ten inhabitants (perhaps households were meant). Maurice Beresford describes this area as 'a classical district of ruined churches and lost village sites'. It is clear, however, that most of the deserted villages had been small before the Black Death. The five depopulated townships of Haverstoe wapentake which appear in the 1334 list had an average quota of only 15s. 6d., and only three of the nine cases in Hill wapentake had quotas in excess of £1 11s. 4d.

Six of the twenty deserted villages of the Yorkshire Wolds which appear in the 1334 list had quotas of £1 or less, and only one exceeded £1 14s. That was Eastburn, at £3 10s. (therefore assessment £52 10s.), showing that a moderate size was no guarantee of survival. Most of the deserted medieval villages occupied sites which were clearly less favoured than the main valley settlements, including the three already mentioned, Arras and Gardham on the hilltop east of Market Weighton, and Bartindale and Argam, on exposed sites between Burton Fleming and the coast. There were exceptions. Caythorpe is in the valley of the Gypsey Race, although in terms of the 1334 assessment it was less than half the size of either of its neighbours along the valley.

The process of sheep depopulation in the Cotswolds in the fifteenth century has been discussed by Harold Fox in the first section of our joint chapter. In the Yorkshire Wolds the process of depopulation, for whatever motive, went on over a long period. Caythorpe (where twenty people had been evicted and 300 acres of arable converted to pasture), Thirkelby, and Wharram Percy were reported to the commission of 1517. Other townships of which there is no demographic record after this date had probably been deserted, before the cut-off date of the commission's investigation, 1485. Kilnwick Percy had suffered a similar fate by c.1600. Eastburn, however, was not finally depopulated until c.1670, and Cottam not until the early eighteenth century. Nearby Cowlam still had fourteen occupied houses in 1674, but formed a single farm of 1,900 acres in the 1780s.

Alan Harris has pointed out that removing tenants to make way for sheep, and sometimes rabbit, farming had a very different effect upon the landscape from piecemeal or parliamentary enclosure. In fact it fossilized the open rural landscape until the ploughs of the improvers were put to work from the middle of the eighteenth century, and the appropriate field divisions made. The word 'enclosure' in the context of depopulation can be a misnomer. In 1698 the land of Eastburn and its neighbour Battleburn (another depopulation, of unknown date), totalling about 1,200 acres, lay 'open for Sheepe Walks'. Even a ring fence was not essential for sheep, as they could be 'hoofed' on to the ground, i.e. herded by dogs until familiarity kept them from straying. Few fences were found at township boundaries. A grass baulk or sod wall often sufficed. Rabbit warrens were surrounded by sod walls. Those at Cowlam may have served a dual purpose, as a flock of 600–800 sheep was kept within the warren of *c*.1500 acres, but away from the burrows.

Writing about four Cotswold townships enclosed between 1767 and 1822, H. L. Gray commented, 'That only half of [the] arable was cultivated yearly after the middle of the eighteenth century may seem incredible …' In fact there could be more flexibility in the cropping system than Gray realized, which is one reason for its long survival. Similarly, the landscape had more variety than a simple division between growing crops and idle fallow. One method used was laying down sections of the arable field to grass for a period of years (leys). Another was the periodic cultivation of part of the common pastures (outfield). New crops were sometimes introduced into a section of the fallow field.

Leys, that is strips temporarily under grass, were found in the common fields as early as the fifteenth century. Ley farming became common thereafter, helping to maintain fertility as well as adjusting the balance between arable and pasture. It was practised at Stanton on-the-Wolds (Nottinghamshire) in 1475, at Bloxham (Oxfordshire) in 1513, and in the open fields of the Lincolnshire Wolds in the sixteenth and seventeenth centuries. At Weaverthorpe (Yorkshire) in 1800, leys made up 22 per cent of the area of the common arable and 44 per cent of all recorded grassland.

Outfield cultivation was widely practised in the Yorkshire Wolds. Durand Hotham described the system in his response to the Georgical Enquiries in 1664–5: 'they have in many towns seven fields, and the swarth of one is every year broken for oats and let lie fallow until its turn at seven years' end, and these seven are outfields.' Of course, the practice was much more varied than this. It was said that the outfield on Bishop Wilton Wold was cropped no more than once in ten or twelve years in the eighteenth century. Isaac Leatham (1794) quotes intervals of three to six years for different places. In 1770 it was estimated that of the 5,000 acres of high wold belonging to the Ganton estate, about 500 acres were under the plough at any one time. This did not mean that all, or even most, of the common pasture was treated as outfield. Strickland (1812) thought that a quarter of the open pastures of the wolds was occasionally cultivated. In favourable conditions the exploitation could be much more intensive. About 70 per cent of the open grassland of Weaverthorpe was ploughed up at some time during the eighteenth century. The outfields of Weaverthorpe (over 500 acres) and neighbouring Helperthorpe (nearly 200 acres), as designated in 1800, were cultivated every second year, but the infields were then cropped more frequently than one year in two.

There were other variations on the same theme. The tenant of a newly-formed rabbit warren-cum-sheepwalk at Warter (Yorkshire) had permission in 1749 to plough 30 acres, take three crops of oats or other grain, and then put the land back to grass.

The fallow field was not simply at rest. As already explained it formed part of the pasture for sheep—in some townships nearly all of it apart from the leys. Durand Hotham asserted that for this reason no winter corn was grown in the Yorkshire Wolds, although he mentioned wheat at Kilham as an exception. 'Their design being sheep, the winter corn would "straighten" the herbage for them from October till March.' In fact inventories of the seventeenth and early eighteenth centuries show that wheat was grown in many valley communities, although barley (spring sown) was dominant on the higher land.

The townships with little or no permanent pasture, including Butterwick, Langtoft, and Kilham, had to provide the grazing needed for cattle, and any required for sheep beyond the stubble and fallow, through the management of the open arable fields. This could go well beyond the use of leys. There are medieval references to a demesne pasture at Kilham, but a survey of 1729 shows all of the land, except for a small village green and the garths behind the farmsteads, divided into forty-seven open-field furlongs. The by-laws of the manor, which survive from 1620, reveal that these furlongs were not simply grouped into two or three cropping divisions. Each year the six by-lawmen, and six other men chosen by them, decided which furlongs were to be cultivated and for which crops, and which were to be laid down to, or left as, grass. The two cow pastures were to be marked out at the end of March, so that they could be given a month's respite for the grass to grow before the cattle were turned out. These and the horse pasture were presumably moved around the furlongs, but some of the poorer ground was left as grass for a period of years. The Kilham system delivered convertible husbandry on a community basis.

Kilham was enclosed under an Act of 1771. Taking the wolds as a whole, parliamentary enclosure was the most important process in creating the modern landscape of farm and field. Early enclosure, piecemeal or by agreement, had progressed furthest in the textile districts of the Gloucestershire Cotswolds, leaving the rest of that region largely open; and in Lincolnshire, only 41 per cent of the wold villages had unenclosed land in the middle of the eighteenth century, and some of these contained early enclosures beyond the crofts of the farms. Frank Emery calculated that 15 per cent of the Oxfordshire Cotswolds had been enclosed before 1540, and a further 21 per cent by 1730, leaving 64 per cent for the enclosure commissioners. In the Redlands 87 per cent was enclosed after 1730, the corresponding proportion for the Yorkshire Wolds being about two-thirds. In assessing the effect upon the landscape, we can add to the parliamentary acres the lands which had been subject to depopulating 'enclosure', the use and appearance of which were transformed after the middle of the eighteenth century.

An inscription on the village well at Sledmere (East Riding, Yorkshire) proclaims that Sir Christopher Sykes, during his tenure of the Sledmere estate from 1700 to 1801, 'by assiduity and perseverance in building and planting and enclosing on the Yorkshire Wolds, in the short space of thirty years, set such an example to other owners of land, as has caused what was once a bleak and barren tract of country to become now one of the most productive … districts in the County of York'. Barbara English has pointed out that land purchase and improvement had been objectives of the Sykes family before Sir Christopher was born, and that other leading landowners were similarly engaged before 1770. In fact, the transformation of the wolds landscapes, although profound, was not rapid, for two reasons. The enactment of enclosure was spread out over a long period; and the physical changes following the issue of the enclosure award might not be completed for decades. Large allotments of land had to be ring-fenced within a specified period (for allotments in lieu of tithe this was normally done at the expense of the landowners in general), but subdivision into fields of a convenient size was at the discretion of the owners. Improvement did not always follow quickly.

The length of the process of transformation is reflected in an account of the farming of Lincolnshire published in 1851. With an exaggeration worthy of Arthur Young, J. A. Clarke described the wolds in the mid-eighteenth century as 'a succession of rabbit-warrens from south to north'. By c.1800 extensive enclosure of the open arable fields had taken place, but warrens and gorse still dominated 'the loftier hills'. All the open fields were now gone, 'a great part having been enclosed within the last thirty years … the highest points are all in tillage, and the whole length of the Wolds is intersected by neat white-thorn hedges …'.

Writing at the same period, Sir James Caird described the Yorkshire Wolds as

> all enclosed, generally by thorn hedges; … plantations, everywhere grouped over its surface, add beauty to the outline, while they shelter the fields from the cutting blasts of winter and spring. Green pasture fields are occasionally intermixed with corn, or more frequently surround the … homestead. Large and numerous corn ricks give an air of warmth and plenty, whilst the turnip fields, crowded with sheep, make up a cheerful and animated picture.

His description of the Cotswolds was similar, if less poetic:

> the grass-lands have now been nearly all brought under the plough, the richer pastures in some of the valleys being the only portions left untilled. The fields are large, and are enclosed either by hedgerows or dry stone walls.

'The fields are large …' This was a general feature of the wolds, particularly on the higher parts. Caird describes the cornfields on the Yorkshire chalk as being 30 to 70 acres in size. In Lincolnshire they were 'generally … 30 to 100 acres, presenting to the eye of the stranger the aspect of open-field lands, the fences being often concealed' by undulations. Large fields on all the wolds went with large farms, most said to range between 300 and 1,300 acres in extent.

The labour required to create productive cornfields on the previously thin soils of the sheep-walks and outfields is not now obvious to the eye. An intensive process of dressing with chalk (effective and still practised in recent years, despite the apparent absurdity of putting chalk on to a chalk soil) and bones, and manuring by sheep, was necessary. In Clarke's words, 'the farmers were first obliged to *make* a soil …'.

It also took a long time to create the plantations which Caird so admired. William Marshall, praising the beauty of the Yorkshire Wolds in 1788, observed, 'Wood and water would render it most beautiful. Water is forbidden; but wood may be had at will; and it is extraordinary that the spirit of planting should have broken out so late. Utility, as well as ornament, calls loudly for this obvious

Sledmere, Yorkshire Wolds, from the north-east

Shown here is Sledmere House (begun 1751), and the park, the making of which (Capability Brown, 1777) involved the demolition of the old village on the east side of the church. Part of the new village can be seen in the foreground

improvement.' Mrs Finberg noted that by 1800 little planting had taken place in the Cotswolds: 'new plantations, such as Lord Chedworth's great wall of beech at Stowell [near Northleach], now a landmark for miles around, were in ungainly infancy'. More progress may have been made in Yorkshire. Henry Strickland, writing in 1812, referred to extensive plantations made on the family estate at Boynton by his father Sir George (1729–1808), who had been allocated all of the land in the township, except for eight acres of glebe, at the enclosure in 1783. He observed that the Sykes family had planted about 2,000 acres in Sledmere and district. John Bigland wrote, also in 1812, that the farms and fields of Sledmere, together with 'the numerous and extensive plantations skirting the slopes of the hills, and the superb mansion with its ornamental grounds … form a magnificent and luxuriant assemblage, little to be expected in a country like the Wolds …'

Long-established natural woodlands can be found on the wold escarpments, in some steep-sided valleys, and more extensively in the Cotswolds, e.g. in small groves adjacent to the villages. Most of the woodlands, however, are artificial creations, some functional, others partly or mainly aesthetic. In the first category come the linear plantations, the shelter belts, which seem to stand on every skyline in the rolling landscape, and the more modest windbreaks wrapping round three sides of the post-enclosure farm buildings, easily identifiable as such because they are surrounded by rectilinear fields. The more extensive stands of timber, such as those at Stowell and Sledmere, were exploited economically but were created also to beautify the surroundings of country houses and their dependent villages. Conversely, villages not blessed with the presence of planting squires tend to be relatively treeless, apart from gardens and the self-sown trees in hedgerows and along lanes.

There may be other correlations. An academic assignment to investigate the relationship between the presence of trees and the absence of dissenting chapels might seem like something out of *1066 and All That*, but as buildings are part of the landscape the proposition has its serious side. Kilham, a large village with no 'big house', which has not had a resident lord of the manor since the Norman Conquest, had in the nineteenth century four public houses, three nonconformist chapels, but relatively few trees. A few miles away the well-wooded Boynton, firmly controlled by the Strickland family, had Methodists but no chapel, and drinkers but no pub. Both had to seek refreshment outside the village.

The devout and the thirsty were amongst the figures in the landscape. Another characteristic grouping consisted of the ten or twelve labourers typically employed in the middle of the nineteenth century on a large wolds farm, which is today worked by the farmer and his family with perhaps one extra pair of hands. Alongside the farmworkers were large numbers of farm horses. In 1901 Painswick had 188 (in addition to those involved in the transport of stone from the quarries), Binbrook 151, and Kilham 385. The most characteristic figures in the moving landscape were, of course, the sheep in the turnip field.

In the Sledmere–Kilham area of the Yorkshire Wolds, of the acreage devoted to corn, pulses, and roots in the period 1870–1901, roots (nearly all turnips and swedes) took up about 25 per cent. Selected townships in the Cotswolds—Painswick, Bourton-on-the-Hill, Milton-under-Wychwood, Burford, and Wigginton—average out at a similar figure, but at Bloxham and Weston-sub-Edge a different regime was followed, with the acreage devoted to beans and peas greatly exceeding that of roots. The 1801 crop returns from 101 parishes in the Lincolnshire Wolds gave roots an average, of the land growing corn, pulses, and roots, of 32 per cent. Almost exactly the same figure is produced by taking the average of three selected townships—Binbrook, Wold Newton, and Welton-le-Wold—for the years 1874–1901. The Lincolnshire Wolds would seem to have been the stronghold of the turnip.

The agricultural statistics, collected from 1866, need careful use. Some of the earlier ones are unreliable because of the large number of estimates made in lieu of unreturned forms, and it is necessary to look out for administrative changes. An increase of 150 per cent in the sheep population of Sledmere was the result of a decision to include the returns from neighbouring Cowlam and Cottam. Nevertheless, it is possible to trace the effects of the agricultural depression on the farms of the wolds. One general feature, which did not directly affect the landscape, was a switch from wheat, the price of which halved between the mid-1870s and the mid-1890s, to barley and oats, which suffered falls of 37 per cent and 34 per cent respectively in the same period. Of seven selected townships in the Cotswolds, five showed a significant fall in both arable acreage and numbers of sheep and lambs, without, on average, a compensating rise in the number of cattle. Painswick bucked the trend, with virtually no fall in the arable acreage between 1879 and 1901, and a substantial increase in its sheep flock. Numbers of cattle and pigs also rose. In the Cotswolds in general rents fell by 50 per cent, and some marginal land was abandoned.

In both the Cotswolds and Lincolnshire it was noted that the size of farms, needing a considerable input of labour at a time when agricultural wages were rising and children were being excluded from the labour market, made adaptation difficult. Splitting farms into smaller units would have required new buildings and water supplies. A survey of eight townships in the Sledmere–Kilham area of Yorkshire shows no major change from the pattern of the 'high farming' years, apart from the shift out of wheat and an increase in cattle numbers which was high in percentage terms, but from a low base. Three neighbouring townships—Burton Fleming, Rudston, and Kilham—saw the area devoted to corn, pulses, and roots increase slightly between 1870 and 1901; the roots acreage decrease by 4 per cent; the numbers of sheep and lambs remain steady at nearly 15,000; the cattle population rise from 644 to 924; and the numbers of pigs increase by a third to 1,191.

The reasons for the contrasting fortunes of different wold regions lie beyond the scope of a book on the landscape. In all of the areas a shortage of funds caused a neglect of maintenance which gave many buildings and fences a bedraggled look. Economic conditions improved slowly from the early years of the twentieth century, and the frenetic activity of the First World War brought many fields back under the plough.

In the meantime the seeds had been sown of a revival in the Cotswolds which gave the countryside a new role. The pioneers of the Arts and Crafts movement discovered the Cotswolds in the late Victorian period, and prepared the way for a small army of working craftsmen and artists. By the 1930s the most attractive areas were becoming retirement havens for the affluent, as well as tourist targets for a broader social range. Today the Cotswolds are collectively one of the leading holiday resorts of the country, drawing in foreign as well as British tourists. The Lincolnshire and Yorkshire Wolds have not had a similar experience. This is partly a matter of accessibility to the main centres of wealth and population. The Yorkshire Wolds could be described as half-known to the people from the Yorkshire conurbations who pass through on their way to the coast, but turn for rural recreation to the more muscular landscapes of the Pennines and the Lake District. The gentle scenery of the Lincolnshire Wolds is still a closely guarded secret.

A factor which is more significant than location is housing, both materials and style. Indeed these two are closely related. The availability, in areas such as the Cotswolds, of high-quality building stone which lent itself to the carving of fine detail, encouraged the development of distinctive and attractive styles, so that even yeomen whose instinctive choice was 'tin in the buttons but gold in the pocket' eventually succumbed to the pressures of social aspiration. Today strict planning controls, informed by an appreciation of the economic as well as aesthetic value of the Cotswold style, preserve the feature which defines the region in the eyes of most visitors. Buildings such as a Little Chef and Travelodge can therefore be seen occupying either conversions or new buildings in the local style, even if, to avoid using the scarce and expensive labour of the stonemason, the stone used is often 'reconstituted' through the concreting of crushed oolite.

The 'great rebuilding' in the Cotswolds began in the sixteenth century. By contrast few of the listed domestic buildings in the Lincolnshire or Yorkshire Wolds appear to be older than the early eighteenth century. According to Edward Anderson, East Yorkshire's McGonagall (c.1792), the housing of the wolds was greatly improved after enclosure.

> Since I came home, as I this country view,
> The towns, the fields, now everything looks new;
> The old thatched cottages have ta'en their flight
> And new tiled houses now appear in sight;

Anderson's 'towns' are, of course, now called villages.

The older houses in the Lincolnshire and Yorkshire Wolds which have attracted listing fall into two stylistic groups. The first could be labelled 'polite Georgian', brick-built to designs copied from the towns through the pattern books used by local builders. The second, the 'vernacular', is not only more varied, sometimes incorporating visible or concealed chalk, but has often been treated as flexible in the absence of strong stylistic conventions. A close examination of the brickwork and fenestration of some of the houses reveals a succession of changes, e.g. the introduction of dormer windows where previously the base of the window rested on the floor of the chamber. There is much to interest the student of domestic architecture in a cluster of 'polite' and 'vernacular' houses, e.g. the contrasting positions of the main chimney stack; and, in Yorkshire, the horizontal-sliding-sash windows in the gables which appear in both styles. Judged as a harmonious composition, however, few of these villages could aspire to be more than an agreeable mixture with a touch of the picturesque, and some fail to achieve even a decent mediocrity. It is significant that the most prominent example of 'traditional' housing in the Yorkshire Wolds, the thatched terrace at Warter, was built in the 1930s, a product of nostalgia rather than heritage.

This is not to decry either the aesthetic quality of the landscapes on the chalk in Lincolnshire and Yorkshire, or their potential value for discerning tourism, or for residence as more people work from home in a village of their own choice, merely to suggest that domestic buildings are not the most attractive feature of

Twentieth-century
thatched houses in
Warter, Yorkshire Wolds

either area. The visitor to the Cotswolds from other wold areas may perhaps be forgiven if admiration is occasionally tempered by resentment at what appears to be the smugness of Broadway, Chipping Campden, or Castle Combe, as if the oolitic stones themselves were exuding an 'easy consciousness of effortless super-iority'. A more appropriate ending to a chapter written in the eastern wolds of Yorkshire, in an old farmhouse built partly of chalk in a 'rambling vernacular' style, is provided by the description by William Marshall (1788) of 'the most magnificent assemblage of chalky hills the Island affords. The features are large; the surface is billowy, but not broken; the swells resembling Biscayan waves half-pacified. The ground in general is particularly graceful ...' This demi-paradise? The Yorkshire Wolds.

3 Lowland Vales

Christopher Dyer, David Hey, and Joan Thirsk

Introduction

Since the vale lands of England spread over such a large area of the country, it is not surprising that they are conventionally taken to represent the English landscape as a whole. In our postcard views of Scotland or Wales, mountains, hills, and pasture dominate the scene. But in England the characteristic scenery shows vales. They are prominent across the centre of the kingdom, occupying large parts of the counties of Buckinghamshire, Bedfordshire, Oxfordshire, Leicestershire, Rutland, Northamptonshire, Lincolnshire, Nottinghamshire, and Warwickshire. Still more patches lie in Hertfordshire, Essex, and East Anglia, in central Yorkshire, and in all four northern counties. In southern England a fine stretch runs through the Berkshire vale, finally emphasizing the fact that vales are scattered everywhere, for they are river plains, in this case of the river Kennet.

Writing of the largest river plain of all, in the East Midlands, watered by several rivers, and in fact the countryside where he was born, H. E. Bates described it as 'probably the dullest plain in England', lacking any impressive landscape features, 'a pudding pattern of elm and grass and hedges'. Yet he saw it as 'the basis on which the entire English countryside is built'. Ash, elm, and willow were staple trees in the landscape; in his childhood, he knew only these. Beech trees were a rarity, alder and sweet chestnut were never seen. The fields he described as 'parallels of steel clay that would later be roots and corn'. 'I knew no other hedgerow except the rule-straight line of laid hawthorn, or the high, cow-rubbed umbrella-shaped variation of it.' By the age of 20 he had come to dislike 'that plain pudding country'. But later when he had travelled more widely, he warmed to it again; he experienced a sense of friendliness and comfort whenever he recognized it elsewhere, in counties as far apart as Cheshire and Berkshire. So he was tempted to call the clay vales 'the real England', but then checked himself, knowing that a remarkable feature of the kingdom as a whole is its capacity for variation in a small space. In contrast with much larger countries where one kind of scenery stretches over long distances and becomes monotonous, we are extremely lucky to have the contrasts and distinctiveness of many landscapes clustering close to one another.

The fact remains that when we generalize about the scenery of England as a whole, we choose the image of the vales and are always likely to do so. Whether they were some of the oldest areas to be settled by man may still be debated. Our downlands preserve more spectacular prehistoric monuments, but the river valleys had many attractions for the first comers.

Early Settlement Patterns

The claylands took on their distinctive form in the early Middle Ages, between about 850 and 1100. But long before then, in the prehistoric and Roman periods, and in the five centuries after the end of the Roman province, they had been settled and farmed. This is clear from the aerial photographs showing ditched fields and the sites of farms, and also from the pottery and flints found scattered over the modern ploughed fields. The lighter, gravelly soils nearest the rivers were settled most intensively, but there are enough indications from the heavier lands between the river and stream valleys to show that they were being exploited for more than a thousand years before the great upheaval of 850–1100. Until that period country people had lived mainly in hamlets and scattered farms, but then began the great movement for peasants to gather into large villages, and for the surrounding land to be organized into great open arable fields. So a countryside which is often seen as 'timeless', sleepy, and enduring was the scene of one of the great revolutions in landscape history. In about 800, when the Midlands were ruled by the kings of Mercia, the traveller would have seen a countryside of great complexity, having groups of houses scattered among small fields, with an occasional centre of royal or lordly power to which the peasants paid their dues. Around important churches, like Breedon-on-the-Hill (Leicestershire) or Kings Sutton (Northamptonshire) rather larger concentrations of people may have dwelt. By 1100 the later visitor would have been struck by the rather monotonous succession of villages, each with rows or clusters of rather similar houses, the church or manor house providing the only break in architectural uniformity. The villages stood in an expanse of cultivated land, mainly brown from ploughing in the winter, entirely green by May as the crops grew, yellow from the ripening corn, and shorn down to prickly stubble in August and September.

Leland and the other sixteenth-century commentators on the landscape called this country 'champion', indicating that expanses of open field were their most prominent characteristic. The English word based on the same idea is Feldon, and this was used particularly to describe the countryside of villages and unenclosed corn fields in south-east Warwickshire. Of course, in the thousand years since the formation of the villages they have changed in their plans and their architecture, a minority of them have disappeared, and the fields have been enclosed. But the landscapes of the lowland vales described by Leland in the sixteenth century, and by H. E. Bates in the twentieth, were formed in their essentials in the two centuries after 850.

As the 'village revolution' belongs to a period of few documents, we know from the physical evidence that it happened, but have to make difficult judgements on how and why the changes were made. It could have been carried out in a single year: we can imagine, after the harvest, houses being built around a church and manor house, as people left their hamlets and farms, and the new fields were laid

● ● ● ●

out by groups of men carrying measuring poles, ropes, and pegs to mark the boundaries. As evidence for such careful and coherent planning, we need only look at the regular layout of some villages, with identical plots to contain the houses, arranged in rows along a street or around a rectangular green, and the orderly plan of the fields recorded in later documents, as in the coherent pattern at Mears Ashby in Northamptonshire. But, more plausibly, the reorganization

Plan of the village and fields of Mears Ashby, Northamptonshire, reconstructed from documents of the 16th and 17th centuries

Hall Field was the lord's demesne, and in the tenant fields each furlong contained 40 lands or strips, held in consistent order by the tenants of 40 yardlands. The arable covered most of the parish, except for a small wood and meadows beside the streams.

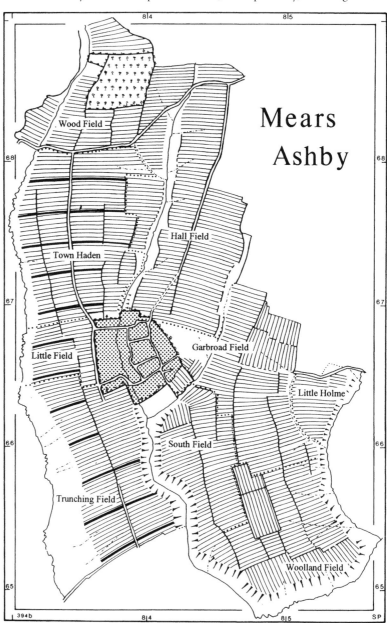

Mears Ashby

Wood Field

Hall Field

Town Haden

Little Field

Garbroad Field

Little Holme

South Field

Trunching Field

Woolland Field

stretched over a number of years, as the houses of the peasants moved gradually into the centre, and the open fields were laid out in sections. Evidence for this more evolutionary view comes from the so-called 'polyfocal' villages, like Hook Norton (Oxfordshire) in which the large straggling settlement consists of loosely connected hamlets, as if the final stage of forming a single large village was never quite completed. Sometimes we find among the nucleated villages a single farm or very small hamlet which appears to have survived the process. More evidence that the procedure was a rather protracted one, and not always completed, comes from the territorial anomalies, when villages which appear to stand in the middle of a compact territory are known to have used land in a detached parcel in another parish, or had common rights in some distant wood or pasture, like the communities in south Oxfordshire which grazed their animals on the Chilterns.

We cannot identify clearly who created the villages. The temptation must be to assume that someone in high authority, anxious to bring some order to the countryside, initiated or encouraged the process. Perhaps kings, especially in the period when the English state was formed under Alfred and his successors, were responsible? Or did the Danes, when they were sharing out their territory after the conquest of the late ninth century, impose a new coherent plan? Neither explanation convinces because the village and field forms on close inspection are not as regimented as we would expect from state planning. If some superior power was responsible, it must have been the local lords. We know that at the time the villages were being created manors were established for a more numerous group of aristocrats, the thegns, and for newly founded monasteries. The countryside was being broken down into smaller compartments, which were to become the parishes and the village territories, often in the region of a thousand acres. Lords were building manor houses, bringing the peasants under close local control, and building new parish churches. This process was set in motion rather later in the Yorkshire vales, where the new Norman lords were developing their estates in the late eleventh and twelfth centuries. But if lords, either in the Midlands or in the north, were also planning the villages, they were doing it in a very piecemeal fashion since, on a single estate, we find striking variety in the shape of villages and the layout of fields. It could well be that the peasants themselves played an important role, as we know that in later centuries communities were capable of governing themselves, and in particular of regulating, modifying, and even remodelling the field system.

Evolving Communities after 1200

The form of the villages and their fields reveals some of the thinking behind the 'village revolution'. We can reconstruct the medieval landscape from the plans of villages which are still inhabited, and the ridge and furrow which preserve the form of the fields before enclosure, as well as from documents of the twelfth century onwards. The whole structure of the village and its territory reminds us of a great medieval church, springing from a single coherent and simple conception, but constructed of hundreds of interlocking pieces. The village itself was formed from a combination of collective and private space. The individual houses were enclosed by ditches and fences, and gave on to shared space such as a street or green; the community was focused on the parish church. Most of the land was under the plough, in strips (selions, ridges, lands were the words used at the time), which typically measured about 10 yards by 120 yards, ideally suited in their narrow shape for cultivation by an ox-drawn plough which turned round with difficulty. The strips were gathered into furlongs—between twenty and fifty together—and the furlongs were grouped into two or three great fields. Individual tenants were responsible for the cultivation of their own strips, but the management of the fields, deciding the time when the fields were fenced from grazing animals, for example, was agreed by all cultivators collectively. The villagers had to strike a balance between corn and livestock, so large sections of the arable—a third or more often a half—were each year left fallow and available for grazing. In addition, the villagers shared some hay meadow, permanent pasture, and wood.

Modern observers find it difficult to envisage how such a complicated method of cultivation worked. For example, we define boundaries with fences and hedges, but in a field divided into thousands of parcels the only markers were stones or stakes set at the end of the long strips. The scattering of strips is also hard for us to comprehend, as a peasant with a yardland of about 30 acres would need to travel to at least 60, and more likely 100 strips, dispersed over the fields, some of them more than a mile from his home. The arrangements had an internal logic, however. The scattering ensured that everyone had a share of good and bad land. Every cultivator was bound into the collective scheme, which was concerned with working a great quantity of arable land, while ensuring that farming was sustainable in terms of maintaining some livestock. This method of managing the countryside was evidently born out of a desire to increase grain production above all, and to protect the interests of the community, while still leaving scope for some individual initiative.

To us the emphasis on the cultivation of grain seems dangerously specialized. Surely the land would eventually fail from repetitive cropping? The proportion of arable in the lowland vales, which was already high at the time of Domesday in 1086, increased further in the next two centuries, and the fields were by then supporting a village population that had often increased by 50 per cent or even more. In some village territories by 1300 more than nine-tenths of the land lay in arable fields. The peasants and their lords were aware of the potential difficulties, and took sensible measures to prevent ecological problems. Many villages had divided their territories into two fields, fallowing the land in alternate years. They were aware that some villages worked three fields, and thus cropped their land more intensively, but they resisted the temptation to convert their fields, fearing that their yields would be reduced and they would gain no advantage. They made efforts to improve the fertility of the land by growing alongside the usual wheat, barley, and rye (oats were a very small crop on the clays), an increasing acreage of peas and beans which put nitrogen into the soil, and when fed to animals helped to increase the supply of manure. Disastrous crop failures occurred in the early fourteenth century (the worst years being the famine of 1315–17) and land was being abandoned by 1340, notably in the champion country of Bedfordshire and Buckinghamshire, but we cannot be sure that this was entirely the result of the failure of the farming system.

We look at the modern villages of the lowland vales with approval, and indeed wish to live in them. They seem to have all the qualities that are lacking in modern cities; they are clean, quiet, harmonious, neighbourly, and co-operative. Surely in the Middle Ages these were communities at peace with themselves? As with any generalization, the society of the vales did accord with part of this rather idealized picture, but in important respects they were discordant and troubled by conflict. Village society was in some ways egalitarian. The village of Gonalston in the Trent valley (Nottinghamshire) in 1299 can stand as an example. Here were fourteen tenants, each with an oxgang of arable land—about 15 acres. They, together with four other tenants having two oxgangs apiece, cultivated most of the tenant land in the village. About half of the land was held as the home farm or

Map of Laxton, Nottinghamshire, in 1635

The map shows clearly the long, narrow house plots, on which were built houses presenting their gables to the main street. The same layout was still visible in 1935, though the spaces in between had been filled with additional farm buildings.

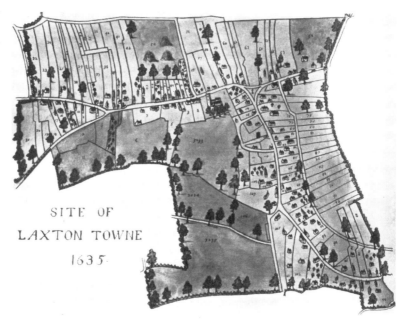

SITE OF
LAXTON TOWNE
1635

demesne by the lord of the manor, a knight called John de Heriz. The holdings of these eighteen peasants were large enough to feed their families and to provide a surplus which would allow them to pay rents to the lord and dues to the parish church. The community was dominated by these people, one of whom, William the reeve, managed the lord's demesne, while others acquired influence as jurors in the manor court. Other inhabitants of Gonalston were fifteen cottagers and smallholders, whose plots were too small to grow enough food for a family; they must therefore have earned wages as employees of the lord and of the tenants of larger holdings, especially those with two oxgangs. They were inferior in wealth and status, but essential to the working of the demesne and peasant holdings, and so the different sections of the village were bound closely together. Gonalston had a strong community organization to manage the fields, and also to maintain law and order through the lord's court, and to deal with the demands of the royal government, especially for taxes.

Villages were not as egalitarian and cohesive as they first appear, however. The growth of a market for farm produce in the twelfth and thirteenth centuries gave at least a few peasants the opportunity to accumulate modest wealth—the Gonalston people visited the market at Southwell, for example. When we see a list of tax assessments compiled in 1327 or 1332 of those people who, in theory, had equal resources of land, we find that some of them had goods worth two or three times those of their neighbours, presumably reflecting their greater talent, luck, or ruthlessness in producing and selling, The lords made the lives of a large section of the peasantry more difficult by imposing servile dues. Thirteen of the more substantial Gonalston peasants held their land in bondage, and so paid at least 6 shillings each year to the lord in rent, and owed labour services.

The unfree peasants' resentment at their condition was expressed in agitations when they attempted to prove before the royal courts that they were free, or at least that they were entitled to protection from increased services because they had once been tenants of the king—this happened at Mickleover in Derbyshire and King's Ripton in Huntingdonshire in the late thirteenth century. Here the organization of the village community was being used not to co-operate with the lords but to make claims against them. Tensions within the community also came to the surface, as the divisions between rich and poor became more pronounced. The court records at this time give evidence of much quarrelling and violence, as well as petty theft and such unsociable acts as listening to conversations by lurking under the eaves of neighbours' houses.

Frictions and divisions, some old, but others new, became more acute in the century and a half after the Black Death and consequent upon long-term decline in village populations. The better-off peasants accumulated more land as neighbours died and holdings could be acquired cheaply. By 1400 or 1450 many peasants held 50 or 60 acres rather than the 15–30 acres which were common before the plague. But life was not easy for those who gained extra land because the cottagers had either died in the plague, or had acquired land of their own, so that labour was scarce and expensive. They resolved these problems by converting land from arable to pasture, which reduced the costs of production (arable farming required many more workers than pasture farming) and the high demand for meat and for wool for the cloth industry made pastoral specialities more profitable. All of this caused much tension within villages, as the number of animals kept by a wealthy minority increased, threatening the common grazing

Aerial photograph of the deserted medieval village of Downtown in Stanford-on-the-Avon, Northamptonshire

A wide main street is marked by a holloway running away from the camera, and on either side banks and ditches define the rectangular tofts which once contained houses. Ridge and furrow of the village fields can be seen to the right. The site is an example of a village that has been totally deserted, but it also shows the regular planning of a settlement, in the twelfth century or earlier. A modern canal cuts through the site.

of the whole community, and as the more ambitious cultivators broke the rules of field management by putting strips down to grass, and enclosing them in blocks. Villagers were becoming more mobile than ever. Lords could not prevent the departure of serfs, and everyone saw opportunities in other villages or towns as the low population made land and jobs readily available. Emigration became an acute problem in some villages, especially those in which a few wealthy yeomen were disrupting the fields which lay at the very base of the community's livelihood. Villages commonly shrank in size, and some were totally abandoned.

An example where we can trace the stages of terminal decline is Compton Verney in south Warwickshire. In 1280 forty-four peasants, half of them yardlanders with 40 acres each, lived in this apparently healthy village. Their lords, the Mur-

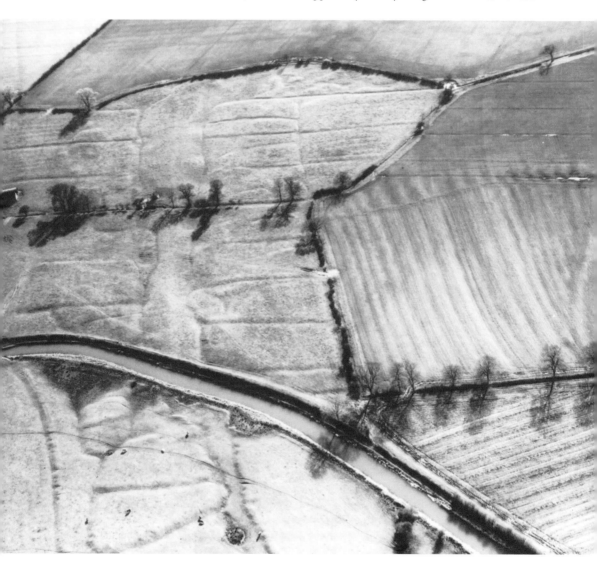

dak family, were not especially oppressive or demanding, and the tax records of 1327 show that many of the yardlanders enjoyed considerable prosperity. By 1400 almost all the old village families had disappeared, and the newcomers were changing the whole structure of the fields. They consolidated blocks of strips which had previously been scattered, and converted land to pasture. But for reasons that were unclear, they did not enclose their land, and there were many problems of trespass, both within the village and from nearby Combrook and Lighthorne. Peasants like Robert Cele held not the yardland of his predecessor but fragments of holdings, bits and pieces acquired from other tenants. Occupiers were allowing their buildings to fall into ruin. The lord scarcely knew what was happening to his once orderly manor, and, more important, the leading villagers seem to have taken no action to remedy this lack of control. Peasants left for other villages such as Tachbrook, and one serf's son became a schoolmaster in Warwick. By 1450 the lord's court could muster a jury of only six tenants instead of the customary twelve, and by 1461 a new lord, Richard Verney, had a rental drawn up which showed that most of the village fields were being leased out as pasture closes. Eventually the lord profited from the demise of the village, and so did the graziers who leased the newly enclosed pastures, and it is tempting to blame them for the desertion of the village. But while they may have hastened the departure of the last tenants, the community had collapsed from within. The site of the village has been obliterated by a park, landscaped in the eighteenth century, but excavation shows that the walls and floors of the medieval village are still lying under the surface of a field called 'Old Town'. The fields are once again under crops, but fragments of medieval ridge and furrow can still be seen, and some of the modern hedges, which were planted by Verney along the headlands of the decaying field system, still preserve the outline of the medieval furlongs. The new breed of graziers put together remarkable new estates from the pastures on deserted village sites, the most enterprising of these being the two John Spencers, uncle and nephew, of Hodnell in Warwickshire and later of Althorp in Northamptonshire.

New Circumstances and Solutions after 1500

A principal virtue of the clay lands, from the farmers' point of view, is their flexibility of land use. They can equally well grow arable crops or be put under grass and feed livestock. Some of the clays, of course, are much heavier than others. H. E. Bates described the worst in winter as 'a pudding of stodge that pulls the guts out of a man; in summer it turns into a land of concrete'. This kind never offers strong inducements to the ploughman. But for the most part vale soils are a mixture of light and heavy clays with varying amounts of sand and gravel. They have allowed for a mixed farming system under which the balance can be tilted towards more arable or more pasture as the economics of the situation command.

The flexibility of the clays had allowed farmers after the Black Death to turn much arable to pasture. New circumstances after about 1500 persuaded them to turn pasture once more into arable. The population had begun to rise significantly, a hungry population needed bread before all else, and so cereal prices after long decades of low rewards began to climb again. Strong objections against ploughing were overcome, and people passed judgement on the landscape

Crimscote in the Feldon half of Warwickshire

The photograph shows the curving strips of former open arable fields, fossilized in pasture. Note the scrub in the grassland, suggesting that it was taken during the years of agricultural depression in the 1930s. One well-to-do Crimscote husbandman in 1545 had no cereals growing when he died (though he had wheat, barley, and peas in store), but he had 4 cows, 15 oxen, 13 horses, and 200 sheep. Probably he was already, first and foremost, a grazier.

with very different eyes. Sir Thomas More, for example, denounced the sheep roundly; those seemingly meek, tame, and small eaters had devoured men, fields, and houses. While every device was now needed to increase cereal yields, the history of agricultural change after the Black Death was inseparably associated in people's memories with enclosure and the conversion of arable to pasture, and so Tudor governments concentrated their attention now on preventing any more. They also prohibited the amalgamation of small farms into larger units, another trend that always accompanies a boom in arable farming. They wanted their villages well populated with husbandmen, especially since strong-muscled ploughmen made the best soldiers; 'shepherds be but ill archers', was an adage of the age.

We see in these government statements how the archetypal image of England in the minds of politicians consisted of villages in vales; alien to their thinking was a landscape of hamlets or scattered farms, filled with a farming population of specialized cattle breeders, sheepkeepers, or dairymen, such as those found in pastoral country. Politicians legislated always for nucleated villages and common-field farming, and expected a manorial lord to be living in his manor house at the centre of things, next to the church.

Farming experience as a result of the Black Death taught one important lesson, however, which eventually modified the anti-enclosure policies of the sixteenth century. It showed the benefits of leys, short or long periods when the arable was left under grass, which enabled it to recover its fertility. The government continued to legislate fitfully against enclosure until the 1630s, but it learned from the countrymen's protests that arable yields were indeed much improved if fields were laid to grass for a few years and allowed a period of rest. A change in the routine of vale farming gradually took hold, as husbandmen put strips or whole furlongs down to grass for a period of years even while adhering to the system of common-field farming. They were no longer accused by officialdom of converting arable permanently to pasture and robbing labourers of work. In another respect the landscape underwent small changes under the pressure of an expanding demand for grain. Farmers strenuously improved the fertility of their soils by putting marl on the fields. Many marlpits were newly dug, or else those that had been neglected for long spells were renovated for further use. They dot the countryside to this day, often overgrown with shrubs and trees or else turned into ponds.

Government exhortations and threats against enclosure did not prevent it from continuing piecemeal throughout the sixteenth and seventeenth centuries. Some calculations for Leicestershire, so much of which, apart from the Charnwood and Leicester forests and the wolds of the north-east, sits in the Midland clay vale, show how phases of vigorous enclosure came and went. Before 1550 36 per cent of parishes had been enclosed, the Black Death having given the strongest stimulus. Between 1550 and 1650 44 per cent of parishes were enclosed; high cereal and wool prices were then the goad, though wool prices flagged after 1580 and grain became the more rewarding objective. Between 1650 and 1750 enthusiasm for enclosure subsided as cereal prices declined, and only 28 per cent of parishes succumbed. The next revival of interest had to wait on the industrial revolution, which would make fresh demands on the farmers after about 1750.

Changes wrought by enclosure in the landscape of vale parishes were radical in the early modern period. Hedges were planted to enclose small fields, and contemporaries defended them for giving more shelter to livestock and supplying much-needed fuel. Open areas of common pasture were subdivided, and old paths blocked or redirected. But across the larger scene, a patchwork of enclosed and open areas resulted since the pressure to enclose varied greatly between districts. In some parishes their small endowment of land gave little room to accommodate a growing population. When parish boundaries were defined between the tenth and twelfth centuries, congested populations in some areas had resulted in relatively small acreages of land being apportioned per parish. By the end of the sixteenth century, if not earlier, residents had pushed cultivation to the boundaries and had little or no spare land for commons and waste. If a determined encloser arrived in their midst, he deprived the whole community of precious common grazing after harvest. Some villages simmered with discontent on this account, while others became hotbeds of rebellion. Anger burst out in 1607 in the Midland Revolt, when hedges were pulled down and gates thrown open in certain parishes in the counties of Warwickshire, Leicestershire, Northamptonshire, and Bedfordshire. It left a particularly disturbing memory in the area between Rugby, Kettering, and Market Harborough.

More generally the bitter antagonism towards enclosure was assuaged in course of time, as people themselves found their own solution, devising local agreements to enclose. Manorial lords sometimes made peace with their tenants by giving gratuities or favourable leases to obtain their consent to enclosure; such a lease was given to the parson of Shearsby, Leicestershire, 'to stop his mouth', as another churchman bluntly reported in 1633–4. The opposition of ordinary farmers was to some extent worn down by the fact that they themselves became enclosers in a modest way, seeing its practical advantages, and presumably recognizing, as the momentum of enclosure built up, that change was under way which they could never entirely stop. So negotiated agreements to enclose became fairly usual by 1660. But long before then the landscape of the vales became a mixed scene of closes and common fields, sometimes mingling the two in single parishes (Lubenham in Leicestershire, for example, was partly enclosed in 1602, but not finally enclosed until 1766), sometimes separating the two farming systems between neighbouring parishes. Not surprisingly, some shrewd farmers exploited the opportunity to occupy both enclosed and common land in adjoining parishes for both systems had merits if you were supplying the market and wanted, say, to produce meat for the butcher in early spring (from grass springing early in your closes) and late autumn (when the stubble of the common arable fields was grazed). This farming strategy of the most commercially minded farmers has been well documented in the good grazing country around Lutterworth.

Tensions between the proponents and enemies of enclosure smouldered through the middle seventeenth century, and were heightened by the Civil War between Royalists and Parliamentarians. They make a vivid human story in Leicestershire where, between 1653 and 1656, two parsons, Joseph Lee at Cotesbach and John Moore at Knaptoft, became locked in public argument for and against enclosure. Moore knew villages that had been depopulated since 1600 by enclosure, while Lee, equally forcefully, denied depopulation, denied the loss of ploughland, and pointed to the run-down condition of the commons, and the greedy men who put out excessive numbers of animals to graze on them; he saw no future in common-field husbandry. Parliamentarians were trying hard at this time to improve the efficiency of farming, and so Major-General Whalley, administering affairs in the Midland counties, tried to find a middle way, suggesting legislation that would make enclosure acceptable and fair to all classes of farmers. Parliament, however, paid no heed to his sober proposals. Its MPs were gentlemen who feared to lose control over their property. So would-be enclosers had to find their own way, making peace locally, if they could, with their neighbours.

Fields in the vales, whether enclosed or open, continued to be much taken up with the growing of wheat, the most desired crop at the market, and barley, usually though not always the second choice of cereal since it supplied both beer and bread. Beans and peas in the third year of a three-course rotation fed livestock. But a new crop arrived in the vales in the seventeenth century, making something of the same impact as rapeseed in the 1970s. This was woad, bringing a fresh yellow colour to the fields when it flowered in May, and yielding a blue dye from its leaves that was essential for dyeing cloth blue, black, purple, and green. Whenever an occupier wanted to plough land after some years under grass, he did well to choose woad for two or three years, since it checked wireworm before cereals were planted. First, however, he had to negotiate with woadmen, who would manage the crop for him, grind the leaves, and prepare the woad balls for the dyer. Certain families travelled around the Midlands offering these skills. They set up their portable woadmills, and lived in the parish for two, three, or four years, before moving on to another place where woad growing was planned. Certain surnames are associated with woad people, for example, Pickering, Jeacock, Noble, and Vines, for the job ran in families. They are found in one parish register marrying, and bearing children, and then they are gone. But family historians find them in another place and yet another; pursuit of them has become a fascinating detective story. Coventry was a noted dyeing centre where the woad people could be found, and a cropping agreement concluded. Their temporary cabins of turf and wood, lasting only a few years, seen through the summer, were a familiar feature of the vale landscape, especially in Northamptonshire, a county which was said by one early eighteenth-century historian to have been 'woaded most'. In the early nineteenth century, people were still familiar with the 'wadders' huts', and even today seeds of woad long buried are liable to sprout if disturbed by deep excavations. The plant may become much more easily recognized hereafter if the research at present under way is successful in making woad into an ink for bubble-jet printers.

The pastures of the vales accommodated a miscellany of animals: sheep yielding meat, which was said in the sixteenth century to be winning favour over beef, and a second grade wool, less fine than that produced on the downlands, but entirely satisfactory for many varied, new textiles, and certainly good for knitting. Cattle produced good beef, and although the vales in the eastern half of England were not usually prime dairying country, some dairying spread there in the second half of the seventeenth century when once that speciality became a more deliberate commercial operation. Many farmers bred a few horses for extra income (a nicely matching pair for the carriage could pay them well), and in any case horses increasingly replaced oxen as draught animals. Pigs were nowhere seen in great numbers in the vales for they had to be fed on kitchen waste, but most countrymen down to the level of the cottager expected to have a pig or two, and Leicestershire farmers fed them with beans; hence their rallying cry for 'beans and bacon, food of kings!'

Creeping enclosure in the vales altered the landscape, but it also contributed to changes in the social structure of many villages, and this too would have struck travellers forcibly. In a village where one lord reigned supreme over his tenants and was, moreover, resident, he expected to command the villagers' lives and was likely to be restlessly watchful. Sometimes his paternal concern was benevolent and employment in his house and on his home farm was an advantages to the whole community. The village usually betrayed this circumstance in its layout and the lifestyle of its tenants. Such a place was described as a 'close' village, for the lord regulated settlement and often did not encourage incomers since they might end up as a burden on the poor rates of the parish. In villages where the lord was not resident and was negligent, or in villages where several lords shared authority between them, more freedom prevailed among the tenants and the village was labelled 'open'. Sometimes authority passed to the freeholders, and they even encouraged incomers, building cottages for them, since the rents yielded an extra income. Such villages tended to grow in size whereas close villages stayed small. The difference between the two kinds could be even more obvious if the residents in open places took up industrial work, combined with their farming. Industries had usually first taken root in forest and pastoral areas, interspersed between the lands of the vales, but if successful, they were likely to spill over into open villages beyond. Thus in Leicestershire and Northamptonshire a knitting industry started in the forest areas, and then spread into the vales. In Leicestershire it expanded into two sectors after the mid-seventeenth century: some stockingers knitted by hand while others used the new stocking frame. In Buckinghamshire a lace industry took hold. In central and west Hertfordshire and in the Bedfordshire vale lands the plaiting of wheat straw for hats made another occupation for the poor: we catch a glimpse of its life-saving value in the possessions of Mary Young, spinster of Redbourn in Hertfordshire, who lodged in a single room in 1691 and had a stock of straw and straw hats when she died that was twice the value of her furnishings. We can readily picture her, like other industrial workers, sitting on a fine day at the door of her house in the village street, pursuing her craft in a good light and exchanging chat with her neighbours. Varied work in village-based industries relieved some of the poverty in the vales as enclosure steadily deprived country folk of land and work in the fields.

Agricultural and Industrial Revolutions after 1750

When a renewed burst of enclosure began between 1755 and 1780, the vale lands of the Midlands were again deeply affected, though the procedures were now changed. Parishes wishing to enclose sought individually a private parliamentary act and if this was seen to represent the wishes, not of the majority of the landholders, but of those holding the majority of the land, then it was likely to be approved. Commissioners were appointed to draw up an agreement and enclosure went ahead. About 30 per cent of Midland parishes were transformed by these parliamentary enclosures, and in Northamptonshire the figure rose to 50 per cent. All open-field arable systems which had survived till then, whether whole or in part, were now swept away, only the minority holding out until the nineteenth century; the last in Leicestershire was Medbourne, enclosed in 1842. Laxton alone, in Nottinghamshire, has miraculously managed to cling to the old methods of management, and survives to this day.

Everywhere else in the clay vales the huge common arable fields which had served many parishes since the Middle Ages passed out of sight. A more geometrical pattern resulted of small, square, or rectangular fields enclosed by hawthorn hedges; and similarities in layout between one parliamentary enclosure and the next reflected the fact that the change was completed in a relatively short period, mostly within a generation, and that the same commissioners arbitrated over many enclosures in the same district. At the local level, young people grew up to see the hedgerows mature and red-brick farmhouses become established in fields away from the village on sites which had never been inhabited before. This planned Georgian landscape gave a new face to the clay vales, different in many of its features not only from the older common-field landscape but from vale countryside that had been enclosed slowly and piecemeal in earlier centuries.

The long, straight lanes authorized by the enclosure awards cut across the vale lands in a way that had not been attempted since the Romans. They were laid out at standard widths, often of 40 feet, and formed a pattern which is readily recognized wherever it occurs. It is thoroughly different from the winding, sunken lanes of old enclosed countryside. The metalled strips which now run down their middle are twentieth-century improvements for cars and tractors; they do not need to be as wide as the original unmetalled surfaces which had to offer a route even when part of the surface was worn into deep ruts. The resulting wide verges are covered in tall grasses and Queen Anne's lace (or, if you prefer a drab name for the same plant, cow parsley!) and brambles are tucked into the hedgerows. Sometimes enclosure commissioners simply improved existing tracks, but most of the lanes were newly made, forming junctions at sharp angles with other lanes that now provided access to the new Georgian farmhouses. These last were often given topical or fanciful names, such as Quebec House, Hanover Farm, and Belle Isle, all found in the one Leicestershire parish of Sileby.

The new, tall, symmetrical farmhouses were usually double-depth in plan, very different from the older farmhouses of the villages, though here too Georgian farmhouses were soon erected. They were warm and comfortable, with rooms leading off each side of a central hallway. The bricks were hand-made and of various textures and colours, for it was the usual practice to clamp-burn them locally for each job. Local tiles, notably those from the famous Swithland quarries, were used for roofing. So each house looked different, despite their standard design. Very many of these Georgian farmhouses survive, albeit frequently with Welsh-slate roofs, but their visual impact has now often been reduced by drab, modern outbuildings and high silos. Subsidies can be obtained for farm buildings but not for farmhouses, and the new erections are not subject to planning regulations. Not all farmhouses that were built at enclosure were as grand as these, of course. Humbler buildings and field walls continued to be constructed of mud walling, 2 feet thick, which was reckoned to last 150–200 years. A photograph of Naseby (Northamptonshire), taken in 1855, shows many houses and cottages that were still built with mud walls and roofed with thatch. The Reverend John Martin described them in his *History and Antiquities of Naseby* (1792) as being 'exceedingly firm and strong, and, if kept dry, are said to be more durable than if built with stone or indifferent bricks'. The walls were coated with cow dung once a year, to protect the mud and to allow the dung to dry before it could be used for household fires.

The Poor Law commissioners of the nineteenth century became thoroughly conversant with the division of villages into the two types: 'open' and 'close', with a large number that were not quite one or the other. The ones that were closed to the poor included numerous estate villages: in Northamptonshire two out of every three villages were of this kind, owned either by a squire or by a small group of proprietors. In Leicestershire, in contrast, a county with an ancient tradition of numerous small freeholders, villages were more evenly divided: some 134 in the nineteenth century were closed to the poor and 174 were open. Since the latter were free of supervision by a squire, they attracted ever more poor immigrants, and were often rambling and untidy. Their inhabitants, moreover, were far more radical in politics and religion than the tenants of estate villages who had the consolation of superior housing.

As the national population grew enormously in the nineteenth century, more people in the open villages turned to crafts for their living, and what had been part-time crafts turned into miserable cottage industries. Hosiery became wholly dependent on middlemen, known as 'bag hosiers', who were based in the towns of Leicestershire, south Nottinghamshire, and south Derbyshire. Red-brick terraces of cottages and small workshops came to dominate many of the villages in these areas. Further south another industry using the clay to industrial advantage persisted quietly for centuries and then grew into a major enterprise. Small brick kilns had long lain scattered among the fields of south Bedfordshire. But digging for clay surged greatly in the Fletton area, so that now the London Brick Company provides up to a half of the country's brick production, and its huge clay pits cannot fail to catch the eye of train travellers nearing Bedford. Elsewhere in the Midland vale small-scale industry, perhaps in the form of a brewery or an ironworks, a tannery or a corn-milling business, altered the appearance of country market towns, many of which took advantage of canals, turnpike roads, and railways to prosper in a modest way.

Naseby, Northampton-shire, in 1855

This early photograph shows that most of the houses, cottages, and field walls were constructed of mud and the buildings were roofed with thatch. In 1792 the local clergyman wrote of the village: 'If we except a few of the modern and best houses, it is built principally with a kind of kealy earth dug near it; excellent in its kind, and the best calculated for building I ever saw.'

Most vale villages remained completely agricultural, however. In *Lark Rise to Candleford* (1939) Flora Thompson described the hamlet of Juniper Hill 'in the flat, wheat-growing north-east corner of Oxfordshire' as she remembered it as a child in the 1880s:

> All around, from every quarter, the stiff, clayey soil of the arable fields crept up: bare, brown and windswept for eight months out of the twelve . . . only for a few weeks in later summer had the landscape real beauty. Then the ripened cornfields rippled up to the doorsteps of the cottages and the hamlet became an island in a sea of dark gold.

This was former furze or common land which had come under the plough after enclosure. Elsewhere, most of the heavy arable land was soon converted to pasture.

Naseby
Sep. 25 1855.
WL

The square or rectangular fields approved by the enclosure commissioners varied in extent according to the size of the previous farms, that is, in proportion to the value of a farmer's former common rights over the open fields and common pastures. The new plan was drawn up by a surveyor working at his desk and then imposed upon the landscape with little regard for former features. Older patterns of ridge and furrow can often be seen, even at ground level, but especially from aerial photographs, continuing beyond the hedges of the new enclosures and bearing little or no relation to the new boundaries. John Clare looked with sadness at the visual effects of enclosure in his native Helpston, between Stamford and Peterborough, in his poem 'The Mores':

> Fence now meets fence in owners little bounds
> Of field and meadow large as garden grounds
> In little parcels little minds to please . . .

Where the vales were characterized by numerous small freehold farms, the new allotments were usually partitioned in fields of 5 to 10 acres, though all parishes also had larger farms of up to 50 or 60 acres. In grazing districts these larger farms were soon subdivided into a number of small fields of about 10 acres each. Some of the much older enclosures in the Midland vale were also further subdivided in the early nineteenth century. Cattle moved from one 10-acre field to another, so that they were always eating fresh grass.

The enclosure apportionment defined lots that usually had to be fenced within a year of the award, so we can date many of the existing hedgerows precisely. The internal divisions of the larger pieces came a little later. New fences were commonly made of hawthorn, known as 'quickset', planted on a bank with a shallow ditch on one or both sides, so that the undulating countryside was soon divided by miles upon miles of white may-blossom. Proverbial advice to 'ne'er cast a clout till may is out' almost certainly refers not to the month but to the heavily scented blossom. These endless hawthorn hedges make a sharp contrast to the multi-specied, ancient hedges in the countryside of old enclosure. They are dotted with ash trees, and sometimes elm, spaced out at wide intervals, which, together with the willows which were planted along the banks of the streams, give the vale land of the Midlands a more wooded appearance than it really possesses. When they came to full growth, the hedgerows also added enormously to the bird population.

The comparative scarcity of trees in the hedgerows—especially in Northamptonshire and Leicestershire—may be the result of the passion for fox-hunting. The strict rules of the organized fox-hunt had been made by Hugo Meynell, the first master of the Quorn hunt (Leicestershire) in the second half of the eighteenth century. At first, fox-hunting had attracted supporters from across the social spectrum. But by mid-Victorian times it had become an expensive sport that was restricted to the upper and middle classes. Conversion to pasture after enclosure allowed long gallops over miles of rolling grassland with hedges that could be jumped, but enclosure and the improvement of heaths and commons had reduced the extent of the natural gorse patches where foxes could hide.

In the late eighteenth and early nineteenth centuries the fox-hunters, therefore, planted gorse coverts and spinneys, between 2 and 20 acres in size, usually on the edges of parishes and sometimes using former pieces of common land. Ordnance Survey maps in fox-hunting country are marked with numerous small splashes of green, with such redolent names as Botany Bay, Fox Holes Spinneys, Norton Gorse, and Lord Morton's Covert, all of which can be found to the east of Leicester. Fox coverts can be readily distinguished from ancient deciduous woodland by their names, sizes, and shapes. Such plantations are usually the only clumps of trees to be found in the clay vales. Their use is well illustrated by Anthony Trollope, who, in his final Palliser novel, *The Duke's Children*, depicts a character who grumbles endlessly about the lack of a gamekeeper for Trumpington Wood on the Duke of Omnium's estate. 'Nobody shoots there', said Lord Chiltern, 'because there is nothing to shoot. There isn't a keeper. Every scamp is allowed to go where he pleases, and of course there isn't a fox in the whole place.'

The old open arable fields had depended on communal co-operation in their management, but when turned into privately owned fields they allowed enterprising landholders to tackle their serious drainage problems. Lack of good drainage was a bugbear everywhere. Peter Bigmore, writing about the clay vale south of Bedford, wonders if John Bunyan, born in Elstow, near Bedford, perchance had this district in mind when he described the Slough of Despond, rather than Tempsford on the Great North Road with which his despondency is usually associated. Nineteenth-century writers returned constantly, though in more prosaic terms, to the mud of the clays in winter, all 'muck and misery': in 1850 Philip Pusey put the need for drainage at the top of his list—'if there be any land which requires improvement, it is our heavy clays', he wrote; the following year James Caird chose three adjectives to describe clayland farms in north-west Wiltshire—'wet, filthy, and depressed'.

The earliest method of draining fields had involved the cutting of either parallel or herringbone patterns of ditches across enclosures. These had then been packed with brushwood or stones, and filled up with soil. A less popular method was to use horses or a windlass and cable to drag a metal plug or 'mole' beneath the surface of the fields. Both types of drains tended to get crushed by trampling hoofs, so that farms on the heavy clays had remained notoriously underproductive. Underdrainage schemes became much more effective from the 1840s when cheap tiles and drainpipes were first manufactured in the Potteries. Government grants for underdrainage became available in 1846, and large areas of heavy land, especially those on the estates of the wealthier landowners, were soon improved. A vivid picture of the effects is embedded in an unusual book, entitled *Talpa: Or the Chronicles of a Clay Farm* (1852), written by C.W.H. (Chandos Wren Hoskyns), farming at Wroxall (Warwickshire)—in the Arden rather than the Feldon country. Drainage pitched at the correct depth was a consuming interest of the author, while his turnip crops, which the drains made possible, were another. This serious book is permeated with a strong sense of fun because of the drawings by George Cruikshank.

Wherever former arable land on the heavy clays was converted to permanent pasture, the ridge and furrow patterns of open-field farming became fossilized. Even now, despite modern deep ploughing, the fields in large stretches of the clay vales still have prominent curving ridges decked with buttercups in the spring and deep furrows that are often filled with rainwater or late snow. The conversion to grass occurred within a generation or two of enclosure. When Wigston Magna (Leicestershire) was enclosed in 1764–6, the parish had about 600 acres of permanent grass amongst the 2,887 acres of fields and commons; by 1832 the grassland amounted to 2,000 acres. The new pastures were improved by a variety of manures, notably guano, which was imported from South America in the 1840s, and bone meal which was widely used by the mid-nineteenth century. Government grants for manures were available from 1856 and all market towns had retail suppliers of artificial fertilizer by the 1860s. Selective stock breeding had begun in the previous century, bringing much success and publicity to Robert Bakewell of Dishley, near Loughborough, and more thought was now put into the management of land, crops, and stock. For example, by feeding cattle with more grains and oilcake, farmers became less dependent upon hay, so that meadowland could be turned into pasture and the increased numbers of livestock yielded more manure. The use of artificial fertilizers meant that less arable was needed to produce the same yields, so fewer working horses were required and therefore less fodder. More land could be used for dairying to supply milk to the towns.

During the period of 'High Farming' from 1846 to 1873, when national output increased by about 60 per cent, farmers of vale land were tempted to put more land under the plough and the proportion of permanent pasture actually declined. The agricultural depression starting in the 1870s hit them hard. Imported cereals from the American prairies forced down wheat prices by one-half and those of barley and oats by one-third. At the same time, Australian imports caused wool prices to fall sharply. Corn-and-sheep farmers were badly affected and their plight was made worse by a series of wet summers between 1878 and 1882. Those farmers who concentrated on supplying milk and other dairy products to large towns, via the railways, continued to prosper.

In 1866 about 60 per cent of Leicestershire's farmland already lay under permanent grass, but from 1880 onwards, during the long depression that lasted until 1939, that acreage continued to rise, suffering only a temporary reversal during the First World War. In 1938 Leicestershire was reckoned to have 86 per cent of its farmland under grass, the highest proportion ever known in its history.

The Second World War forced another dramatic change on the landscape, however, for farmers in the vales had to plough as much land as they could to produce home-grown food, and the 'plough up' campaign was so successful that by 1943 the county of Leicestershire, as one example, had 51 per cent of its farmland under the plough. The same demand for home-grown foodstuffs persisted after the war, so that by 1979 the county's arable amounted to 62 per cent, and its grassland 38 per cent, the same as the national average. At present the clay vales are again mixed farming districts but with a difference; cereals and oilseed rape are grown in rotation with intensively managed temporary grassland. But machinery and the drive to achieve economies of scale have led to the absorption of smaller farms into larger units, and have drastically changed the appearance of the landscape. Hedgerows have been grubbed out to make larger fields and deep ploughing has eradicated much of the ridge and furrow. The wide open spaces of Laxton's common fields, which still survive though much buffeted since 1945 by changing times and a changing outlook, no longer present such an exceptional scene. Moreover, the change of routine, whereby farmers plough and sow immediately after harvest, so that the fields are green before the end of the year, means that there are no more stubbles and that the countryside is no longer bare during winter as it was when Flora Thompson was young.

Little Moreton Hall in Cheshire
This late fifteenth- and sixteenth-century house was built on a courtyard plan by successive members of the Moreton family. The abundant use of timber for construction and ornament is characteristic of the north-west Midlands. Many houses of this type still survive, reflecting the wealth and social ambitions of gentry families. The moat surrounding the house, a feature which is especially characteristic of the woodlands, probably originated in the thirteenth century.

Green Man

When the cloister was rebuilt at Norwich Cathedral between 1297 and c.1430, nearly 400 bosses were incorporated, including several in the foliate head or 'Tête de Feuilles' tradition from 13th-century France. The interaction of humans or gods with woodland is symbolized here in this wonderful mask of hawthorn (may) leaves, removed from its original pagan symbolism to take its place in Christian iconography or decoration in the east walk of the cloisters.

Halvergate Marshes from the air

At the bottom, the long, curving line of a lost tidal watercourse—the Halvergate Fleet—is still marked by the line of dykes and tracks. To the north, the water-filled dykes are of varied origin. Serpentine watercourses are survivors from the original system of salt-marsh drainage; innumerable straight dykes represent piecemeal modifications to this pattern.

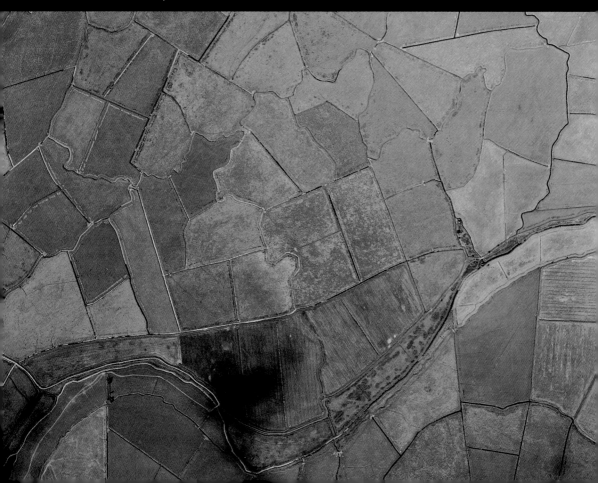

Remains of St. Benedict's Abbey Henry Bright (1810–1873)

St. Benedict's or St. Benet's Abbey is the most well-known sight on the Broadland marshes, the subject of innumerable sketches, paintings, and photographs. The ruined medieval gate house was used, in the 1740s, as the base for a drainage windmill. Other mills can be seen far away across the marshes.

View from Hayeswater Gill, looking west towards Helvellyn, over the great fell-commons of Deepdale, Patterdale, and Glenridding in the Lake District. Here the ancient wastes survive, but they have always been used as sheep-pasture and for many other purposes. The hamlet of Hartsop, in this valley, was formerly noted for its lead mines, and Patterdale for its slate quarries.

The southern English end of the Debateable Land
The parish of Kirkandrews-on-Esk had belonged to the diocese of Glasgow, and therefore to Scotland in the thirteenth century. In the early sixteenth, this neutral ground was illegally settled by Grahams (and others), one of whose tower houses is shown.

Detail from *Dedham Vale* by John Constable (1828)
Looking east towards the Stour estuary, the picture shows how the valley opens out at this stage, with the extensive north flank on the left, but with scrubland on the still pronounced scarp to the right. By this period, Dedham, the only major settlement on the Essex side, was considered to be a town.

Stonor, from a painting of c.1690

This shows the formal E-shaped Tudor façade, built of warm local brick, behind which many older, medieval elements remained. In the eighteenth century the Tudor forecourt, wall, and gatelodge were removed, sash windows replaced the mullions, and the gables were changed into a hipped roof with dormer windows, but the 1690 picture remains quite recognizable today. The ancient Catholic chapel is on the right of the picture.

THE
Healthy Future

FOR YOU
AND YOUR
FAMILY

VEALE
GILCHRIST
STUDIOS

LAND SETTLEMENT ASSOCIATION

The Healthy Future
The booklet published *c.*1938 advertised the LSA as an organization that provide
capital and knowledge for those who had ambitions to go on the land but woul
normally be unable to do so. Some LSA holdings were available for unemployed me
outside the designated Special Areas. Smallholdings at Rookery Farm, Wyboston, par
of the Chawston Estate in Bedfordshire, were reserved for London unemployed

4 Woodlands and Wood-Pasture in Western England

Christopher Dyer

Here we will examine the landscapes which account for a high proportion of the countryside of western England, from Cheshire to Dorset. These landscapes are taken for granted, and we do not have a single agreed term to describe them. They are commonly regarded as ordinary and unremarkable, though in fact they have a distinctive personality and well-established character. The early topographical writers sometimes used the word 'woodland', not meaning a piece of territory covered completely by trees, but a country with a mixture of pasture, woods, arable, and heaths, in which fields were often hedged. When Leland travelled through western counties in the early sixteenth century he noted the main characteristics of the land on either side of the road, and used phrases like 'metely well wooded', or 'woody and enclosed', or 'corn, pasture and wood' when he passed through north Warwickshire and Worcestershire, southern Shropshire, southern Staffordshire, and central and eastern Herefordshire. On his journey from Bradford-on-Avon to Trowbridge and Frome, through north and west Wiltshire, he saw 'stoney and woody ground' and 'woody ground and pasture'. The custom has developed among modern landscape historians to make a distinction between woodland and champion, or woodland and feldon, the terms 'champion' and 'feldon' being old usages to describe a countryside of common fields. When landscape types lie close together, as they do in Warwickshire, a frontier, coinciding roughly with the valley of the river Avon, can be defined between the feldon of the south-east and the wooded Arden of the north-west. In Worcestershire the champion of the vale of Evesham and the adjacent Lias clay lands in the centre of the county are in sharp contrast to the woodlands in the west and north. In the south-west the clay vale of Wiltshire, which specialized in dairying, has been dubbed 'cheese' country, the very opposite of the 'chalk' of the centre of the county, where sheep and corn husbandry was practised. When modern travellers cross these frontiers from the south or east into the woodlands they leave countries with large nucleated villages and large rectangular fields made by parliamentary enclosure, and enter a land of straggling half-formed villages, hamlets, and fields often of irregular shape defined by thick and ragged hedges. The sharpness of the divide between the chalk country of Dorset and the Vale of Blackmoor in the north of the county was very apparent in the 1880s to Thomas Hardy, describing the setting of the home village of his character, Tess Durbeyfield: 'The traveller from the coast, who, after plodding over … downs and corn-lands, suddenly reaches the verge of one of these escarpments, is surprised and delighted to behold, extended like a map beneath him, a country differing absolutely from that which he has passed through.' In contrast with the chalk downs' large fields, low hedges, and 'colourless' atmosphere, 'in the valley the world seems to be constructed upon a smaller and more delicate scale; the fields are mere paddocks … Arable lands are few and limited; … the prospect is a broad mass of grass and trees …' In Leland's day the boundaries were even clearer; as the champion, feldon, or chalk countries were largely unenclosed with much arable, and the settlements of the woodlands were smaller and more scattered. But 'woodland' is rather an ambiguous term. Part of our western landscapes lay in royal forests in the twelfth and thirteenth centuries, such as Shirlet and Morfe in Shropshire, Feckenham in Worcestershire, and Selwood and Blackmoor in the south-west, and while the land under forest law included many settlements and much agricultural land, there were some tracts of wood with continuous tree cover. In north Warwickshire lay Arden, often called a 'forest', but never under royal forest law. It contained some large woods in the eleventh century, many of which survived into the later Middle Ages and beyond. On the other hand the woods were sometimes absent, and Leland remarked, in central Herefordshire for example, that there was 'no great wood', and he noted the combination of arable and pasture. In such areas now, and for many centuries, the main tree growth has been in the hedgerows, which can be plentiful enough to give a wooded appearance from a distance.

An alternative way of describing these landscapes is to call them 'wood pasture' or 'pastoral lowlands', which derives from the way in which the land was used in the early modern period, and indeed these descriptions can be applied to other periods, such as the later Middle Ages. The use of the term 'lowland', or sometimes 'pastoral vales' is useful for drawing a distinction between our midland landscapes and the nearby higher ground, with its very different economy and culture, including a much more pronounced specialization in animal husbandry. The hills which lie adjacent to woodlands or wood pastures on the extreme west of Herefordshire and Shropshire, or the ridges that run through Shropshire such as the Long Mynd and Wenlock Edge, or the western slopes of the Pennines and the Peak District which take up much of eastern Cheshire and north Staffordshire are not our concern here. Instead most of our landscapes lie below 500 feet, many of them in the valleys of lesser rivers such as the Cheshire Weaver, the Corve in Shropshire, the Lugg, Frome, and Teme in Herefordshire, and the Stour in north Dorset. The land rises to form a plateau in north Warwickshire, but neither here nor on the similarly rather high ground in parts of central Shropshire could the terrain be described as very hilly.

For the modern visitor, as Thomas Hardy remarked, these are landscapes on a small scale. They lack the wide spaces and grand vistas that we appreciate in the moorlands or the chalk hills. Today, with the enlargement of fields, the country is more open than at any time in its recent history, but even now the view is constrained by hedges and the undulations of the terrain. Often the visitor is travelling along lanes with little visibility beyond the road itself, because the carriageway lies at the bottom of a holloway, and the earth banks flanking the side of the road are surmounted by high hedges. These lanes wind a good deal, because these landscapes, having escaped parliamentary enclosure, also missed the new network of straight roads laid out with an eighteenth-century surveyor's ruler alongside the neat new field hedges. Houses are rarely far away, and settlements are encountered scattered along the roads. Often we see a single cottage, tucked into the roadside, with an elongated garden, which had once clearly been part of the verge. Or the houses stand in groups, and sometimes even a large village is encountered. These villages are often associated with smaller hamlets, so they are not the only settlements in their territory, and in some districts although the modern village has no urban pretensions, a former market place and neat rows of regularly spaced plots facing the main street betray its medieval founder's ambition to create a town.

This countryside derives from its western situation a relatively high rainfall and mild climate. It suffers less from summer droughts than the eastern side of England, and there is no shortage of ponds and small streams. The soils vary, but tend to be quite heavy, with clays, and in the northern counties, marls, called Mercian mudstones, which are capable of bearing crops but are also often used for pastures. The land under arable has been extended considerably since the Second World War, but while visitors will now see many large cornfields created by grubbing out hedges and merging three or four former 'crofts', it is still possible to find sheep and cattle grazing in the shade of stagheaded oak trees in old pastures enclosed by high hedges, in much the way that their predecessors have done over the last five centuries.

The woodland landscapes may not offer great scenic attractions or even leave the visitor with any very striking impression, but we must regard them as an important section of the English countryside, covering a large area and providing a living for many people. The density of population that they supported between the eleventh and the seventeenth centuries was near to the average for the whole country, and in the last two centuries in parts of eastern Shropshire, south Staffordshire, and north-west Warwickshire industrialization has attracted very large concentrations of inhabitants.

The Origins of the Woodlands, up to 1000

The obvious first question that presses itself upon us is, 'How old are these landscapes—when did the lanes, hedges, and hamlets first emerge?' The traditional interpretation emphasized their late development. The more obvious prehistoric monuments, such as barrows, were rare, and no great number of objects datable to the Roman or prehistoric periods from the woodland landscapes were preserved in museum collections. As the soils were generally rather heavy and damp, it was presumed that they were not extensively cleared and settled until the Middle Ages. Until the 1970s it was still possible to say of Shropshire that 'The area was composed partly of Glacial and Triassic clays of a type abhorred by early cultivators, partly of dense and damp oak woodland which effectively resisted every attempt at penetration', and another writer described south-west Worcestershire

Aerial view of Romano-British farm site, Pendock, Worcestershire

Settlement and agriculture in the woodland landscape of Malvern chase has ancient roots. The dark crop-marks in the field in the centre reveal the existence of buried ditches. The rectangular enclosure surrounded a Romano-British farmstead, and the other marks represent field boundaries of the same period. To the right is ridge and furrow, part of the West Field which existed from the Middle Ages until the early nineteenth century. To the left is a road of medieval origin, and beyond that the M50 motorway.

before the Middle Ages thus: 'On present evidence it would appear that a considerable extent of genuine primeval forest still survived … and that permanent settlements were, as yet, few in number.' The river valleys were seen as exceptions to this rule, because of their lighter soils, but even then their role as routeways rather than zones of permanent settlement was emphasized—for example, occasional caches of tools, weapons, and scrap metal dating from the Bronze Age were thought to have been lost or hidden by itinerant metalworkers on their passage to some more important focus of occupation.

Now we know that woodland landscapes contained a considerable amount of settlement in the Neolithic and Bronze Age, and supported quite high populations in the Iron Age and Roman periods. This complete reversal in the assessment of human activity in such a large section of the English countryside has been made possible by the more systematic use of new evidence. Aerial photography shows abundant indications of early occupation in the major river valleys, such as the Severn or Wye, which is not surprising, but a growing number of sites have been found in the valleys of minor streams and indeed on the heavier soils which in theory repelled early settlers. Characteristic monuments of the Bronze Age are ring ditches, dug in the course of building round barrows, which were used for burials and also for more general ritual functions—an impressive group of twenty has been found at the confluence of the rivers Onny, Corve, and Teme at Bromfield in Shropshire. Few settlements are known from this period, but the burials and ritual sites must have been associated with a local permanent population, who were exploiting the heavier soils beyond the narrow confines of the gravels of the river valleys where the ring ditches are observed. Such settlements are revealed more directly by finds of flint implements on the clay and marl soils when they are subjected to systematic field-walking, that is when the ploughed surface of modern arable fields is examined closely for archaeological evidence. An indirect sign of extensive cultivation in the Bronze Age comes from the layers of soil found in the banks of major rivers such as the Severn, which began to accumulate around the beginning of the first millennium BC. Large quantities of earth were being washed by rain into the river and its tributaries from newly cultivated fields, and the silts were deposited by the water much further downstream.

In the Iron Age and Roman periods isolated farms were established in some numbers in the river valleys, and judging from a scatter recorded on aerial photographs and from field-walking, also on the heavier land. They often consisted of timber buildings surrounded by a ditched enclosure, taking in no more than an acre, and sometimes linked with rectangular, ditched field systems. These farms sometimes originated in the Iron Age and continued into the Roman period, and others were founded in the first or second centuries AD, so that by the height of rural development in the Roman period the exploitation of land can be described as intensive. In sample areas of north and west Worcestershire, for example, on typically 'inhospitable' soils, Roman farms have been recorded at a density of one for every square mile, and this must be a minimum figure given the incomplete nature of our evidence. In fact, a sizeable farming population could have been predicted from the presence of sites with large numbers of people who were not growing their own food. In the Iron Age, hill forts were especially numerous in Herefordshire and Shropshire on the high ground overlooking the clay vales and minor river valleys. Excavations on a number of these show that they contained a large and apparently permanent population living in small, closely packed huts. The building and maintenance of the elaborate fortifications and the provision of the inhabitants with food, fuel, and raw materials for building and crafts must imply that extensive farmlands were being worked nearby. The same argument can be applied with greater certainty for the Roman period, when large cities such as Viroconium (Wroxeter) in Shropshire, the great fort and settlement at Chester, and smaller towns like Kenchester in Herefordshire would have depended on a developed rural hinterland for their food, and would in turn have supplied the farms with manufactured goods. The Roman towns at Middlewich and Nantwich in Cheshire, and at Droitwich in Worcestershire specialized in salt extraction (continuing an Iron Age tradition), and ironworking was practised on a large scale at the Herefordshire town of Ariconium (Weston under Penyard). Such places were especially dependent on a

rural workforce which must have been employed in cutting and carting large quantities of fuel, and the local woods would have been kept under careful management to produce crops of underwood for fuel or charcoal. Industrial activities, such as pottery and tile-making, were also sited away from towns, presumably because these fuel-hungry activities were most conveniently sited near good supplies of wood and clay. The location of these kilns in western Worcestershire and north Warwickshire shows that we should not take our doubts about the 'oak wood' depiction of the pre-medieval rural scene too far— in the Iron Age and Roman period large areas were still covered by trees, but we must suspect that many of these woods were not 'genuine primeval forest' but managed coppices at no great distance from the settlements that exploited them. They also would have been associated with areas of grazing land.

The people who lived in our woodland landscapes in the Roman period do not seem to have achieved a very high level of wealth or sophistication. Most of the farms had wooden buildings, and the inhabitants used local pottery and limited quantities of metalwork. Some lived in villas, but in the northern woodlands in particular these were few and not very opulent, like that at Eaton by Tarporley in Cheshire, which was built to a simple plan and of modest size. The peasants who presumably occupied the majority of the farms may have been subordinate to some estate, but if there was a rural aristocracy their residences cannot be traced in the countryside, so they may have been based in the towns, or in some cases the land was subject to the administration of the imperial state.

We can conclude that the woodlands had been inhabited and cultivated throughout prehistory, and that when Roman imperial government ended, there was an extensive man-made landscape, both in terms of settlements and the division of the land into fields and managed woods. The inhabitants were served by a network of roads, both the older system of routeways, and the imperial roads established for military purposes but also used for trade, such as those radiating from Middlewich and Droitwich along which salt was carried. But is all of this evidence for the early landscape relevant for discovering the origins of the medieval and modern pattern of fields, settlements, and estates? Surely the break with the Roman past in the fifth and sixth centuries was so sharp that in effect a fresh start was made in the early Middle Ages? Most Roman sites were abandoned, so that few medieval hamlets or farms lie directly over a Roman predecessor, and indeed the countryside was transformed to such a degree that land containing Roman sites later became wooded, as trees regenerated over abandoned settlements and cornfields.

This rupture in the post-Roman period cannot be the whole story, because there is also evidence for the continuation of early field boundaries into medieval and later times. Around the south Staffordshire town of Lichfield a regressive technique has been used to reconstruct an early pattern of rectilinear fields, and as a Roman road seems to cut across the field-system, ignoring the orientation of the boundaries, it is presumed that the layout predates the construction of the road. If 'topographical logic' is correct, then the line of ditches and hedges of the Iron Age continued to influence the organization of that part of the countryside throughout the upheaval of the early Middle Ages, and the land was never abandoned. Some of the hedges and lanes that we see and use today, even if in their present form they are relatively recent creations, follow lines first set out 2,000 or more years ago. This topographical deduction is in harmony with the known history of Lichfield, which grew in the territory of the small Roman town of Wall. The focal point of the local population shifted from Wall, which in the fifth century was sited rather dangerously (in an age of raids from the west and internal disorder as government deteriorated) on the main road, Watling Street, to the hilltop now dominated by the very old church of St Michael. It then moved again down into the valley immediately to the west with the foundation of the cathedral in the seventh century. Here is an excellent example of movement of administrative centres, but within the context of a countryside which remained inhabited and cultivated.

The continuation of territorial units and 'estates' helps to explain the transition from the Roman to the medieval countryside, and some of the best examples have been identified in the woodland landscapes. For example the early medieval estate based on Sherborne in Dorset is thought to have been in existence by the seventh century, and the place seems to have been important in the Roman period. The medieval estates around Wroxeter (Shropshire) and Worcester have also been plausibly argued to have been formed out of territories attached to those important Roman towns. Some of the early estates are known to have been based on royal tuns, or to have had 'minster' churches, or indeed to have been distinguished as focal points of both secular and religious government. Many of these early estate centres later became medieval market towns, and one can see how these 'central places' have continued to exercise an influence over their surrounding countryside until the present day, though of course performing different functions in successive phases of their history.

We are prevented by a lack of evidence from examining in any detail changes in the landscape and its inhabitants in the period between the fifth and the tenth century, which is all the more frustrating because we know that it must have been a crucially formative period, when many of the characteristics of the woodland landscape crystallized. To know that territories may have remained as units of administration is valuable information, but we are still very uncertain about activities within the boundaries. Even if land was still managed in a way that preserved old fence and hedge lines, one suspects that the fields were used in new ways. For example, in the Roman period a good deal of surplus produce would have been consumed away from the place where it was grown, because grain, cattle, cheese, and other goods were paid in taxes and exchanged in markets, whereas in the fifth, sixth, and seventh centuries rulers may still have demanded taxes or tributes, but the collapse of markets and the imperial coinage would have left most country dwellers producing mainly for their own consumption. The peasants of south Shropshire might have specialized in the period of Roman rule in producing cattle for the market, as we know that the citizens of Viroconium (Wroxeter) ate great quantities of beef, but after about 400, though remaining the residence of a local chief and his entourage, the city shrank drastically in size, and the nearby rural population would have been mainly concerned with mixed farming to feed their own households.

A break with the old order is certainly implied by the eventual abandonment of the Roman farms, the transformation of the political structures, and the ethnic changes implied by the adoption of a new language. Further east—indeed on the eastern frontiers of the woodlands, along the edge of the Peak District, the Trent valley, Warwickshire Avon, and the escarpment of the chalk downs in Wiltshire, strong Germanic cultural influences are exhibited in cemeteries in which burials were accompanied by weapons, jewellery, and pottery of a distinctive 'Anglo-Saxon' type. These are almost entirely absent from the woodlands, and the gaps cannot be explained by deficiencies in modern methods of discovery, as the bulk of the river valley cemetery sites were found by accident when the earth was disturbed in recent times by the building of houses, railways, and roads—activities which are as common in the woodlands as elsewhere. There must have been strong Germanic influences on our part of the countryside, because the place names are overwhelmingly Old English, coined in the sixth to tenth centuries. Perhaps the newcomers consisted of a small number of powerful aristocrats, who took over key places such as Penkridge (Staffordshire) from native British rulers, so that a former Roman town (*Pennocrucium*) became the centre of an administrative district. Or conceivably settlers arrived in larger groups but they quickly came under the influence of the British church, and abandoned their custom of burial with grave goods. A small cemetery at Bromfield in south Shropshire appears to have been used by a group of people in the seventh century, a few of whom were buried with objects—a bead, brooch, and knife—similar to those found in Anglo-Saxon cemeteries. If they were not Christians, they were certainly following some Christian burial practices, as most were not accompanied with grave goods, and the graves were oriented east–west. We have strong suspicions based on evidence such as place names, dedications, or later administrative arrangements that British churches persisted throughout the post-Roman period at Eccleshall in Staffordshire and Sherborne in Dorset. Indeed a number of churches that are still in use in the district of Archenfield in Herefordshire are likely to have been founded before 600 by British clergy.

This is all relevant to solving the problem of 'How old is the woodland landscape?' because we are drawn to the conclusion that there must have been a considerable survival of both British institutions and British people, admittedly in greatly changed political circumstances, which makes it more likely that the Middle Ages inherited a great deal from the Roman and prehistoric past in terms of occupied and cultivated land, roads, and other physical features. Important clues to the extent of the debt comes from place names, which tell us something about the use of land in the period when other sources are not available. One of the most widespread place-name elements is '-ley' or '-leah' in such names as Bromley, found in Staffordshire. 'Ley' can mean 'wood', and on that basis one could build up a picture of thinly populated 'natural' landscape, returning to the 'oak wood' and 'primeval forest' conception already mentioned as colouring earlier views of the countryside in prehistoric times. However in most cases '-ley' seems to be used as a settlement name, referring to a clearing in or near woods, and its relatively frequent use, probably in the period c.750–950, indicates a rural patchwork of interspersed woods and open country. This does not mean that woods were being cleared at that time, because the name describes a location, not a process, and in many cases the '-leys' could have had their trees removed centuries or millennia before they were given their new Anglo-Saxon name. Both this place-name element, and other names referring to enclosures, settlements, and cultivation reflect the features of a varied landscape under human exploitation. Some place names which derive from woods, like '-hurst' and '-grove', imply that woods were relatively small features which helped to identify a particular place—Nuthurst (Warwickshire) was for example a wood containing nut trees—and they do not suggest that the trees formed a continuous expanse. The foundation in the seventh and eighth centuries of major churches in our woodland landscapes, and their endowments with local estates which were clearly valuable assets, encourage us further not to underestimate the densities of the population who provided rents and labour. The bishoprics of Hereford, Lichfield, and Worcester, and the monasteries of Much Wenlock (Shropshire), Malmesbury (Wiltshire), and Shaftesbury (Dorset) all did well out of woodland assets, though they also held lands of other types to provide them with a balanced portfolio of resources.

We can glimpse sections of landscape of the ninth, tenth, and eleventh centuries described in the boundary clauses of charters. Those for north and west Worcestershire use as boundary marks, as well as the inevitable streams and ponds, areas of cultivation (headlands and 'acres', strips of ploughland, were most commonly mentioned) and with distinctive frequency hedges, crofts (small enclosed fields), trees, and the edges of woods. At Upper Arley on the boundary of Worcestershire and Staffordshire at the end of the tenth century the edge of the estate was defined by means of thorn trees, Wulfsige's horse croft, an alder shaw (a small wood), Eadwulf's croft, a flax clearing, a swine pit, an acre yard or acre fence, and Winna's tree. The references to individual tenants or owners clearly imply that some of the land was held outside common fields and pastures. The remarkable conclusion is that the tenth-century woodland landscape seems to contain all of the main elements which gave it its special character five centuries later. John Leland's comments in *c.*1540, together with terriers, surveys, deeds, and court rolls of his time, depict mixed land use with pasture, crops such as flax, fields enclosed with hedges, and much land held in severalty. The features of the woodland landscape were clearly in place by the tenth century, and they were not novelties then. The boundaries will sometimes describe a feature as 'new', but they also use boundary points such as an 'old dyke'.

Hedge on the northern boundary of Upper Arley, Worcestershire

This multi-specied hedge line is mentioned in a charter boundary of 994, when it formed part of the enclosure round 'Wulfsige's horse croft'. Now, as then, it is a field boundary for the control of livestock, a boundary for the estate and parish of Upper Arley, the county boundary between Worcestershire and Staffordshire, and the boundary between the dioceses of Worcester and Lichfield.

Expanding within a Framework, 1000–1300

When we trace changes in the woodlands through the centuries after 1000 we are clearly not dealing with its formative pioneering phases, but with developments within an already established framework. Domesday Book in 1086 describes a countryside with a density of ploughs and people rather lower than those recorded for the adjacent champion or chalkland regions: the number of people usually falls between three and seven per square mile (that is, 15–35, allowing for whole families), with fewer in Staffordshire and more in north Dorset. There were gaps in the settlement pattern, but sometimes these can be shown to be the result of administrative arrangements whereby areas of woodland were attached to remote estates largely given over to arable, which relied on their outliers for timber, fuel, and grazing. So a large section of the Forest of Arden in Warwick-shire, later named as Tanworth-in-Arden, belonged to Brailes 20 miles to the south, and its assets were described under the heading of the parent estate. These included many trees, and no doubt extensive pastures, but there were at least a few people as well. In other ways Domesday may understate the resources and population of the woodlands, in that the survey was focused on the arable land of manors, and the peasants who provided the labour to work it. Woods are recorded quite systematically, but meadow and pasture is only mentioned if it belonged specifically to the lord of the manor, so large areas of common grazing were in effect omitted. This type of countryside was more likely to contain people who, because they did not fit into the conventional categories of peasant tenants, were not enumerated in the survey. These included specialist workers involved in crafts or pastoral agriculture, such as woodcutters, charcoal-burners, and swineherds, and tenants paying rents in cash.

There was still plenty of scope for expansion, and this was the predominant tendency up to the early fourteenth century. The units of landholding increased in number and became smaller, continuing the fragmentation of the old great estates that can be traced back to the ninth and tenth centuries. Some enormous manors persisted at the time of Domesday, which describes estates with hundreds of tenants at Bromsgrove in north Worcestershire and Leominster in Herefordshire. Indeed, the shadow of the large estates became permanently fossilized in the parish boundaries which were fixed in the twelfth century. While many parishes to the east contained only 1,000–2,000 acres—the territory of a single village—many woodland parishes covered 5,000 acres or more, and commonly included in late medieval and modern times a number of townships and manors. This suggests that the fragmentation of secular landholding came later than in other landscapes, and the twelfth-century rectors had been able to resist the foundation of new parish churches that would have reflected the new manorial structure. The lords and peasants within these large parishes could only provide for their separate religious needs with chapels dependent on the parish church. An example of the proliferation of small manors is Hanbury in Worcestershire, still today a parish of 8,000 acres, which is 5 miles long. It had been carved out of an even larger parish of the seventh century, as in 1086 it consisted of two large manors, one belonging to the bishops of Worcester and the other to the Crown. By 1300 ten units of landholding had been established, two granges belonging to the Cistercian monastery of Bordesley Abbey, a manor held by the Knights Templar, and most of the rest of the manors in the hands of gentry families—the de Hamburys, the Webbes, and so on.

Behind this spread of smaller manors lay a number of important social and religious movements, felt in every part of the countryside, but having their greatest impact in the woodlands. Great lords were creating new holdings for their followers in order to obtain military services from their new tenants, but also to use them as administrators. The de Hamburys, for example, served the bishops of Worcester for generations as bailiffs, stewards, and in other useful offices. At the same time these lesser aristocrats could take the initiative and create their own manors, above all by buying property. Their lands were also on occasion divided by inheritance, or by gift, or by sale. During the twelfth and thirteenth centuries monasteries belonging to the new monastic orders were being founded, especially in those regions not already dominated by wealthy Benedictine houses. In Staffordshire, Shropshire, and Herefordshire, where relatively few monasteries are recorded in 1086, the Cistercians sought out 'deserts' where houses could be established in relative solitude at places like Abbey Dore in Herefordshire and Buildwas in Shropshire. The Augustinians attracted patrons who founded houses at Haughmond and Lilleshall in Shropshire, Wigmore in Herefordshire, and Stone in Staffordshire. In Warwickshire and Worcestershire the Benedictine presence was concentrated towards the south, and so the Cistercians spread in the forests of Arden and Feckenham at Bordesley, Merevale, and Stoneleigh, and the Augustinians at Kenilworth. These were the larger establishments, but dozens of smaller houses were also founded, including a number of nunneries. Stanley was another Cistercian house which filled a gap in the monastic settlement of north Wiltshire.

The new lords who were multiplying in the two centuries after 1086 were changing the landscape not just by building new manor houses and monasteries, but also by reorganizing the countryside to aid their production. The monastic granges set up by the Cistercians are well known for the systematic way in which by exchange and purchase property was consolidated, and existing tenants moved away, so that the land could be cultivated in a single block. The Augustinian canons of Haughmond did likewise, using gifts of land and their rights over the church of Stokesay to create a grange at Newton in Stokesay parish, which survived into modern times as a farm on the edge of Craven Arms. In the early thirteenth century at Caludon near Coventry (Warwickshire) a gentry family, the Segraves, similarly set about forming a manor with a large grain-growing demesne and a park. But most gentry, and indeed many of the manors belonging to smaller church landlords, relied more on the rents paid by peasant tenants than efficient compact demesne farms. If they cultivated a demesne they needed a local supply of labour, which sometimes came from peasant labour service, but more often was obtained by hiring smallholders or the families of more substantial peasants. They would usually acquire land on which some peasants were already established, and their aim would be not just to gain control of their existing tenants, but also to attract others, so the proliferation of new manors was also accompanied by an expansion in the numbers of settlements.

We have very little direct evidence for the settlement pattern of our landscapes in the eleventh and earlier centuries. The remains of a compact group of farms, of which no more than six may have been inhabited at the same time, and which was abandoned soon after 900, has been revealed at Catholme in south Staffordshire, but most excavations of rural settlements of the ninth–eleventh centuries, like that at Trowbridge (Wiltshire) reveal no more than fragmentary building plans consisting of rows of post-holes, which probably belonged to a hamlet. Small settlements, consisting of hamlets or isolated farmsteads, apparently dating from the twelfth and thirteenth centuries, can still be seen reflected in the modern landscape. These might consist of a group of houses, often not with any regularity of plan, standing around a road junction or apparently established near a focus such as a manor house, church, or chapel. More often the houses stand in a row along a lane, not in a continuous line, but each separated by a field or paddock from its neighbour. The line sometimes extends along a green, perhaps on the edge of a wood, as if the houses represent a frontier of colonization which eventually reached limits beyond which no more encroachment on the waste or incursion into the woods could be allowed. In the great majority of cases the buildings are of modern brick, so we might doubt if the settlements were very old, but references in early documents, or scatters of pottery found nearby show that they were occupied in the Middle Ages, and early maps confirm that they were in existence in the sixteenth and seventeenth centuries. They can have distinctive place names indicating their origin, with a wide variety of names deriving from Old English or Middle English, like those ending in -cot for example, meaning cottage or cottages, and most characteristic of all, names ending in End or Green. These hamlets proliferated, so that a large parish, which, as we have seen, might be split up into ten smaller manors, will often contain Ends, Greens, and small groups or rows of houses to the number of twenty and more.

The fields associated with these settlements were also distinctive. Areas of open field, on which some form of crop rotation was practised, bear a superficial resemblance to the fields found in champion country attached to a single large nucleated village. But on closer inspection we often find that the open fields were restricted in area, and account for less than a half of the total agricultural land, and in any case were divided into many more than the two or three fields found in a conventional field system. For example, the arable land of Wybunbury in Cheshire, is recorded in 1297 as lying in fourteen divisions of arable, only two of which were called 'fields', while another piece was described as a 'furlong'. The remainder were identified just by their names, which ended in 'croft', implying an enclosed parcel, or 'ruding' suggesting their origin in the clearing of new land, as does the name 'Newacre'. Now these fields, crofts, and rudings belonged to three seasons, implying that they could be cropped in successive years in a sequence of winter corn, spring corn, and fallow, so they resembled in organization an orthodox field-system. But there similarity ends, because in most woodland townships a great deal of land lay in 'severalty', that is in enclosed crofts and closes, each containing a few acres. Sometimes the crofts survive in the modern landscape: they are roughly rectangular, though rarely with any geometric regularity, or straight boundaries. They can bear the marks of early cultivation in the form of ridge and furrow, and the medieval documents confirm that land held in crofts was ploughed in ridges. Individual peasants in such a settlement would hold part of their land in the open field, and part in closes. They would be restricted in their choice of crops in the open field by communal decisions about rotations, and they would be expected to allow their neighbours' animals to graze in some or all of their closes in the 'open time' when common pasture was extended throughout the fields. But they could choose how to use their land in the closes—as arable or pasture, or for crops such as flax, or for orchards. Like medieval peasants

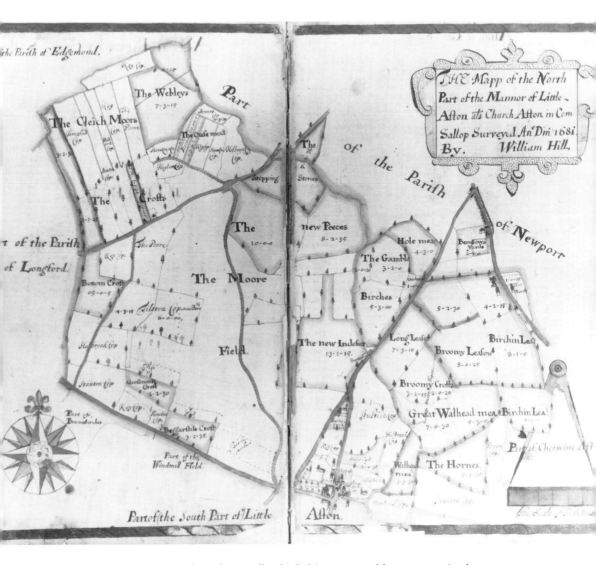

everywhere the woodland inhabitants were able to graze animals on commons and collect fuel in woods, but these assets would often be more extensive than those available to the people of champion country.

A distinctive type of settlement found in woodland landscapes was the moated site, often a residence occupying a rectangular platform of no more than a quarter- or half-acre, surrounded by a broad and shallow ditch filled with water. These sites have often been abandoned, and the moats have degenerated into stagnant ponds, though some moats surround houses which are still inhabited. They were commonly constructed in the period between 1180 and 1320, and are especially numerous in some large parishes with a pattern of dispersed settlements. Solihull in the Warwickshire Arden, for example, with six 'greens' and three 'ends' has no less than twenty-eight moated sites. Many moats surrounded manor houses, rectories, granges, and other aristocratic dwellings, but others were dug by wealthy or aspiring peasants. They had a useful defensive function, which would deter thieves or small raiding parties, but they were at least as important as symbols of social exclusiveness: neighbours would respect the wealth and standing of someone who could surround his house with such a visible barrier.

The moat's construction often involved elaborate water management schemes, by which a valley bottom was adapted not just as a site for a house but was also used for ponds for fish and to power mills. Even these practical and profitable features may have conveyed messages about their owner's status, and, together with the moat, added to the visual attractions of the residence. The same mingled motives can be seen in the creation of parks, also a characteristic activity of the thirteenth-century woodlands. More than a hundred are known in Staffordshire, for example, and most of the ninety identified in Wiltshire were concentrated in the 'cheese' country in the north-west of the county. These were exclusively associated with the landed aristocracy, and only those with a considerable estate could have afforded to assign so much land to a hunting ground. Again we can appreciate that they had practical uses, not just for keeping deer, but because the enclosed grassland could feed cattle and be occupied by horse studs; the trees in the parks provided fuel and timber as well as cover for the deer. More important, the park advertised the social standing of its owner with a prominent pale which left no doubt of his resources and privileges. The venison itself was prized as high-status food, which would serve to impress honoured guests, and could be sent as gifts to other aristocrats, or to inferiors, to win or confirm ties of loyalty and friendship.

We can visualize the appearance of woodland landscapes by piecing together the elements that still survive, together with those parts which have to be reconstructed from documentary and archaeological research. Only then can the combination of the dispersed settlement, enclosed fields, woods and pastures, moats, ponds, and parks be appreciated in their entirety. A contemporary glimpse of this countryside can be found in the poem *Piers Plowman* written in the late fourteenth century by William Langland, who probably came from the Herefordshire–Worcestershire border. At one stage directions are given by the honest peasant, Piers Plowman, for finding the house of his virtuous master, Truth. Needless to say the language is allegorical, but the landmarks are based on those of a real woodland landscape—the traveller proceeds by a brook, 'then shalt thou come by a croft', and past two tree stumps and a hill with a fence to a house surrounded by a moat. Piers Plowman is depicted in a famous scene ploughing a half-acre, which is located not in an open field, but in a croft, a small enclosure.

Some of the features of the woodland landscape were already established before 1100, and others developed in the twelfth and thirteenth centuries. The moats, and most of the parks, came relatively late. Settlement spread at the same time, and the population density increased. Large-scale encroachments on heaths and wastes, together with clearance (assarting) of woodland, increased the proportion of land under the plough, and the amount of enclosed land. Large areas were brought under more intensive use, like the 1,569 acres recorded in 1213 as asserted in the Golden Valley in Herefordshire. Even when we lack precise records of this kind, the number of fields still called in modern times rudding, stocking, or 'newland', and declaring their origin by their small size and irregular shape, provide ample evidence of the extension of cultivation at the expense of wood and scrub. Thousands of new houses were built to accommodate a population which more than tripled.

The impact of these changes on one place can be seen at King's Bromley in south Staffordshire, where fifteen peasants were recorded in 1086, whose holdings together with the lord's demesne accounted for perhaps 500 acres of arable. The woodland covered a similar area, and the pastures would have been more extensive. By 1300 the number of peasant families had increased fivefold, to more than eighty, and they were cultivating well over 1,000 acres of arable. The woodland held by the lord had shrunk to 20 acres. Most tenants held land measured in the standard units of virgates or yardlands (c.30 acres of arable), representing the old cultivated lands going back to the eleventh century and earlier, together with parcels of 'waste newly appropriated'. Thirty peasants were holding new land for rents, some as low as ¼d., ½d., and 1d., suggesting the small size of the parcels. King's Bromley was an attractive place to live, as it had opportunities for acquiring assarted land, and the movement of immigrants is reflected in the surnames of the peasants deriving from nearby villages like Barr and Brewood, and also the name 'Newman'. Many of the assarts were probably too small to feed a family, but surnames indicate that the tenants supplemented their income from land by following a craft, like John le Colyere (charcoal-burner) and Thomas le Couper (barrel-maker). Assarting could be carried out by lords, or at least they seem to have organized peasants into clearing new land by renting out parcels of waste and recruiting settlers. But this is very unlikely to have happened at King's Bromley, and indeed assarting in most places was the result of peasant initiatives, reflected in the numerous piecemeal clearances. The peasants at Bromley were not burdened by heavy obligations to their lord, as they owed no labour services, and paid modest cash rents. The manor belonged to the royal demesne, so, as was often the case, the tenants were all free. The manor may have been unusual in the complete absence of serfdom and labour obligations, but woodland peasants were generally less subject to servile burdens than their counterparts in the champion, wold, or downland landscapes.

Was this growth in people, settlements, and developed land just a matter of quantity, or did the character of the landscape change? Although the farming system became more heavily tilted towards arable, animal husbandry often still retained an importance in the peasant economy. The new arable was not always land of the highest quality, and in the wetter and colder districts oats occupied a high proportion of the land under the plough, which meant that the cultivators had to be content with a crop which yielded rather badly, had a low market value, and provided much less edible grain per acre than other cereals. Even in the kind environment of central Herefordshire, where much wheat was grown, pastoral activities retained their importance. For example at Woolhope in 1308 46 peasants are found to have been feeding 387 pigs, which is much more than they needed for their own consumption, so they were clearly destined for the market. The shift from grazing land and woodland to cultivation was not a revolutionary change, but involved extending arable fields which went back centuries. The inhabitants took conservation measures lest the advance of arable be taken too far. That was one of the motives for lords enclosing areas of pasture and wood in parks, and for putting hedges round woods and setting them under the supervision of woodwards. Individual peasants might secure access to wood and timber by owning small enclosed groves, and communities could take collective action to protect remaining commons, for example by paying their lord a rent to keep an area as part of the village pasture. The dispersed settlement pattern was not transformed by growth, but instead seems to have accommodated expansion as assarts and encroachments on former open pastures added new farmsteads along roads and on the edge of greens. A scatter of larger nucleated settlements appeared at this time, and a high proportion of them were either founded as boroughs, or the village houses were grouped around a market for which the lord obtained a royal charter, hoping to stimulate urban growth. When we cannot discover any records of an official market grant, a market place can sometimes still be identified in the village plan, as at Tanworth-in-Arden in Warwickshire, so the settlement seems to have been at least a minor centre for trade.

These commercial influences played an important part in allowing the multiplication of smallholdings in dispersed as well as nucleated settlements. Families could survive on a few acres by keeping animals on the common pastures, by earning wages in agriculture or industry, by exploiting the raw materials of woods and wastes in pottery-making, charcoal-burning, coal-mining, and other crafts, and by participating in small-scale retail trade. Peasants with larger holdings needed access to markets, as they paid a high proportion of their rents in cash, and the produce of their pastoral husbandry needed to be sold. This helps to explain why small boroughs, which provided local centres for exchange, were numerous, notably in Shropshire and Staffordshire.

By the thirteenth century we can detect a spirited and independent mentality among the inhabitants of the woodlands, born out of the absence of close control by their lords, their freedom from the routines of cereal cultivation in common fields, and their involvement in the market. Some of the most prolonged and dogged struggles with lords who were attempting to impose servility were located at such places as Darnhall in Cheshire and Halesowen in Worcestershire. Much more frequent were the battles fought over common rights in woods and pastures, leading at their most extreme to crowds gathering to destroy fences and hedges that encroached on dwindling areas of common grazing. In an incident at Alvechurch in Worcestershire in 1273 enclosures were destroyed by people from adjoining King's Norton and at Lydlynch in Dorset in 1279 a bank and ditch were removed by 'certain malefactors', 'with force of arms by night'.

Weathering the Crisis, 1300–1500

The woodlands were as hard hit by the late medieval crisis as any other landscape, and yet they emerged with surprisingly limited damage to their fabric. The great famine of 1315–18 took its toll—at Halesowen in Worcestershire for example mortality reached 15 per cent, and in the aftermath, in 1322, Staffordshire peasant holdings lay vacant and land uncultivated on the estates of the earls of Lancaster. When royal tax officials were gathering taxes in Shropshire in 1340 they heard many complaints that land was infertile and uncultivated, and that peasants were either poor or had migrated. The Black Death (1348–9) and successive plagues caused as severe a mortality in the hamlets and farms as in

Aerial view of the deserted hamlet of Sidnall in Ditton Priors, Shropshire

The parish of Ditton Priors, north of Clee Hill, contained at least seven hamlets in the Middle Ages, a characteristic of woodland settlement patterns. Sidnall probably consisted of six households in the thirteenth century, but had shrunk to only one by 1510. The photograph shows, next to the modern farm, the earthworks of ditched enclosures and holloways marking the deserted hamlet site.

nucleated villages to the east. But when we look for evidence of deserted villages, or of whole parishes transformed into pastures managed by a few shepherds, we search in vain. There were plenty of parishes in which a third or a half or an even higher fraction of holdings had been left vacant and their houses had fallen into ruin. If we search carefully we can see the tell-tale signs of grassed over holloways, boundary ditches, and house platforms from abandoned farmsteads and hamlets, but usually only parts of the settlements were lost, and the number of households in each place was reduced. A common pattern was for hamlets which had previously contained a number of houses to be thinned down to one or two. As neighbours left, or sold up, those who remained took over their land and let the redundant buildings fall into ruin. But this process stopped eventually, leaving a rump of tenants to continue to use the land—these settlements were not afflicted by progressive collapse, like some declining nucleated villages.

The woodlands saw a shift from arable to pasture, reversing the trends before 1300, which in the Warwickshire Arden reduced the proportion of land under the plough from two-thirds to one-third. The increase in livestock led to stresses

among neighbours, with complaints that selfish individuals were keeping excessive numbers of animals on the common pastures, or were refusing common grazing in their closes. But in general the inhabitants of the woodlands weathered the storm, adjusted their farming methods gradually, and survived as functioning farming communities. They benefited in particular from the rise in consumption of beef, and made profits by breeding their own animals, or by taking in stock from Wales and fattening them for the butchers.

Enclosures extended further in the fourteenth and fifteenth centuries. In most parishes a proportion of the land lay in closes and crofts already. The management of these was changed by their tenants asserting private control and denying common rights, and in the open fields individuals exchanged land and fenced off their strips, and groups agreed among themselves to enclose a larger area. Lords were involved by granting licences to enclose, or by encouraging enclosure by agreement. This process has left its mark on the modern countryside because the boundaries of the consolidated groups of open-field strips followed the curved lines of the lands and headlands, and hedges still in use echo the layout of the defunct system. Enclosure, like the shrinkage of settlements and the conversion to pasture was a matter of shifts and tendencies rather than complete transformations, though it is possible to find townships, like Bordesley in north Warwickshire which had been totally enclosed by *c.*1500. More often the common arable was enclosed by a consortium of tenants, as happened at Sambourne, also in the Forest of Arden, but the extensive pastures remained open and common until the eighteenth century. Some common wastes and heaths lost acres to enclosures on their edges, either by hedges erected by acquisitive lords, or by smallholders like those at Sedgley in Staffordshire who in the late fifteenth century were encroaching by building new cottages.

This period brought some prosperity to parts of the woodlands. Peasants who could take advantage of the market for beef expanded their operations, and the surviving smallholders could earn good wages. The people of the industrial areas, such as south Staffordshire, which was involved in ironworking, and the clothing districts of west Wiltshire, were doubly advantaged, in that profits could be made from crafts, and the demand from the workers stimulated food production. The best physical indication of wealth in the woodlands comes from building activity. Dozens of peasant houses still survive, especially in Shropshire and Herefordshire, built distinctively with substantial frames based on crucks (pairs of curved vertical timbers made by splitting a whole tree) and founded on low stone footings. Analysis of the tree rings of their timbers shows that these houses were being built between 1380 and 1550, with a high proportion in the middle of the fifteenth century. Church building reflects the fifteenth-century prosperity of west Wiltshire, and many churches in the Warwickshire Arden were rebuilt or at least refurbished at this time.

The lords were rebuilding too, judging from the number of manor houses of high-quality timber frames, though increasingly they used stone. It is hard to believe that many of them were facing financial difficulties, though their rent rolls were bound to suffer from the fall in the number of tenants and the value of land. Some families found ways of profiting from woodland resources, notably by going into large-scale beef production. Those who felt a pinch cut their costs, for example by reducing the number of their residences, so many moated sites fell into disuse. The greater lords often withdrew from direct agricultural management, and leased out their demesnes to ambitious peasants or gentry who believed that they could make a profit from agriculture. This led to a further shift in the settlement pattern, because in the woodlands the demesnes often lay in scattered parcels. The new lessees (farmers) found it convenient to rent not the whole demesne but separate blocks of land, so that what had been one large agricultural unit was now split into two, three, or more pieces. The farmers built houses on their new leaseholds away from the manor houses, and fresh sites were being developed in a period when settlements as a whole were contracting.

New Pressures and Opportunities, 1500–2000

In the centuries after 1500 some of the woodland landscapes were adapted to new circumstances, while others were transformed by powerful economic and social changes. To begin with the peasants and farmers who managed the countryside, we would expect to find that they were greatly reduced in numbers as commercial pressures, and the policies of their landlords, tended to increase the size of agricultural holdings. We can observe these forces at work in the small north Worcestershire parish of Rushock, where in 1572 twelve of the nineteen tenants held 50 acres or less, and only two had more than 100 acres. In the late sixteenth century the landlord's agent was demanding from new tenants high entry fines which were claimed to be preventing heirs from succeeding fathers in their copyhold lands. In the seventeenth century the lord continued to pursue a policy of taking into the demesne as much of the tenants' land as possible, and then letting it out on lease. The free tenants who in theory could not be subjected to pressure from the lord found that their common rights were threatened in the 1650s by the lord's policy of grazing a large herd of cattle on the pasture. By 1755 four landholders at Rushock each had acquired a holding of between 120 and 236 acres, which together accounted for almost two-thirds of the land in the parish.

While the Rushock case reveals one important trend, we cannot fail to be impressed by the capacity generally of small farms to survive. First, the extent of freeholds helped to protect many communities from ruthless lords. And secondly pastoral agriculture gave some farmers the ability to prosper on small acreages, and this is especially noticeable in the dairying districts. Commercial growth in the sixteenth and seventeenth centuries encouraged farmers to specialize, and none more so than in Cheshire and north Shropshire, where the high-quality pastures were ideally suited to milk production. In the early eighteenth century Defoe was commenting on the richness of the Cheshire grass which 'disposes the creatures to give a great quantity of milk, and that very sweet and good'. The cheese was sold in great quantities in the London market, where it commanded a high price. A dozen animals could give their owner a living, and many herds contained thirty or more by the 1690s.

In north and west Wiltshire, the 'cheese' country earned its name by producing large quantities for the London trade, which in the nineteenth century was dyed to give it the colour of beeswax, apparently a selling point for metropolitan consumers. Another type of cheese came from the vale of Blackmoor in Dorset, made from the skim, as the cream was used for butter. Production was organized

Cheese market at Chippenham, Wiltshire, 1850

The pastoral agriculture of the woodlands produced for the market, and tended to greater specialization as commercial and transport facilities developed. The dairy industry of north Wiltshire expanded to supply London consumers. The cheeses were brought into market by farm carts, shown here in large numbers, and were then carried to the capital by rail.

through the 'Dorset system' by which farmers would rent cattle and pasture to a dairyman, who would pay a fixed sum per cow each year—it could be as much as £6 in the late eighteenth century. The system gave the farmer a fixed return, and provided opportunities to entrepreneurs without much capital to manage a herd and, if they could develop the skills, market a high-quality product.

Dairying used land intensively—in the vale of Blackmoor for example thirty cows could be kept on 60 acres, and there were profitable by-products, like the surplus calves to be sold, and the pigs which were fed on the whey. So we find that smaller holdings survived in the Wiltshire 'cheese' country into the eighteenth century, and in Cheshire as late as 1919 46 per cent of farms contained between 5 and 50 acres. They could not last forever, and although the dairy industry was able to survive the depression of the 1870s, the advent of the railways enabled farmers to sell fresh milk and abandon cheesemaking. Even in Cheshire the large farms eventually triumphed, and between 1920 and 1971 the number of holdings halved, from c.12,000 to c.6,000.

Horticulture was another type of intensive cultivation well suited to the woodlands. Herefordshire, and adjoining parts of Shropshire and Worcestershire, from the seventeenth century became involved in hop cultivation. Orchard fruit was also grown, in these counties, and Herefordshire already

Hop picking in Herefordshire, 1940

The woodlands were well suited to various forms of horticulture, including market gardening and the specialized production of fruit. Herefordshire has since the seventeenth century been famous for its cider apples and its hop cultivation. Harvesting requires much seasonal labour, often provided by women.

enjoyed before 1700 a reputation for cider which was to become its most generally recognized product in the late twentieth century. Flax and hemp, grown on specially designated plots or on newly broken pasture, were also well-established crops in the woodlands in the sixteenth and seventeenth centuries. The growth of large towns has encouraged market gardening in their vicinity in the last two centuries, and specialisms have developed where the soil and climate are most suitable at Wombourne in Staffordshire (serving Wolverhampton), Bromsgrove in Worcestershire, adjoining Birmingham, and Altrincham (Cheshire) on the edge of Manchester.

Enclosure continued throughout the sixteenth, seventeenth, and eighteenth centuries, the result of agreement rather than parliamentary acts. The latter were used in the late eighteenth century to bring the surviving areas of commons and wastes into private control. Such land often accounted for a relatively small pro-

portion of the parishes affected, although in Staffordshire a quarter of the county consisted of heath and waste according to a seventeenth-century commentator. William Barnes, the dialect poet who was brought up in the vale of Blackmoor, commented in the late nineteenth century on the loss of amenity resulting from the enclosure of commons:

> Girt banks do shut up ev'ry drong [= narrow way],
> An' stratch wi' thorny backs along
> Where we did use to run among
> The fuzzen [= furze] and the broom.

A more serious protest was caused in the twenty years before the Civil War by the disafforestation of the royal forests and their division into new farms, with a loss of common rights to grazing and fuel which had been traditionally enjoyed by the local population. Riots were provoked in Selwood in Wiltshire, Gillingham in Dorset, and Feckenham in Worcestershire. The first stage of the enclosure of Needwood in Staffordshire, in Uttoxeter ward, caused such a disturbance in 1636–9 that the rest of the chase survived until large scale disafforestation in 1803–11. This last act of destruction of a forest has left a landscape of straight roads and neat, geometric fields to the south of Tutbury which is so unlike any other part of the woodlands that the visitor is made immediately aware of its untypical history.

The continued process of quiet and relatively uncontentious division of the common fields meant that the agricultural improvers had no need to advocate enclosure. Some of the principal writers on agriculture in the sixteenth and seventeenth centuries had contacts with the midland woodland landscapes—Walter Blith came from Allesley in Warwickshire, where he would have seen the convertible husbandry he preached as part of the normal practices of the Forest of Arden. John Beale of Herefordshire played an important part in improving apple-growing in his county, and Andrew Yarranton of Astley (Worcestershire) together with Blith and Richard Gardiner of Shrewsbury proselytized on behalf of horticulture and, especially in the case of the first two, advanced the cause of dye plants such as woad. One of the great projects of this period was advocated by Rowland Vaughan for turning the Golden Valley in Herefordshire into a version of Lombardy by an irrigation system, traces of which can still be seen. Water-meadows are still visible in the woodlands, but the main surviving remains of agricultural improvement are the ubiquitous marl pits. Before the advent of chemical fertilizers, and when liming could be very expensive because of the cost of transport, farmers of every type dug deep into the subsoil for the calcareous earths which could be spread over the arable fields, combating acidity and improving soil texture. Now the disused pits remain one of the enduring legacies of pre-industrial agriculture—seventy have been counted in one Worcestershire parish.

Recent developments in agricultural technology have in some districts seriously damaged the distinctive features of the former landscape. Farms have been thrown together into large units, and hedges removed in order to give access to great machines which have turned the land into a single expanse of cereals. In the growth of ruthless land management since the 1960s all traces of ridge and furrow, ponds, and other physical remains of early agriculture have sometimes been removed, even though these remnants of the past had managed to survive all the other agrarian changes over the last three or four centuries.

The changes in holdings and farming methods have had a number of influences on the pattern of settlement. The reduction in the number of holdings, as at Rushock, had thinned out the dispersed straggle of houses in a continuation of the shrinkage of the later Middle Ages. Some former peasant farms were reduced to labourers' cottages, but often the labourers were housed in new locations, either near the surviving farms, or in new nucleated settlements which grew up in the nineteenth century where none had existed before. Nucleation has accelerated in the twentieth century with the building of groups of council houses and private housing estates near existing nuclei, in a move encouraged by the mistaken belief among planners that the nucleated village was and should be the normal rural settlement. But in opposition to nucleation, various types of squatter settlement developed along roadsides, and on the edge of commons and wastes. This had been an almost continuous process since the later Middle Ages, and renewed the trend towards a dispersal of cottages across the landscape. The new dwellings, mostly of wage-earners and the poor, have a volatile history. They can appear on maps, such as those drawn in connection with the tithe commutation around 1840, soon after they were built, and have since disappeared. Some have been given a new lease of life by modern commuters who have transformed small cottages into substantial houses, which betray their origin by their rather precarious situation, squeezed onto the edge of the road. The shifts and movements in the location of houses may be continuing a long tradition, but the houses since about 1700 have been built almost entirely of brick, ending abruptly the timber framing which reached a high level of technical and aesthetic achievement between about 1380 and 1650.

The social character and mentality of woodland landscapes again shows a continuation of earlier developments. On the commons the 'cottage economy' was practised, by which the relatively poor inhabitants lived by makeshifts, seeking jobs in agriculture or industry as they became available, and taking advantage of the resources of the commons for grazing a few animals, and for gathering fuel and rushes, bracken, or furze that could be used or sold. That way of life was endangered and even extinguished with the enclosure of the commons in the late eighteenth and early nineteenth century. There had always been elements of distrust in the relations between the more substantial inhabitants of the older settlements and the shiftless (as they saw it) population on the commons. The culture of the 'cheese' regions of Wiltshire in the seventeenth century as in other wood-pasture districts, contrasted with the villages of the 'chalk'. The inhabitants tended to puritanism and frowned on time-wasting and drunken sports and pastimes. They collected money for community projects by means of the rational rating system rather than the convivial and potentially sinful church ale, and when they played games, they preferred the individualistic 'stool ball' to football, which was boisterous and disorderly.

A powerful influence on the modern landscape of the woodlands has come from the great estate owners. Old families, like the Talbots of Shropshire, survived from the Middle Ages, but in the inevitable pattern old families became extinct and were replaced by new blood. In the sixteenth century newcomers appeared who took over monastic estates, such as the Holcrofts at Vale Royal in Cheshire, and in subsequent centuries those who made their money elsewhere acquired land and built great houses. The proximity of the woodlands to industrial and urban centres meant that capitalists made an important impact on landed society,

Ironbridge, Shropshire, 1963

The expansion of rural industry in the eighteenth and nineteenth centuries created this sprawling and incoherent collection of houses, workshops, mines and waste tips. The combination of local mineral resources and easy transport on the River Severn (in the foreground) attracted investment and workers. A road crossing of the river at this difficult point was provided by the bridge in 1777–80, made at the local ironworks of the Darby family.

like the iron masters, the Foleys, who built a mansion at Great Witley in western Worcestershire, or the France-Hayhursts who put their wealth gained in Liverpool shipping into Bostock and other estates in central Cheshire. In the sixteenth century the gentry were still making new deer parks, as at Longleat in western Wiltshire created by the Thynnes, but the later landscape parks required the removal of settlements, or the conversion of farm land. The woodlands were suited to these ventures, as new parks could be based on earlier deer parks, and an enclosed and pastoral landscape was more easily adapted to this new use. Some of the novel ideas in the development of parks and gardens originated in the woodlands, like the 'picturesque' movement led by William Shenstone, based on his remodelling of the Leasowes near Halesowen (Worcestershire), and Uvedale Price who modified the wooded landscape on his estate at Foxley in Herefordshire.

The influence of the large estates extended far beyond their country houses and grounds. Parts of the woodlands were dominated by great landowners. Shropshire contained such vast accumulations as Lord Forester's estate, and in Staffordshire, a third of the land belonged to eight families in 1873. The estates invested in agricultural improvement, and their works still survive in the impressive brick farm buildings, from the home farm of the 1760s at Acton Scott (Shropshire) to the functional, but solidly constructed farms on the Lilleshall estate. In the same spirit they built workers' cottages, and even contributed to the modern nucleation of settlement by founding model villages like those at Eastnor in Herefordshire and Shugborough in Staffordshire. But that era of the building of country houses and model farms came to an end in the late nineteenth and early twentieth century as the agricultural depression, debt, and death duties began to break up the estates, and the farmers were able to buy their freeholds.

Industry, as well as providing some of the wealthy landlords with the means to buy land, had a more direct impact by spreading through the countryside. In the sixteenth and seventeenth centuries craft work was combined with farming to make 'dual occupations'. The woodlands were well suited for these developments because the population had spare time for employment, as pastoral farming did not absorb all of the family's time. In some places raw materials, water supplies, and transport (down the River Severn) were available. The main districts affected were western Wiltshire, where woollen cloth was made, south Staffordshire and south-east Shropshire, where coal was mined and iron smelted, and parts of north Worcestershire and western Warwickshire, the locations for the specialized manufacture of iron goods—scythe blades in some villages, and in others nails and needles. Glass was made in Staffordshire and north Worcestershire. These early industries were accompanied by growth in the density of settlement, and in particular an increase in cottages on wastes, and in the more intensively developed parts of Shropshire and Staffordshire the proliferation of mines, furnaces, mill ponds, and other industrial structures. Large-scale ironworks, such as furnaces and slitting mills, were also built in places away from the main focal points of the industrial revolution, such as Charlcott and Cleobury Mortimer (Shropshire).

These early rural industries did not always survive. With the intensification of production from the eighteenth century some places were affected by 'de-industrialization', and the artisans making blades or needles moved to the larger centres, leaving villages like Belbroughton (Worcestershire) or Sambourne in the Warwickshire Arden to adjust to a mainly agricultural economy once more, as did the villages which had been involved in woollen cloth-making in Wiltshire. In the eighteenth and nineteenth centuries industrial growth became so intense that it attracted a huge workforce—Staffordshire's population increased tenfold between 1701 and 1901. Aided by improvements in communication by canal and railway, districts like the Black Country between Wolverhampton and Walsall (Staffordshire), and the Ironbridge Gorge in Shropshire filled with industrial buildings, coal pits, and slag heaps, with groups of houses scattered almost haphazardly over a landscape which was still not totally urbanized. The new settlements reveal their origin by their names—in Madeley (Shropshire) we find Tea Kettle Row, Bedlam, and New Buildings. In spite of all of the economic changes names of hamlets in the Black Country like Netherend and Lyde Green remind us that all of this modern activity was ultimately based on the framework of a medieval landscape.

5 Forests and Wood-Pasture in Lowland England

Brian Short

> Perhaps of all species of landscape, there is none, which so univer-
> sally captivates mankind, as forest-scenery: and our prepossession in
> favour of it appears in nothing more, than in this; that the inhabi-
> tants of bleak countries, totally destitute of wood, are generally con-
> sidered, from the natural feelings of mankind, as the objects of pity.
>
> (William Gilpin, *Remarks on Forest Scenery* (1791), 269)

Stretching across the lowland clay vales of southern and eastern England by the time of the Norman Conquest were the remains of ancient woodlands, dense and impenetrable in a few places, open, pastoral and savannah-like in many others. Between the calcareous uplands of limestone and of chalk, and from the sandstones of Devon to the coastline of East Anglia and the Fens, the wood-pasture countrysides occurred locally, yielding valuable resources to the communities within them and on their margins. In some cases the regions of wood and intermingled pasture were large—the Anglo-Saxon Chronicle for the year 893 referred to the 760,000-acre Weald as 'the great forest which we call Andred' stretching for nearly 100 miles east–west and 40 miles north–south at its outer limits. Elsewhere the woodlands were small as with the 'sour woodsere land', described by John Aubrey in western Wiltshire between the Cotswolds and Salisbury Plain as 'very natural for the production of oaks'. This woodland pattern was itself the product of generations of landscape change: at times there would be woodland clearance, at other times regeneration to give a secondary woodland cover once more, but one which never again reached the habitat-rich standards of the 'wildwood' with its many herbaceous plants. This generic type of countryside was complex, much of it settled early and unplanned, but much also relatively late in being colonized for a variety of physical, political, and cultural reasons.

In what follows, the initial concentration will be upon the greater and lesser landscape features of forests and wood-pasture country, before turning to a chronological narrative of landscape evolution, and a final section which emphasizes the fact that such landscapes are both the outcome of, and the setting for, cultural display and social and economic endeavour.

Wooded Landscapes

The landscapes being considered here contain greater or lesser amounts of woodland. Such landscapes were complex for several reasons: first, they were subjected to great human modification through time, so that the precise amounts of woodland and clearings varied according to the period under consideration. Second, the physical presence of woodland in the landscape must be distinguished from the legal use of the term 'forest' which need not refer to woodland at all but rather to an area subject to superimposed forest law from the early medieval period onwards. Forests were areas of deer ranching and hunting, reserved to the crown or its lessees, but there were many non-royal equivalents in chases, as well as areas of woodland which had once been royal forest but were so no longer, and were referred to as 'purlieus'. There were also many deer parks and cony warrens which might be inside or outside forests or chases. Extra layers of complexity were added to the landscapes so created because most were subject to ongoing use rights from surrounding communities, who constantly exercised their gathering or grazing rights until such rights might be extinguished at enclosure. One effect of the creation of the deer parks was to fossilize prehistoric, Roman, and Anglo-Saxon landscapes beneath them, since the marks upon the landscape of these earlier settlers would be protected to some degree from later ploughing and habitation, and since in most instances they were superimposed upon countrysides which were already operating within the Saxon or early medieval local agrarian economies. Only with disparking could the processes of agrarian change fully work upon the landscape again. There were also straightforwardly private woodlands, and interweaving between such areas there were large stretches of wood-pasture, specializing in pasture farming or having mixed farming economies, integrated to varying degrees with local markets and with woodland by-employments. Clearly, we have to consider the intertwining of social, cultural, and ecological evolution to understand fully the landscapes which have resulted.

Lowland England's woodlands and wood-pasture were not confined to one geological or topographical base. In general the woods might cling tenaciously to the heavy, sour lowland clays of the south Midlands or the great south-western Jurassic clay vale, the Weald, or the heavy soils of Essex and Suffolk. In such areas, wood-pasture economies were overlain by forests, chases, and parks which in many cases thereby effectively prohibited cultivation and settlement. The Jurassic clay vale had by the thirteenth century a group of royal forests: Blackmoor, Gillingham, Selwood, and Chippenham, for example. The wooded midland clay plain contained the forests of Whittlewood, Salcey, and Rockingham, and also Shotover, Bernwood, and Wychwood.

By contrast, infertile sandy or gravelly soils formed the basis for oak or birchwood forests and gorse-covered heathlands in Sherwood and on the metamorphosed Precambrian rocks of Charnwood. Similarly in the south were those of the New Forest, Alice Holt, Woolmer, and Windsor. By contrast again, and outside the scope of this chapter, were those in upland Britain on Dartmoor and Exmoor, on the limestones of the Peak District, and on the sandstone plateau of the North York Moors at Pickering and Scalby.

The landscape features of such areas are both broad and detailed. Some are obvious to the most casual traveller, others require more laborious fieldwork in

The High Weald of Sussex

This general view northwards towards the Dudwell valley illustrates many of the components of a poor wood-pasture landscape. Pastures are small and weed-strewn, with shaws (hedges) wide and unkempt and with much hedgerow timber. Here, near Kipling's home at Batemans, Burwash, one sees his 'dim, blue goodness of the Weald'.

dense and obscuring undergrowth. Perhaps the most obvious landscape features were those of the vegetation cover: woodland, open moorland, heath and under-wood, although the trees might be scattered among grassland or shrubs rather than occurring in denser stands, the vegetation cover depending upon the environment within which the forest was located. Oak, birch, and alder dominated the medieval forest woodlands, with ash, maple, and lime of secondary importance, together with beech in southern England. The amount of woodland cover in such areas is disputed, but Rackham calculates that only about one-fifth of the legal forests were wooded, although this is probably too low, since many were in areas which were already heavily wooded. The trees themselves give character to individual woodlands: the last surviving woods of Sherwood, at Birklands and Bilhaugh near Edwinstowe (Nottinghamshire), incorporate several hundred late-medieval oaks whose curved and sharp-pointed dead upper branches form the essence of Gilpin's 'blasted oak, ragged, scathed, and leafless; shooting its peeled, white branches athwart the gathering blackness of some rising storm'. The introduction of Scots pine to the New Forest in 1776 has probably changed the landscape of that forest more than any other single event, whilst the importation of conifers from North America, begun by Douglas in 1827, has similarly made a great impact on modern forestry techniques.

Ancient coppice woodlands, with longer and quite rigidly defined felling cycles of the growth from the stool of between five and thirty years, yielded diverse shrub, grass, and herb elements resulting from grazing amongst their standard oaks, with ash and hazel, as in Rockingham Forest. Some large stools of ash are estimated to be 1,000 years old, and coppicing was well established by the early Norman period. Clearly, the advantages of coppicing in order to yield a reliable supply of timber for building, wattle and daub, or hurdles and fences from the forests were appreciated. The Tudor forests were certainly also seen as much as reserves for timber for naval uses as for venison. In other areas pollarded trees gave a quite different habitat, the trees being cut between 7 and 14 feet from the ground to prevent animal grazing on the regrowth. Old pollarded trees are a feature of the English landscape and are probably better represented in England than anywhere in Europe, typified by the 500-year-old oaks of Sherwood and the even older pollards of Windsor Forest (Berkshire).

Relict areas of ancient wood-pasture also support rich assemblages of flora and fauna and the biodiversity of such areas is now coming under closer scrutiny for conservation purposes. Similarly, the surviving ancient oaks of Sherwood are now classed as Sites of Special Scientific Interest (SSSIs) for they afford valuable dead wood habitats, and their conservation is rendered the more important by their enormous attraction to tourists, now encouraged by the Sherwood Forest Visitor Centre. Conservation legislation and planned change in the woodlands were to some extent overwhelmed by natural forces in the Great Storm of October 1987 when anything up to 15 million lowland trees were flattened overnight. At Knole Park (Kent), for example, granted to Lord Sackville by Elizabeth I in 1566, half the ancient trees were destroyed together with one and a half miles of deer fencing. Avenues of seventeenth-century oak and chestnut were lost, as were all but 25 of 10,000 young trees planted in 1958 and half of a 75-acre beech wood. Altogether 10,000 tons of timber had to be disposed of. Similar storms have

Pollarded hornbeams in Epping Forest

The hornbeam woods of the London region and southern Essex were prolific, and successive crops of wood were produced from the bolling, or permanent trunk, out of reach of cattle and deer.

weeded out older trees in previous centuries, as in 1222 or 1703 for example, but nothing on the scale of the 1987 storm has been so well recorded in England. And we should also remember the devastating impact on hedgerow timber of Dutch elm disease, calculated to have killed 25 million elms in the last thirty years.

Another conspicuous feature of parks and some chases as well, though rarely of royal forests was the deer-leap fence (*saltatorium, salteries*, or *deer lopes*) and pale, the former designed as a bank with a ditch or multiple ditches on its inner perimeter to prevent deer from straying beyond the landowner's confines, but also designed so that deer jumping into the park from neighbouring lands could not then easily escape back again. In Salcey Forest in Northamptonshire, Britton and Brayley described the 'creeps and deer leaps, made in the fences' around the underwood to allow deer into the enclosures but to exclude commoners' cattle. Both John Norden and Christopher Saxton used the pale to distinguish parks on their maps, although by the 1570s such parks were becoming less important. That such features predated the Conquest is attested by the reference in a tenth-century Somerset charter for Weston, near Bath, to a *hlipgete* which allowed deer to leave a wood but not to enter it. Such earthworks, especially dating from the medieval period, were massive and would be prohibitively expensive to initiate and keep in repair around the large perimeter of a forest, and some chases only had such features in the absence of easily distinguishable topographies for purposes of perambulation. By the nineteenth century wood margins were of a lesser size and tended to be straight with a small ditch or bank with a hawthorn row. Parks were usually enclosed by a ditch and pale, with the bank topped by cleft oak stake palisades, although the royal parks at Devizes (Wiltshire) and Woodstock (Oxfordshire) had stone walls, as did the smaller Forest of

A deer-leap fence at Wolseley Park, Staffordshire

The moderated height of the pale allowed deer to jump into the Park, but they were then prevented from escaping by the steep upward slope in the ditch inside the park boundary. Such contrivances were sometimes presented to the Forest Eyres as nuisances, and they could be removed if deemed to be too close to the forest.

Wychwood (Oxfordshire). Empaling was repeated as the pales rotted and this might be undertaken as a labour service. The custumal of Hatfield Broadoak (Essex) stipulated that one tenant 'shall make and keep 32 perches of pale round the park of Hatfield from the lord's timber in the park, and shall have the old timber of the said pale when it is not worth putting back in the pale'.

To gain access to the pale for inspection and repair, there was often an open area referred to as the freeboard. Other boundaries depended on local topographic circumstances, and at Sonning and Hampstead Marshall parks (Berkshire) the northern boundaries were the rivers Thames and Kennet respectively. Where private individuals gained permission to develop parks within a royal forest, they had to ensure that no deer-leap fences were erected. Where routeways cut through such pales there would be hatches, gates designed to cover the openings in the pale to prevent animal escapees. The forests, according to Sir William Manwood in 1598, should be 'mered and bounded with unmoveable marks, meres and boundaries, either known by matter of record or prescription'. At Whittlewood Forest, Britton and Brayley reported that the central part was demarcated by 'a ring mound, which has been its boundary beyond the memory of the oldest man', with purlieu woods outside the ring. But in general there were recurrent disputes concerning forest boundaries, and the strength of feeling against forest law was such that repeated perambulations to determine the exact boundaries were held. With the revival of forest laws under Charles I the boundaries of Dean (Gloucestershire) were disputed in 1634, according to whether certain townships were inside or outside the boundary. The King's representative asserted the findings of the perambulations of 1228 and 1282, but these were rebutted by countervailing claims relating to perambulations of 1298 and 1300 and their confirmation by Act of Parliament in 1336. In return the 1300 perambulation was denounced as 'false and erroneous', and another boundary based on the Forest Eyre roll of 1277 was produced. This latter, much in favour of the King, was agreed by the jury under some duress. With enlargement and shrinkage of the areas of forest, much depended on the interpretation of the perambulations and their dates.

Undisputedly within the forests there were internal boundaries, similarly constructed to the pales, but on a smaller scale (often consisting only of a low wooded bank with an accompanying ditch) while hedged enclosures were known as 'hays'. Such banks probably resembled more closely the older Anglo-Saxon *wyrtruma* or woodbanks which delimited private woodlands before the Conquest. Since woodland was valuable by the medieval period as more clearance took place, rounded banks with outer ditches to keep deer out, rather than in as with park pales, continued to be constructed, typically about 25–40 feet in total width and possibly set with ancient hedges and pollarded trees. Such banks might in fact compartmentalize the woodlands to assist in coppice management. Private woods were treated in the same fashion: thus three miles of new bank were being built around the Norwich Cathedral Priory woods at Hindolveston (Norfolk) in 1297–8, financed by selling the branches and underwood growing along the line of the new earthwork. A hedge was planted on top of this bank, and bridges, gates, and padlocks also supplied. Fencing was expensive but necessary, since fallow deer are more agile than goats and as strong as pigs, and many of the older parks thus have a characteristic rectangular shape with rounded corners to minimize costs.

The woodland might also be intersected by lawns (launds), which were grassy clearings designed to provide hay or grazing for the deer. Salcey Forest had 180 acres of such lawns by the beginning of the nineteenth century, on which the cattle of the warden and keepers also grazed. There would also be keepers' lodges with their surrounding small areas of cultivated land to provide food for the keepers of the king's deer, the land often providing a marked contrast with the surrounding forest landscape. In Salcey the warden lived in the 'Great Lodge' which by the early nineteenth century was a 'well built brick house, having offices, with gardens, and pleasure grounds attached to it'. However, such buildings were frequently merely keepers' workshops used for storage or overnight accommodation in fence months, when the deer required more careful attention. To help this there might also be deer houses which provided shelter in severe weather, together with mangers or feeding troughs which might be covered, for supplementary browse, hay, or oats for the voracious deer. Pools and streams might also be artificially created to offer water.

The forest might also have hunting stations (*tristera*) where hounds were held under cover until the moment they were required to be let loose. There were also 'standings', high points with railed constructions from which the King or Queen might view the hunt and from which they might safely shoot at the deer which were driven past. Thus we find 'Kings Standing' as an important topographical feature on Ashdown Forest (Sussex). An excellent surviving example of such a landscape is to be found in Hatfield Forest (Essex) with its coppices, boundary banks, lawns with pollard trees, cattle and deer, seventeenth-century lodge, rabbit warren, and purlieu woods. The latter surrounded deer parks or forests and had been considered forest by the perambulations of Henry II, Richard I, or John, but were no longer subject to forest law following the Forest Charter of Henry III. We therefore find Dibden Purlieu on the eastern edge of the New Forest. Such ground could be of considerable extent, as on the southern borders of Rockingham Forest, and tenants holding property had variable common rights to pasture livestock, varying from forest to forest. Terminology also varied regionally, and included *outbounds* (Cranborne Chase), *Venville* (Dartmoor), or *outwoods* (Duffield or Clarendon), the latter also therefore having *outlodges*.

Other landscape elements not associated with the hunt were to be found in the forests and chases: stud farms might be important, as in the New Forest. Ironworks were important in the Forest of Dean and lead-smelting in the High Peak. It should be remembered that woodland landscapes also accommodated much proto-industrial development. In 1574 there were 52 furnaces and 58 forges in the Weald, and in 1634 there were 11 furnaces and 11 forges in Dean, together with glass-making sites, and it was feared that the timber resources would soon be exhausted. Consequently the landscape today reveals sinuous hammer ponds, relict bays (dams), spillways and working areas, often half-hidden deep within woodland areas. The diversity of landscape impact is reflected in a 1608 list of men fit for military service from St Briavels in Dean: of the artisan population, 36 per cent were miners, metalworkers, and stone-workers, 30 per cent were in woodworking or animal processing trades and crafts, 25 per cent were clothworkers, and the remainder in miscellaneous crafts.

All manner of minor surface features may also contribute to woodland landscapes: drainage grip systems were initiated from the middle of the eighteenth century; temporary saw-pits and charcoal-pits may have left some trace, as with other innumerable marl-pits and other depressions. And such features as ridge-and-furrow can remind us that a woodland may be secondary, growing over what was previously farmland. Another landscape feature associated with denser areas of woodland is exemplified by the actions of Walter of Huntley in Huntley Wood, Dean, who in the 1270s 'made a clearing or ridding next the public way to prevent misdeeds done there by thieves, and another above his court-house'. Such clearances were commonly made to deny ambush facilities to more nefarious characters, and were recognized as a duty of the holders of such land by the 1285 Statute of Winchester, wherein Edward I ordered that the edges of highways be cleared for 200 feet on either side of a highway, leaving only large trees such as oaks.

Another important feature was the settlement pattern. Many such areas were colonized, before any imposition of Norman forest law, by small groups of Anglo-Saxon settlers or swineherds, whose hamlets or isolated farmsteads might reflect still more ancient patterns of clearance. Such settlements might then be later incorporated wholly into the forest, connected by woodland tracks such as the *wealdenes weg*, 'the forest road' which ran into Neroche (Somerset) according to a charter of 938. The retarded settlement resulted in larger administrative areas in such regions: the virgates and parishes of the Weald are larger than in the surrounding champion countryside, and their churches were generally later in date. The forests themselves, by their legal existence, did not encourage settlement in the medieval period, although monastic houses might obtain royal grants of land, as did Chertsey Abbey (Surrey) within Windsor Forest or Newstead Priory (Nottinghamshire) within Sherwood. The upper echelons of forest administrators who were courtiers and nobles might also obtain grants and build up considerable holdings within the forests, as did the Wardens of the Forest of Dean and Savernake. The keepers' lodges were also supplemented by more imposing hunting lodges for visitors. Chingford was a favourite of Queen Elizabeth, and Woodstock on the edge of Wychwood Forest had a stone wall 7 miles in circumference. In some cases royal forests were 'assigned to the maintenance' of royal castles nearby, and their constables were frequently also appointed as forest wardens, as at Rockingham (Northamptonshire), Porchester (Bere Porchester Forest, Hampshire), Devizes (Chippenham and Melksham Forests, Wiltshire), St Briavels (Forest of Dean, Gloucestershire), or Windsor (Berkshire).

There were perhaps three very significant later waves of settlement. One came during the land hunger of the period up to about 1300 when assarts were created as typically small, often isolated cottages with a few acres of land held in severalty, whose occupants owed allegiance to manorial lords. On the edges of Charnwood Forest (Leicestershire) we therefore find the Domesday settlement of Groby deriving rents from freshly cleared assarts at Newtown Linford, first recorded in 1293. The village is now a loose linear collection of the original farmsteads with many cruck-framed buildings, demonstrating the abundance of woodland in what is otherwise a district rich in stone. Certainly forest status could not hold back the demand for assarts, and licensed and organized assarting was taking place in some Yorkshire forests by the late thirteenth century, with both hunting and pastoral functions being withdrawn to the large parks. There were significant increases in the arable area by 1350 in the Forests of Sowerbyshire and

Knaresborough. In Sherwood Forest Edward I converted many such assarts into ordinary leaseholds with no common rights. Distance might mean that the typical feudal relations of landholding and services might be commuted to a payment in kind or cash. The patterns left around the edges of forest areas by this medieval land reclamation movement can be seen in such areas as Ashdown Forest in Sussex at Nutley where the village itself was first mentioned in this period and where scattered cottage holdings are still visible in the modern landscape. Within the eastern Weald there was a vast clearance as, for example, the monks of Battle Abbey developed their arable lands, as the manor of Rotherfield assarted 6 square miles of woodland between 1086 and 1346, or at nearby Laughton where nearly 1,000 acres were brought into cultivation between 1216 and 1325. The relict ancient woods thus affected thereby have acquired sinuous or fragmented perimeters, or have small islands of settlement within surrounding woodland or heather and bracken-covered slopes.

Second, following a period of general retreat from the margins of settled Europe in the fourteenth century, the early modern period again saw expanding settlement onto the manorial wastes along roadsides, onto commons, and into the woodlands as population increased, especially in areas where alternative agricultural practices, by-employment, or industrial growth were evident. With the expansion of the Wealden iron industry in Sussex, for example, many cottages were erected around commons and forests during the sixteenth and seventeenth centuries in such locations as Ashdown Forest (e.g. Newbridge), St Leonards Forest, or along the Forest Ridges at Heathfield and Burwash. Many of these timber-framed cottages remain, and the more sturdily built houses of the ironmasters are highly desirable residences today. But in many areas the forests suffered at this time, and surveys and accounts of the Forest of Pickering (North Yorkshire) between 1608 and 1651 show neglect and plunder of the tree cover. Few communities could provide sufficient employment or common rights for the many poor who flocked to the woodlands in the seventeenth century from champion England, as agrarian change displaced many from their cottages and work, and impoverished and underemployed families were the result in areas which themselves were coming under intense scrutiny by agricultural improvers and industrialists alike.

Finally, a building phase of great importance occurred during the Victorian and later periods as improved communications opened up the possibilities for another period of colonization, this time by wealthier escapees from the towns whose *fermes ornées* or red-brick villas lined the turnpike roads and clustered near the new railway stations as they sought simultaneous seclusion within newly appreciated picturesque surroundings and accessibility to the urbane world of economy and high culture. This last movement also brought a cultural collision as middle-class Victorians came face to face with the spirited independence of the forest dwellers. In the twentieth century, leaving aside the enormous impact of clearance and destruction caused by road-building or widening schemes, and noting that the M1 motorway cuts the Forest of Charnwood in half between junctions 22 and 23, we might note also that many of the dispersed cottages have been colonized anew. This has frequently been achieved by wealthier residents who have conserved the cottages, landscaped and exoticized their gardens, and converted outhouses or barns to new cottage employments or to garages for estate or four-wheel drive vehicles, which gain access along tarmac roads that snake through bracken or conifer-bordered driveways.

Prehistory and the Early History of Lowland Woodland Landscapes

Woodland history can be glimpsed from the evidence of pollen analysis, backed up by Carbon 14 dating, allowing us to see changes in some of the earliest post-glacial woodlands in England which were emerging by *c*.11,000 BC. Birch, aspen, and sallow groves of woodland were gradually interspersed with, or replaced by, pine and hazel, then oak and alder, then lime and elm, and then holly, ash, beech, hornbeam, and maple as the climate warmed. From *c*.6500 to 4000 BC localized woodlands proliferated in the relatively stable Atlantic climatic period. Woodland dominated by lime, combined in various degrees with oak, hazel, ash, and others such as pine in the Breckland of East Anglia, formed regional patterns across southern England in the mid-Holocene, covering all soils except for areas of natural moorland or grass, or coastal dunes and salt marshes. Groups of lime woods in Essex, south Suffolk, Norfolk, Lincolnshire, and Derbyshire represent the fragmented remains of this woodland cover. Older ideas of a uniform 'mixed oak forest' must be discarded.

The exact impact upon the wildwood of early hunting and gathering communities remains difficult to establish, although modification of the woodland cover through temporary clearance to facilitate hunting by Mesolithic groups is now widely demonstrated. In the south east early woodland clearances can be measured by the analysis of sediments containing tree pollen in the lower river valleys, indicating considerable clearance within the Weald during the Prehistoric period, in contrast to a widely held earlier image of an unbroken woodland zone. Mesolithic populations have also been credited with the initial formation of heaths on sandier soils and an early wood-pasture landscape of glades within woodland may have resulted from these clearances.

However, the greatest initial anthropogenic impact was that of shifting, and later settled, Neolithic cultivators from *c*.4000 BC who selected areas of elm for clearance leading to a synchronous 'Elm Decline' *c*.3100 BC associated with early pottery and tools, and with the pollen of open-land plants, emmer wheat and plantain, stinging nettles, and other weeds of arable land. Through a combination of felling, grazing, and burning, areas such as the Breckland were cleared to become heathland, and the Somerset Levels and the chalklands were also cleared during this period.

In the succeeding millennia of the later Neolithic and Bronze Ages still more of the wildwood was cleared for arable farming and grazing by cattle, sheep, and goats which precluded woodland regeneration to some extent by their trampling and browsing on the younger shoots. Iron Age and Romano-British settlement also actively extended deep into woodland areas, such as the heavy clays of Suffolk (e.g. the Waveney valley) and Essex, Wychwood, which has also revealed Bronze Age and Neolithic sites, or Rockingham Forest and the Weald and Forest of Dean, where ironworking gave rise both to clearances and the requirements for charcoal from coppicing—a practice well understood in the Roman world. It has been estimated that in lowland England half the area ceased to be woodland in about 700 years, yielding a patchwork landscape of woodland and farmland not very different from that which we know today.

Saxon Woodland Landscapes

Within lowland England, the Anglo-Saxon settlement for the most part took over and continued a well-organized and stable agrarian landscape. Here woodlands were required for swine pastures, building materials, fuel, food, and charcoal. The woodland cover varied, and there may well have been some marginal regeneration in the late or post-Roman period. Where colonization or cultural assimilation occurred within the woods we find the evidence of place names, charters, and the Domesday survey indicating the main remaining wooded regions such as Hertfordshire, the Chiltern plateau, north-west Warwickshire, or the largest concentration of woodland of all in the *Andredesweald* of Kent, Surrey, and Sussex, where '-field', '-ley', and '-hurst' place-name elements show clearings either inherited from previous generations or newly won. Destruction of woodland was also implied in place-name elements such as *Roding* (and variants *Ridding*, *Reed*, etc.), *Stocking* or *Stubbing*. Some woods were recognized as ancient even then: Selwood (Wiltshire) was Asser's 'great wood' (Celtic *Coit Maur*) at the end of the ninth century, and a charter of 682 refers to 'that famous wood which is called *Cantocwudu*' (Quantockwood, Somerset). Although the Anglo-Saxon language has many elements denoting woodland (e.g. *wudu*, *graf*, *scaga*, *hangr*, or *holt*), by 1086 much woodland had been cleared: Rackham calculates that out of 12,580 settlements for which adequate particulars are given in Domesday Book, only 6,208 possessed woodland, with large amounts in the Home counties, in north-west Warwickshire and adjacent Worcestershire, in Derbyshire, and in east Somerset. But there was by this time little in Breckland or the Fens, or in a belt from East Yorkshire across north Oxfordshire and the Midlands to Wiltshire. Even by the early eighth century the Laws of Ine carefully regulated the preservation of woodland, suggesting that a plentiful supply was by then already a thing of the past. The Domesday map of woodland is difficult to interpret accurately since much woodland was at a distance from the manorial centres which were the focus of attention. Thus little woodland is noted for any places within the Forest of Neroche (Somerset) or within the Weald since the manorial centres were clustered around the edges of these woodlands. But we can estimate that the Weald was probably about 70 per cent woodland, Worcestershire about 40 per cent, and Somerset about 10 per cent, but that Cambridgeshire, Leicestershire, or Lincolnshire had a woodland cover amounting to less than 5 per cent of their areas.

Some wooded or heathland areas appear to have been designated for hunting in the late Saxon period, with enclosures for the management of deer, and their location clearly influenced the development of the Norman forests in turn. A 'deer-fold' (*derhage*) existed on the wet, acidic soils of London Clay and Boulder Clay at Ongar (Essex) in 1045, and its antiquity is demonstrated by the fact that much of its perimeter has been adopted in turn as a parish boundary. Edward the Confessor had a hunting lodge at Brill in Bernwood Forest (Buckinghamshire), and hunted deer, boar, and hares in his extensive woods of Chippenham (Wiltshire). Almost certainly the nucleus of a forest administration was already here as well as in Dean,

Savernake, Clarendon, Mendip, and Windsor where huntsmen and woodwards (translated by the Normans as 'foresters') held land of the King in the *silvae regis.* There may have been continuity of administration in many forests from the time of the Confessor to William I, although the exclusive forest laws of the Normans had in turn owed much to the Frankish royal forest institutions of the Carolingian Empire, inherited by the Dukes of Normandy in the tenth and eleventh centuries.

Medieval Woodlands

The regional differentiation of woodland cover should be kept in mind when discussing the medieval landscape of lowland England. Overall the woodland cover in the late sixth century has been estimated at about 33 per cent; by 1086 at about 15 per cent, falling to perhaps 10 per cent by 1350. The Anglo-Saxon centuries saw a rate of woodland decline equalled only in the twentieth century; and in the Weald alone about 450,000 acres were grubbed out in around 260 years before 1350. Perhaps the last real wildwood may have been in the Forest of Dean *c.*1150, after which that area too was pastured and coppiced. In truth, medieval England was one of the least well-wooded countries in north-west Europe, and in the open-field areas of the Midlands one might not encounter more than 5 per cent of the area of any community's lands as being wooded. By contrast, settlement in woodland areas tended to be lacking in the cooperation that was necessary between people farming in open fields, and took the form instead of families farming independently in hamlets and on isolated farms.

By the early medieval period it is possible to differentiate more clearly between the varying timbered landscapes of lowland England. On the one hand there were the managed areas of woodland, yielding timber from standard trees—usually oaks—and underwood from coppiced stools around them. Second, there were agrarian systems based primarily on a wood-pasture environment of woodland, often with pollarded trees scattered amongst small, hedged fields or commonable areas which were productive to greater or lesser extents of some corn, cattle, and sheep, combined in different ways to produce regional contrasts in farming patterns. Third, there were the more specialist uses of these two environments as forests, parks, chases, and warrens: the three former reserved primarily for hunting or as enclosures for the 'storing of live meat' in the medieval period, the latter for the production of conies. Such areas as these might have varying amounts of woodland cover, ranging from a fully mature woodland ecosystem to a landscape bare of trees.

The medieval forest resulted from the arbitrary imposition of royal forest laws upon wide stretches of otherwise conventional countryside. As a legal entity, created or destroyed by royal command rather than by any landscape change, the word 'forest' was derived from *foris*, 'outside' or external to prevailing laws. Richard FitzNigel *c.*1179 defined forests as 'a safe refuge for wild beasts: not every kind of beast, but those that live in woods, not in any kind of place, but in selected spots suitable for the purpose'. In practice forests might therefore also include privately owned farmland and settlements, even a town such as Brill in Bernwood Forest (Buckinghamshire) or Colchester (Essex), as well as royal demesne lands, among which the king's deer would move unharmed, and from which the king could expect income for his coffers via forest courts and officials.

133 ● ● ● ●

The main protected animals were the introduced fallow deer, the main beast of the hunt, together with red deer, roe deer (until they were excluded in 1339 as being too numerous and harmful to the other deer), and wild boar (all these being defined as 'venison', which was any beast hunted by the king), but protection might also be extended to include hares, foxes and rabbits, wildfowl and fish. Fallow deer were common in most forests, but red deer were chiefly present in northern forests and roe were significant only in Yorkshire. Forest dwellers were forbidden to possess bows and arrows, to erect boundaries which would impede the free movement of deer onto their crops, and their dogs had to be 'expeditated', having their claws cut close to the ball of the forefeet so that they were unable to chase the game. The 'vert' was also protected as giving shelter and food to the game, so that numerous laws forbade clearance and cultivation, and restricted communal wood gathering rights.

The superimposition of laws designed to protect deer and support the hunt sat uneasily with pre-existing woodcutting, transhumance, communal grazing, and other use rights and settlements. In some areas the forests may have been indeed created at the expense of settlements: in the case of William I's *Nova Foresta* in south-east Hampshire there were exchanges of land but also possibly evictions and depopulations, hinted at in the Domesday Book, together with reductions in the value of affected manors. Perhaps 30 to 40 whole villages were placed under forest law, and parts of another 40 or so, and substantial revenues from the constituent manors were thereby siphoned off into royal coffers, to the growing distaste of the affected landowning elite. On the Welsh border, William of Eu held Wyegate before the Conquest and it was assessed at six hides, but by 1086 it had been incorporated into the Forest of Dean 'by the king's command' and there was now only a 'fishery worth 10s.' for taxation purposes. From 1087 first William Rufus and then Henry I (1100–35) fiercely maintained forest privileges after promising concessions which were later forgotten, and Henry created new areas of forest law in Rutland, Leicestershire, Bedfordshire, and Yorkshire and covered virtually the whole of Essex in this way. Under Henry II these laws were strengthened and extended so that all classes suffered. It was said that:

> He granted to no one dwelling within the Forest metes liberty to gather twigs in his own woods without supervision by the foresters, or to bring even impenetrable thickets into cultivation … He conferred immunity upon the birds of the air, the fish in the rivers, and the wild beasts of the earth, and made poor men's plots their feeding-grounds.

Penalties for offences against forest law were harsh, and the reality of greater leniency, at least under Richard I (1189–99) and Edward I, is currently disputed. Amercement (fining), loss of estates, mutilation, and even death could technically result, although income from the fines imposed was far more important as a source of royal revenue than the exaction of punishment, at least by the end of Henry II's reign. Vert and venison were to be protected, there should be no building of houses, sheepfolds, hedges or ditches, no cultivation, and no right of grazing swine on nuts and roots, and the passage of carts along the forest roads was forbidden during the fence month in June and July when the deer were fawning. Compensation for such restrictions might include the right to 'forest commonage', unstinted commoning over whole forest areas, as in Epping Forest (Essex) and Wychwood. In 1591 this right was claimed by the tenants of Ascot under

Upholding rights of common usage into the twentieth century

On Oak Apple Day the villagers of Wishford, on the edge of Salisbury Plain, take part in the Grovely Oak ceremony, asserting their right to gather dry wood in Grovely, a nearby Wiltshire forest, which had been created by William I, as a wooded area with large numbers of native red deer. It was attached to the larger Clarendon Forest in the early fourteenth century. Stamp and Hoskins in The Common Lands of England and Wales *(1963) refer cautiously to Grovely as containing 1,118 acres 'over which the inhabitants of Wishford enjoy certain rights of estovers, though the land is not regarded as common by the owners'.*

Wychwood, but more normally some degree of limitation was imposed, as for example, for the manor of Charlbury in 1636 adjoining Wychwood which claimed 'common or mast in time of pannage in the wastes and commonable places within the forest for all their hogs and pigs, ringed, levant and couchant in and upon their lands and tenements in Finstock'. In general, there were to be no purprestures (encroachments onto the royal forest), no buildings and enclosures made without licence, and no assarts or any cropping upon them. Thus, at a time of growing population in England between 1086 and 1300, large areas of the countryside were effectively held back from settlement and cultivation, to the obvious discontent of many levels of society. Furthermore, this was exacerbated where important towns lay adjacent to woodland, such as Nottingham to Sherwood or Lancaster to Lancaster Forest, and where woodland resources and hunting privileges were also denied to the townspeople.

Forest laws were administered by a hierarchy of officials who held office in ways which were very similar in both England and Normandy: Chief Justices of the Forest were at the head of the entire forest system (from 1236 two men held the office, dealing with lands north and south of the Trent respectively until the office was abolished in 1817), and they were assisted by the lucrative posts of wardens, the hereditary foresters of fee, of whom there might be several in a large

forest, and other sworn foresters—'walking foresters', 'riding foresters', and sergeants, each with a defined bailiwick or ward to police and who were thus allowed to carry bows and arrows against marauders. The New Forest, for example, was divided into nine bailiwicks, Windsor into 16 walks. And from among the knights of the locality came elected but unpaid officials such as verderers, regarders, and agisters—unpopular positions to hold locally because they had to enforce a system which itself was seen as corrupt. According to chronicler Matthew Paris, the chief Justice Geoffrey of Langley, who held office between 1250 and 1252, performed his duties with cunning, shamelessness, and violence, and extorted huge sums from the northern magnates. The system of courts included forest inquests, and surveys, which led to the Forest Eyre, comparable to the General Eyre throughout the rest of the kingdom in the twelfth and thirteenth centuries. An eyre for the Peak Forest held at Derby in 1285 heard 517 charges of trespass against the vert and venison, but as thirty-four years had elapsed since the previous one, many of the original offenders must have been dead.

Areas which were similarly enclosed for hunting, but which were not according to Manwood the 'highest franchise of princely pleasure' but rather 'inferior franchises', held by landed magnates or ecclesiastics other than the king or queen, were referred to as 'chases': Cranborne Chase (Dorset) was held in the twelfth and thirteenth centuries by the Earls of Gloucester and Hereford. Others such as Cannock (Staffordshire) or Enfield (Middlesex) were examples from about twenty-six such areas in existence during the medieval period. No forest law prevailed here, although the restrictions on local populations differed little in practice from those asserted in royal forests. The Earls and Dukes of Lancaster in the fourteenth century owned five chases (Blackburn, Bowland, Duffield, Leicester, and Needwood) and appointed officials to carry out duties virtually identical to those of the forest administrations of the king. The ecosystems and landscapes of forests and chases were little different.

Many smaller and enclosed hunting parks (Anglo-Saxon *pearroc*, any fenced area, hence the medieval *parrock*) or hays (Anglo-Saxon *haga*, an area enclosed by a hedge or fence, which could also be a deer corral) might also be found in wooded or wood-pasture landscapes, distinguishable from forests and chases by the presence of the boundary. Parks were essentially designed for the rearing and hunting of deer, and with landscapes quite different from the later amenity parks which began to appear from the late fifteenth century as adjuncts to great houses. At various times in the medieval period there were at least 1,900 hunting parks in existence as landscape features of considerable significance. Most were between 100 and 200 acres, although royal parks as at Woodstock, created in 1113, and now preserved within the later eighteenth-century Blenheim Park, might be larger. Not all were completely wooded since pastures were also required for the fallow deer with which most were stocked, and in fact the pressures of browsing deer often reduced the woodland cover to a savannah-like landscape of grass and scattered large trees. Emparkment was frequently inimical to woodland conservation. The parks belonging to the Crown were frequently associated with forests, such as Gillingham (Dorset) or Clarendon (Wiltshire), but there tended to be more parks associated with chases. The Earls of Lancaster, for example, held forty-five parks at various times, primarily in Lancashire, Staffordshire, and Leicestershire. Their chase at Needwood contained no fewer than eleven parks in the fourteenth century. The Dukes of Cornwall held twenty-nine parks, the Earls

Hunting scene

Depictions of early hunting scenes contained stylized landscapes with elements of woodland and forest environments taken from many locations, but with the best surviving examples mainly originating from the European Continent. The subject matter in the English early seventeenth-century Bradford Table Carpet shown here, with coloured silks embroidered on linen canvas, reflects delight in the countryside, and the border of the carpet includes shepherds and country houses as well as the hunting, fishing, and duck shooting shown in this extract.

of Arundel twenty-one, and the Bishop of Winchester twenty-three. During the earlier Norman reigns the creation of forests took precedence, but between 1200 and 1350 the number of parks multiplied as magnates and knightly families demonstrated their status by conspicuous expenditure. The spatial distribution of parks reflected population densities to some extent: there were more in the south-east and Home Counties and in the West Midlands, but fewer in the north or south-west. On the edge of London, for example, Henry VIII created Hyde Park of 600 acres and St James's Park as convenient sites for hunting by courtiers and foreign guests. Previously established woodland was also clearly more of a locational factor than for forests, with well-wooded counties such as Essex and Staffordshire possessing most parks.

Within all the foregoing landscapes the lord of the manor might also hold rabbit warrens (*coneygarths* or *coneries*), which were areas, possibly on poorer soils, and varying in size from a small field to a square mile or so, superintended by warreners. The first certain mention of mainland rabbits is in 1235 at the royal park at Guildford (Surrey), where the first coneygarth was named in 1241. They seem to have rapidly gained popularity and by 1268 we hear of complaints of poaching. Warrens on Ashdown Forest dating from the fourteenth century were enclosed with perimeter banks and ditches and contained elongated 'pillow-

mounds' for the burrows. The planting of gorse, juniper, bramble, and other prickly shrubs afforded food and shelter; and breeding hutches or 'clappers' might also be provided.

Surrounding and interweaving with such areas, especially in the lowland zone of England, would be extensive areas of wood-pasture, intimate mixtures of landscapes with dispersed settlement, deep winding lanes, and small, amorphous closes seemingly carved directly from the encompassing woods, and devoted primarily to grassland for dairy farming or for the rearing or fattening of cattle, as on the Norfolk–Suffolk borders, or for pigs as in the New Forest, the central Hertfordshire clays or the eastern half of the Vale of York. In fact in many areas, such as central Suffolk or pastoral parts of Essex, woodland farmers cultivated their arable in open fields though they had enclosed them early by agreements or exchange of property. In the dairying and grassland regions of Wiltshire or the Warwickshire Arden such enclosures were similarly achieved during the sixteenth and seventeenth centuries. These economies and societies, together with those of similar areas in the Weald or eastern Somerset, were embedded in landscapes which both influenced and reflected the presence of the pastoral family farm, the lack of manorial control, and the rise of cloth-making and other rural by-employment. Sales of tree bark to tanners, ash wood to the coopers, cordwood to the itinerant charcoal burners, and other woodland resources, helped seasonally to boost family incomes in these regions.

Any hard and fast definition of a landscape deriving from the above types is riven with difficulty because these six landscape elements—woodland, forest, chase, park, warren, and wood-pasture—tended in practice to be to some extent interchangeable. Part or whole of a royal forest might pass out of the hands of the Crown and into another landowner's possession, for example, as when Henry III granted the royal Forest of Dartmoor to his brother, the Earl of Cornwall and his heirs in 1239, or Pickering Forest to his second son, Edmund in 1267. Cannock Chase was carved out of Cannock Forest for the Bishop of Coventry and Lichfield in 1290. Alternatively, disafforestation removed the privileges of the royal hunt and returned land into wood-pasture once more. Conversely, there could be changes in the opposite direction. The chases owned by the Lancastrian faction became royal forests as Henry IV ascended to the throne in 1399—thus Leicester moved from being a chase to a forest. Cranborne Chase passed into and out of royal hands, including those of King John, but mostly remained with the de Clares, Earls of Gloucester.

More normally, however, land became forest as farms and other land were technically declared so to be by the Crown, as during the reign of Henry II when large areas of the south-east were so proclaimed, including what is now the entire county of Surrey, giving rise to a maximum 'forest' area by the end of the twelfth century. Between 1178 and 1180 the Chief Justice of the Forest heard forest pleas in no fewer than twenty-nine counties. As a source of royal revenue such an area was important, although Richard I, John, Henry III, and Edward I sought to raise money also by granting disafforestation in return for large fines. For a fine of 200 marks the knights of Surrey succeeded in getting their county disafforested in 1191, except for the north-west corner west of the Wey and north of the Hog's Back (Guildford downs) which remained within the Forest of Windsor, and in 1204 Cornwall was similarly disafforested to help finance John's war with France, as was Essex 'north of Stanestreet', and Devon outside Dartmoor and Exmoor.

Some of the worst excesses were curbed in the Magna Carta and the ensuing Charter of the Forest in 1217, which ensured that no one was to suffer death or mutilation for taking the king's deer, and that land afforested by Henry II and his sons was to be disafforested. But the disputes between the Crown and the barons continued through to the reign of Edward III, by which time the alternative procedure of raising income for the Crown by parliamentary taxation became well established. By the 1330s there were fewer such complaints. Forests such as Allerdale (Cumberland) or Northumberland disappeared, and as royal power was forced to abandon large areas of forest, so great landowners attempted to absorb them into chases and parks, often against local opposition.

Finally, it should be noted that contemporaries themselves were often lax in their use of the word 'forest', and they did not always use the word in its strictest sense. Thus we have Charnwood Forest which was nearly always held privately in the Middle Ages, as was nearby Leicester Forest, and many northern forests passing from royal to baronial possession were still referred to as forests, and, worse still, many hunting grounds with no record of royal ownership, which should have been styled chases, were also often called forests. And we should also note that the word 'forest' has itself changed in meaning and that from c.1650 by historical back-association, it began to assume its modern connotation of a large wooded area, rather than a legal entity, a process assisted by the early twentieth-century creation of a Forestry Commission.

During the reigns of the later Norman and Angevin kings a long-standing struggle for supremacy with the barons centrally involved England's forests, though not necessarily its woodlands. During the civil war of the twelfth century both Stephen and Matilda sought to gain support by grants and charters connected with the forests; and in 1187 Henry II was angered by the actions of Hugh, Bishop of Lincoln, who excommunicated the then Chief Justice of the Forest for enforcing forest laws against clerical offenders. Sequences of afforestation, disafforestation, reafforestation, perambulations to define 'metes and bounds', and royal dissimulation meant that large parts of the English countryside were alternately within, without, or questionably on the legal boundaries of the royal forests. But overall, from the royal high point of Henry II's extension of forest law to cover about one-fifth of the country, a succession of grants and charters, as well as the licensing of assarts and purprestures, gradually eroded the spatial extent of the forests. Especially following the death of Edward III we therefore find great magnates being granted forests such as Dean, Rutland, Savernake, Feckenham, and the Somerset forests. By 1334 the area covered by the royal forests had shrunk to about two-thirds of its 1250 acreage. And for woodland outside the forests the mixtures of oak and brushwood (*bruera*) were gradually cleared as its preservation within a well-organized landscape took second place to the sheer pressure of people needing farmland and building materials, undoubtedly strongly influenced by the process of subinfeudation, creating new manors which were self-contained tenurial units within the wood-pasture regions. Thus in 1288–9 we find monks on the Westminster Abbey estate at Pyrford on the edges of Windsor Forest clearing a demesne wood, felling and shipping cartloads of great oaks (*robura*) and young rafter-standards (*cheveron*) to Westminster; and in the early 1290s the roots were grubbed out and the area sown with 36 quarters 7 bushels of oats. The landscape of elite pleasure was without doubt giving way to a working landscape managed for profit.

Early Modern Forests and Woodland

The Tudor period witnessed the slow decline of medieval forest administration, but certainly not of the vigorous working society of the woodlands. At its inception we find Henry VII letting out Exmoor and Neroche Forests on a 300-year lease, and despite the interests of both Henry VIII and Elizabeth in hunting there seemed little they could do to halt the decline. In 1628 the Attorney-General stated of the Forest of Waltham (modern Epping) that

> The inhabitants were restrained from commoning with sheep as contrary to the forest laws during the reign of the late King Henry VIII…. and of all other preceding Kings who took delight and pleasure in hunting. But afterward that Royall and princelie pleasure being not so much esteemed by the late king Edward the Sixt (by reason of his minoritie) and by the two succeeding Queenes (by reason of their sexe), the lesse care of the due execucon of the forest lawes consequentlie ensued … whereof the inhabitants of the said fforest … taking advantage, did by degrees, especiallie towards the end of the raigne of the late Queene Elizabethe, encroach upon the said fforest laws by commoning with sheepe.

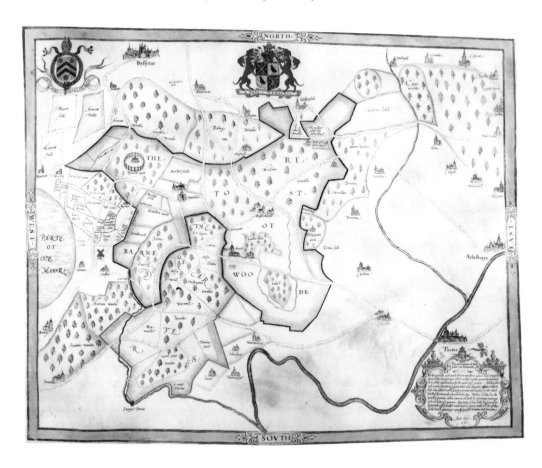

During the reign of James I, enquiries revealed the state of decay. Morfe (Shropshire) had become valueless as a hunting ground through long neglect; and Pickering (North Yorkshire) had little timber left and 'for every redd deare in the fforest, there are 5,000 sheepe'. Forests were leased to improvers to increase royal revenues, as with Lord Cranfield's commission for Bernwood (Buckinghamshire), Feckenham (Worcestershire), and 'other forests and moors' in 1622. In 1620 it was reported, however, that the enclosure of the 'best part' of Galtres Forest (Yorkshire) had inflicted hardship upon the local poor and that the wood required for making the enclosures had further reduced the woodland cover, and the 'tumults arising there from searching for cattle etc.' had 'frightened away most of the deer'. By 1623 Chippenham and Melksham had gone the same way, and Charles I continued in the same vein in 1627 with the Forests of Leicestershire, Neroche, Selwood, and Feckenham. In Selwood (Somerset), the waste and commonable land was to be divided, one-third to the Crown, one-third to the forest landowners, and one-third to those with rights of common pasture. In 1628 Gillingham (Dorset) was disafforested and in 1629 Galtres was granted to the King's creditors. But some tenants in the latter forest townships refused to agree to the proposed terms, and the dispute was still smouldering by 1637 when the lessees stated that 'the people of the country adjacent will not pay any considerable rent for what they say have been their commons'. The land itself was 'wild and unmanured' and Walloon refugees were being imported from Hatfield Chase to cultivate the soil at a reasonable rent. But elsewhere the old woodland cover disappeared: Bernwood was disafforested in 1632 and the tree roots were grubbed up after large-scale felling, hedges and ditches demarcated new closes, and Samuel Hartlib, writing in 1651, admired the improvements. New isolated farmsteads and mixed or pastoral farms had appeared by 1700 and ancient woodlands are thus to be found today only in isolated patches around the edges of the former forest.

Similar schemes for enclosure and improvement were drawn up for the Peak Forest, Chute, Pickering, and Knaresborough, although the latter did not come to fruition. But in Dean the King sold to Sir John Winter 'all his Majesty's Coppices, Wood Ground and Waste Soil of the said Forest (except the Lea Bailey), with the Wood and Timber, and all his Mines of Iron and Coal, and Quarries' for just under £2,000. Employing 500 woodcutters, Winter felled much timber, enclosed about 4,000 acres as coppices, and allotted a similar area to the commoners to compensate them for the loss of their common rights. But the latter land was worthless, provoking violent opposition, and in 1642 Winter's grant was rescinded because of his misappropriation of timber in the forest. The revival by Charles I of ancient forest administration provoked great hostility. Under the direction of the Earl of Holland licences were sold, surveys made, and the forest courts sat again to prosecute offenders. At Dean in June 1634 a swanimote heard 800 presentments of offences against vert and venison stretching back over forty years. The forest justices were also sent out again on eyres, beginning in Waltham Forest in 1630, and attempts to enlarge forests were repeated, as for example at Salcey in 1639 following a new perambulation. Exemptions from the forest laws were sold to landowners within the forests to fill the royal purse. But as opinion within parliament and the country turned against him Charles retreated as former kings had done, and declared the forest boundaries henceforth to be those of 1622, and new perambulations were to be

The Forest of Bernwood, Buckinghamshire, 1590

A map made for a lawsuit in the possession of New College, Oxford shows the clear outline of Bernwood Forest (also the boundary of the parish of Oakley) on the borders of Buckinghamshire and Oxfordshire. The railed area in the west of the forest is Pauncell Rails, with a lodge depicted within it, a royal demesne wood; and Brill, a royal manor and administrative centre of the forest, is indicated by its church in profile. Assarting from the early thirteenth century, and enclosure by the fifteenth century, had affected the forest and the irregular boundaries and the field system to the west bear testimony to this, as does the presence of the twelfth-century newly created village of Boarstall, with its open fields, within the forest boundaries.

made. In 1641 these were certainly made for the Forests of Essex, Windsor, Wychwood, Shotover, Rockingham, and others. In Essex the boundaries were close to those established in a perambulation of 1300, but in fact the forest became restricted to the south-western corner of the county where it remained until the nineteenth century.

During the Civil War the forest laws fell into disuse again, but as the need for money grew in the Commonwealth period, so too did Parliament turn to the forests as kings had done previously. The Parliamentary Surveys of Forests were intended as forerunners of disafforestation and improvement, and 40,000 trees were cut down in Dean alone at this time. The Restoration of the king intervened before further action could be taken, but instead, Charles II granted away some remaining royal forests. In 1664 the last remnant of royal forest in Wiltshire, Clarendon Park, was granted to the Duke of Albemarle, and although some forest courts were held as late as the 1670s, the system was virtually ended at this time. Management of the crown forests now passed instead firmly into the hands of the Surveyors-General of Woods and Forests, whose duties were primarily to procure timber for the navy or for sale, and thereafter to administer customary rights, oversee the deer, and generally protect the forest environments against the depredations of neighbouring populations. This was no easy matter since disproportionate numbers of the poor had come to reside in such areas in the early modern period, descending on land that was seen to be rewarding and awaiting colonization. Such communities relied heavily on forest resources. Thomas Fuller, the Dorset rector of the 1630s and author of the *Worthies of England*, described how New Forest pannage was so important since pigs 'feed in the forest on plenty of acorns … which going out lean, return home fat, without care or cost to their owners'. But the cottagers certainly did care about the threats of forest enclosure, and the 'Skimmington riots' in Wiltshire, Dorset, and Gloucestershire were in large measure provoked by threats to traditional rights. The impact of woodland-consuming developments such as the Wealden iron and glass-making industries was also significant in this period. Thus in 1520 tenants complained that in Ashdown Forest 'much of the king's woods were cut down and coled for the iron mills and the Forest digged for Irne by which man and beast be in jeopardy'.

In general, despite royal attempts at woodland conservation as in the Acts of 1483, 1543, and 1559, and again in the 1580s, and the perceived needs of the industrialists to protect their fuel and raw material supplies by coppicing, little advance in woodland management was made between 1450 and 1700, and despite the interest aroused by Fitzherbert's *Book of Husbandry* (1523) or Evelyn's *Sylva* (1664), there was little to distinguish woodland expertise in 1700 from that of its medieval practitioners.

Forests after 1700

The decline of woodland management seen in the seventeenth century continued throughout the eighteenth century also, and between 1787 and 1793 seventeen reports were published giving the findings of a Royal Commission into the condition of crown woods and forests. In general the picture was one of neglect, with officials unappointed and courts held intermittently or not at all.

North of the Trent there was now just one forest, Sherwood. To the south the New Forest, Alice Holt, and Woolmer and Bere lay in Hampshire; Windsor (Berkshire); Dean (Gloucestershire); Waltham or Epping (Essex); Whittlewood, Salcey, and Rockingham in Northamptonshire; and Wychwood (Oxfordshire).

Felling and selling timber 1717

The frontispiece to the second edition of Moses Cook's The Manner of Raising, Ordering and Improving Forest-Trees, *which had first been published in 1676. It shows timber being felled by axe and measured by rod and dividers, and discussions taking place with a timber merchant. Cook described himself as 'Gardiner to the Earl of Essex', and his book was inspired by Evelyn's* Sylva *twelve years earlier. The emphasis here is on ships' timbers for the 'wooden walls' of English defence at this time.*

By the end of the eighteenth century Arthur Young and fellow improvers, such as Vancouver, were vehemently opposed to the very existence of the forests, which, as they saw it, only required enclosure to make them into cornland. In many forests permission was granted during the eighteenth and nineteenth centuries to fell woodland and to make enclosures: the unenclosed wastes of Epping Forest were reduced from 12,000 acres in 1777 to 3,500 acres by 1871. Enfield Chase (Middlesex) was enclosed in 1777, Needwood (Staffordshire) in 1801, much of Windsor in 1817, Neroche in 1830, Hainault (Essex) in 1851, and Wychwood in 1857. It was reported of the Forest of Dean in 1788 that:

> the Number of the Cottages and Encroachments in the Forest … is nearly doubled since he had known it. The persons who inhabit the Cottages are chiefly poor Labouring People, who are induced to seek Habitations in the Forest for the Advantages of living Rent free, and having the benefit of Pasturage for a cow or a few Sheep, and of keeping Pigs in the Woods; but many Encroachments have been made by People of Substance. The Cattle of the Cottagers are impounded when the forest is driven by the Keepers, as all other Cattle are; and when the Owners take them from the Pound (paying the usual Fees to the Keepers) they turn them again into the Forest, having no other Means of maintaining them … The Cottagers … are detrimental to the Forest, by cutting Wood for Fuel, and for building Huts, and making Fences to the Patches which they inclose from the Forest; by keeping Pigs, Sheep etc. in the Forest all the Year; and by stealing Timber.

Such a description would have answered for most of the remnant forests at that time, indicating the pressures that their environments were under from the growth of squatter settlements. A survey of Salcey in 1783 reported trees fit for naval use reduced to a number which was one-tenth that of 1608. By the late eighteenth century ecological decline in these woodland environments through inadequate crown management, combined with pressures from zealous agrarian interests and over-exploitation by commoners and interlopers, brought matters to a head.

So it was that from the end of the eighteenth century the remaining forest areas were disafforested, enclosed, and 'improved'. In 1795 and 1796 local landowners purchased the walks of Rockingham Forest 'freed … from … the duties and Burthens … of … the Laws and Customs of the Forest'. In 1812 two-thirds of Alice Holt and Woolmer Forests became a nursery for timber, and between 1810 and 1855 most of the other southern forests were disafforested, enclosed, and divided between the Crown, the landowners, and the commoners in lieu of their common rights. The latter group remained frequently unhappy about the loss of common rights, and there were riots in 1831 in Dean. The rump of the forest officials was disbanded, and the remaining deer removed as far as possible, although their numbers were constantly replenished by stock from adjacent woodland areas. As coal replaced wood as fuel even for rural dwellers during the nineteenth century, and as brick replaced timber as the ubiquitous building material, woodlands continued to suffer from neglect.

Inroads of clearance and generations of neglect thus confronted the first serious attempt at central government intervention in woodland management in the twentieth century. Whole woods are actually difficult to eradicate, and most ancient woods survived into the first half of the twentieth century. At that time lowland woods were greatly affected by the zealous attention to conifers associated with the formation of the Forestry Commission in 1919. Subsequently, between a third and a half of all remaining ancient woodland was destroyed in the thirty years after the Second World War, as subsidized and grant-aided agriculture expanded.

Reminders of the medieval forest system now survive in just three locations: the New Forest, Epping, and Dean. In the New Forest former proposals to develop it as a source of timber came to nothing, and plans to disafforest it in 1871 were withdrawn because of strong opposition. Instead, a new sentiment was gaining ground, and in the New Forest Act (1877) commoners' rights were given prominence under the supervision of a reconvened Court of Verderers, together with due provision for safeguarding the scenic qualities of the forest. In 1923 the Forestry (Transfer of Woods) Act transferred the property to the Forestry Commission, thereby taking it finally out of technical royal control. The Commission has become responsible within the New Forest for the management and sale of timber, recreational amenities, and environmental conservation, and as an area of lowland beauty it has been considered on more than one occasion for conversion to a National Park. Epping Forest is the shrunken oak, beech, and hornbeam remnant of the formerly great Essex Forest which also included Waltham and Hainault. The latter was disafforested in 1851 and most of the land enclosed for farming in squared fields after steam ploughs had grubbed up the roots of the ancient trees. But in Epping Forest the rights of the commoners were championed by the City of London, whose corporation held an estate at Little Ilford with appurtenant common rights, and wished to see Epping preserved as an open space for Londoners. During the 1870s various measures were taken to ensure free access to the area, culminating in its disafforestation by the Epping Forest Act (1878), when the Corporation was appointed as Conservators of the Forest with the express duty of conserving Epping for public enjoyment, and working with elected verderers. The deer were henceforth to be regarded as 'objects of adornment' and the Corporation inherited Queen Elizabeth's lodge 'as an object of public and antiquarian interest'. The forest was thrown open to the public by Queen Victoria at a ceremony in 1882. In the Forest of Dean verderers are also elected to oversee common rights and the area is managed, as with the New Forest, by the Forestry Commission. The verderers continued to meet after the Second World War in a room bedecked with antlers in the Speech House but in a forest which was becoming increasingly devoted to conifers (57 per cent by 1958). Iron ore has not been mined since 1945 and only a few free-miners' pits continued to produce coal.

Since the 1970s interest in woodland conservation has revived as part of a wider marriage of agriculture and conservation. The wooded remnants of Wychwood, enclosed in 1853–4, illustrate this well. On enclosure, the second Lord Churchill was granted 1,700 acres in lieu of his office of ranger. This was purchased in 1901 by the Watney family and this in turn became in 1955 a National Nature Reserve (NNR), and now a Site of Special Scientific Interest (SSSI). Woodland now resumes its place in the development of alternative farming enterprises. A Farm Woodland Scheme was introduced in 1988 and subsequently updated to divert farmland into woodland. The plans in the 1990s for the National Forest between Charnwood and Needwood, covering nearly 100,000 acres of open glades and woodland, are indicative of the aspirations of a wider cultural shift away from a purely strategic and productivist rural landscape and towards one which also recognizes the environmental, aesthetic, and recreational value of lowland woods. Although a forest path was opened in the Forest of Dean as early as 1938, it is only since the 1960s that the Forestry Commission has developed a strong recreational policy.

Landscape, Cultures, and Society

In their functional heyday, the settings of the forests, with their hierarchies of officials, were the essential backcloth to the activities of hound, horse, and falcon. The amount of time spent in any one royal forest in hunting is difficult to assess, but the hunting process itself was a noble obsession, a military training, and a space for the mutual exchange of ritual expressions of feudal honour, rank, and servitude. Thus Machiavelli's prince should always be out hunting to inure his body to hardship and to learn the nature of different terrains. Forests were valued perhaps as much for their social and cultural significance as for their economic contributions, although parks and chases might well be seen as deer farms. This apart, poems, paintings and tapestries celebrated hunting occasions, often as allegorical expressions of human nature, and hunting landscapes are sometimes to be seen as background components to some portraits. Thus the Jacobean portrait of Henry Prince of Wales, about to deliver the *coup de grâce* to a stag, has an enclosed park, a town and a castle in the background. The concern with prestige, ritual, etiquette, and the use of the correct terminology ensured that by the end of the Middle Ages hunting symbolism had become complex and subtle. George Turbevile's *The Noble Art of Venerie or Hunting* (1576) incorporates advice on the correct terminology for all manner of animals' droppings to be encountered whilst hunting, and advice too on the correct forms of address to be used upon the presentation of deer 'fewmets' (faeces) to the sovereign, as was the accepted custom. The correct terminology also had to be applied to the condition of the deer, their tracks, and all other landscape features of relevance to the hunters. Thus red deer driven out of hiding were 'unharboured' but fallow deer were 'roused' and boars were 'reared'. The ceremonial 'breaking' of the slain deer also had to be undertaken correctly, and in the sixteenth century it was accepted practice that the animal's tongue, testicles, rectum, and other delights went to the highest-ranking nobleman present, and other parts of the carcass to others who had helped in the hunt.

Although woodland therefore served as a stage upon which feudal and early modern power relations could be made visible, and from which royal largesse could be demonstrated, attitudes towards woods were ambivalent. The mid-thirteenth-century Franciscan friar Bartholomaeus Anglicus noted that

> Woods be wild places, waste and desolate, that many trees grow in without fruit, and also few having fruit. In these woods be oft wild beasts and fowls, therein grow herbs, grass leas, and pasture, and namely medicinal herbs in woods be found … In woods be place of deceit and hunting. Fore therein wild beasts are hunted … There is place of hiding and of lurking, for oft in woods thieves are hid, and oft in their awaits and deceits passing men come, and are spoiled and robbed, and oft slain.

The presence of woodland and game enticed more marginal sectors of society to the forests and wood-pasture areas. Thus William de Curtehope owed service of safe passage as a free tenant 'within the wood' of the manor of South Malling at Wadhurst by which he 'must lodge the bailiffs of Malling and the steward of the Lord Archbishop when they wish to lie at his houses and he must lead the Lord Archbishop and his men through the midst of the Weald (*boscagium*) up to Cranbrooke.'

Where there was also access to unstinted common pastures population growth would be rapid by the sixteenth century. Few areas of forest or former forest in lowland England by this time escaped the attentions of immigrants and poor families searching for land and work. The Northamptonshire forests, close to the depopulating enclosures of the champion areas of the East Midlands, experienced this perhaps more vividly than any others. Some open villages trebled in size between 1524 and 1670. Thus the surveyor of Apethorpe in Rockingham Forest wrote in the 1570s that

> In these fields tenants and cottagers have common of pasture for their horses, oxen, kine, and other great beasts without any rate amongst themselves cessed or appointed, which is a great hindrance to the husbands and a maintaining of the idlers and beggary of the cottagers, for the liberty of the common of pasture and the gentleness that is showed in the forest to the bribers and stealers of woods and hedge-breakers without punishment is the only occasion of the resort of so many naughty and idle persons into that town and others adjoining.

Such complaints were echoed elsewhere, for example in Dean and Kingswood where there were a 'great number of unnecessary cabins and cottages built in the forest by strangers, who are people of very lewd lives and conversation, leaving their own and other countries and taking this place for a shelter as a cloak to their villainies'. The possibilities of earning a living by woodcutting, charcoal making, besom or basket or hoop making, or by the collecting of forest resources either legally or illegally lured many to these areas. The weaving of tammies and shalloons in Rockingham Forest or lace-making in Salcey and Whittlewood were joined in the eighteenth century by hurdle- and ladder-making, wood-turning, leather-working (with deer skins), the provision of oak bark, and tanning. The iron industry attracted many to Dean but since employment was sporadic many newcomers resorted to stealing wood and poaching, and the same industrial structure in the Weald similarly helped to account for far higher population densities by the seventeenth century than in the champion downlands. William Gilpin thought the scenic qualities of the New Forest were spoiled in 1791 by the foresters, 'an indolent race' living by 'forest pilfer ... deer stealing, poaching or purloining timber'. Much depended on the viewpoint. In 1907 Louisa Jebb surveyed the same New Forest and found 'a thrifty, independent race of men who ... might be trusted to work at their own salvation'. Pigs were still widely kept, together with ponies, hardy milking cows and young cattle, and with access to common rights still crucially important.

However, Gilpin's concept of 'forest pilfer' reminds us that here too was the environment giving a home to Robin Hood, symbolizing the free spirit of the greenwood which railed against (Norman) laws in protest against social injustice. Encapsulated also in the fourteenth-century songs and stories of Robin

Hood, but also in less well-known characters such as Hereward the Wake, Fulk Fitzwarin, Gamelyn, Adam Bell of Carlisle, or Johnie Cock, was the theme of woodland resistance. The legend of Robin Hood is first mentioned in William Langland's *Piers Plowman* (*c*.1377) with the implication from the text that the character was already well known. Modern associations with Sherwood Forest are, however, not precisely derived from these early accounts, since the early tales place him accurately in Barnsdale in South Yorkshire (his legendary grave is at Kirklees 30 miles to the west), as well as Sherwood, whilst other names and incidents draw upon the uplands of north-east Lancashire and West Yorkshire. Many other locations in England have been subsequently named after the legend, but much of the association with Sherwood and Nottingham is secondary, derivative, and due to seventeenth-century and later elaborations of the Robin Hood stories, upon which much modern tourism is based in Sherwood.

Here too was the environment of the Green Man—legends of complex mythology which mingled with an older pre-Saxon spirituality. The forest was wilderness, and wilderness invoked spiritualities represented by the antlered stag-god Cernunnos for the Celtic tribes; by animal-guarding figures in Welsh and Irish mythology; by saintly recluses; or by Herne, the antlered bogeyman of Windsor Forest:

> There is an old tale goes that Herne the Hunter,
> Sometime a keeper here in Windsor Forest,
> Doth all the wintertime, at still midnight,
> Walk round about an oak, with great ragg'd horns;
> And there he blasts the tree, and takes the cattle,
> And makes milchkine yield blood, and shakes a chain
> In a most hideous and dreadful manner.

(*The Merry Wives of Windsor*, IV. iv. 26–32)

Here too could lurk danger since the forest was 'fitted by kind for rape and villany' (*Titus Andronicus* II. i) and the deer hunt itself was used from the thirteenth century onwards as a literary metaphor for the tragic fall of a noble victim; and the wounded deer as a trope for lost maidenhood. Raids on the chases and privately enclosed parks and royal forests, together with attacks on the Foresters, were common. However, in time the unsettled woodland came to share its imagery of danger and supernatural awe with the image of natural beauty. The imagery in Beowulf eventually becomes the greenwood setting for Robin Hood, fully sanitized in the post seventeenth-century versions. Silvaticus becomes first 'sauvage/savage' and then 'sylvan'. And ultimately the forest could become benign, as in the work of A. A. Milne and E. H. Shepard whose Winnie the Pooh and friends roamed Ashdown Forest to the continuing delight of tourists today. However, there was no uniform ascent into pleasantry: the lawless reputation of the forest dwellers encouraged Whiggish severity in the eighteenth century equal to any in medieval England as the 'Black Act' (9 Geo. I c.22) reintroduced capital punishment for deer stealing and woodcutting.

Clearly woodland had connotations resonating within successive English cultures. As Keith Thomas notes, trees 'provided a visible symbol of human society' and the woodland landscapes were thus a rich source of symbolism as well as a material resource. By the Georgian period, huge numbers of trees were being planted to accentuate the association of property with power, a developing theme in the work of Capability Brown and others in the pleasure parks which had by then supplanted the hunting parks. At Stourhead (Wiltshire) during the 1790s more than 2,000 acres of woodland were planted. Woodland and maritime strength were also elided in patriotic verse of the 'hearts of oak' strain, and William Marshall believed that 'our existence as a nation depends upon the oak', despite other views that came to see a field of wheat as even more patriotic, and possibly replacing ancient woodland. Oaks were the patriotic emblem, seen as quintessentially English, although conifers were rapidly spreading on newly enclosed heaths and commons. But oaks and other trees could equally be used both as symbols of stability by Gainsborough in his *Mr and Mrs Andrews* (1748) or *John Plampin* (c.1755) and as revolutionary symbols, by Tom Paine.

By about 1200 Southern England had seen its last wolf and beaver, and red deer and wild boar survived only in the royal forests. As woodland retreated over-all so the habitats of associated flora and fauna disappeared. The landscapes of hunting became more jealously guarded privileges for the powerful elite. The impacts on the landscape of William's actions in setting up, as the Anglo-Saxon Chronicle recalled, 'many a deerfrith' over subsequent history have been multi-ple and complex in their unfolding. But the themes of hegemonic control, access to woodland resources, and social inequality have all left their marks on a land-scape which we are struggling to retain and indeed reintroduce as a landscape essence in the twenty-first century. The great Midland National Forest thus car-ries with it not only the aspirations of modern conservationists but the myriad historical and mythical connotations associated in England with woodland as a contested setting for people's living space.

6 Marshes

Anne Reeves and Tom Williamson

Today we often confuse marshes and fens. Before the nineteenth century, most people were well aware of the distinction between the two. Fens were waterlogged, low-lying areas of peat soils from which a variety of resources were harvested—marsh hay, reeds, saw-sedge, rushes, and peat—but which were also grazed for parts of the year. Such areas might be of limited extent, within the valleys of particular rivers, or they might cover vast areas, as in the East Anglian Fenland. In all cases, however, they contained few permanent settlements, except on their margins, or on any large islands which occurred within them. Marshes, in contrast, were much more tamed and settled landscapes. They too were watery lands—most were found beside the coast, and many occupied areas of former estuaries—but they were more thoroughly drained, by networks of dykes, and were usually protected from flooding by 'walls' or embankments. They occupied areas of rich clay and silt soils and were principally exploited for pasture or arable, rather than for the kinds of semi-natural resources offered by the fens. Moreover, unlike fens, most marshes contained permanent settlements from an early date, ranging from isolated marsh farms to sizeable villages. And whereas the majority of fen grounds were common land, most marshes have always been held as private property, in part because land so valuable was not easily yielded up to communal use.

Often these two forms of landscape could be found in close proximity. The area which today we call the Fenland, lying between East Anglia and the Midlands, is in reality two distinct landscapes. The northern section—traditionally referred to, significantly, as Marshland—occupies silts and clays beside the Wash and was first settled in Saxon times. In the Middle Ages it contained numerous large villages. Inland, to the south, lay the Fens proper, a swampy area of peat soils which was only reclaimed and settled in the course of the post-medieval period. In a similar way, on the other side of East Anglia, the area now known as the Norfolk Broads comprises two very different districts. Towards the sea are true marshes, most notably the wide, flat expanse of Halvergate, which occupies what

was once a great estuary into which the rivers Bure, Yare, and Waveney originally discharged. Higher up these rivers, silts and clays give way to peat soils and open grazing marsh to reed beds, alder carr, and those open expanses of water from which the region takes its name, dug to extract peat in medieval times.

Most areas of marsh occur in eastern and southern England. East Anglia has more than its fair share: the marshland of Lincolnshire and Norfolk, the Halvergate Marshes, the marshes of the Suffolk Sandlings. Essex has extensive tracts, especially around the mouth of the Thames. South of the Thames, in Kent and Sussex, some of the largest areas occur: the marshes of north Kent, fringing the Thames and Medway estuaries and the river Swale (47,000 acres); the Wantsum and Stour Levels lying between Thanet and the mainland (17,000 acres); the Romney Marshes (67,000 acres) and the adjoining Rother and Pett levels (7,500 acres); and the great Pevensey Level. Yet although they are a particular feature of the south and east of England, marshes can also be found in parts of the west; most notably in the western section of the Somerset Levels (the eastern section of this vast wetland, once again, occupies an area of peat soils, and is really fen ground rather than marsh).

The quintessential marshland landscape is one of boundless prospects of flat pasture, hedgeless and treeless, dissected by drainage dykes and bounded by high sea walls. Often internal embankments subdivide the marsh into particular drainage areas or 'levels'. Superficially, one area of grazing marsh resembles another; yet each has its own distinctive character. Thus the north Kent marshes, which stretch more or less uninterrupted from London to Seasalter, are wild, windy, and exposed, affording panoramic views over the adjacent areas of unreclaimed mudflats and salt marsh and out over the estuary to the Essex coast beyond. Here a sense of remoteness still prevails in spite of the proximity of oil refineries, paper mills, and power stations. Different again are the Halvergate Marshes, less exposed to the elements, and now quite isolated from the sight or sound of the sea. The level pastures here are scattered with the ruined remains of innumerable brick tower mills, erected to improve drainage in the eighteenth and nineteenth centuries.

On both these marshes, houses are few and far between—a thin scatter of lonely marsh farms in desolate, wind-swept locations. Romney, in contrast, is very different. Although this great tract likewise lies flat and open, exposed to the influence of the sea and every mood of the elements, not only are houses widely scattered but even villages, complete with parish churches, can be found. Marshes, in short, are a very mixed bag.

Indeed, although broad level pastures, and wild creeks and salt marshes, are the classic form of marshland landscape, many areas of arable marsh, where ploughland runs right to the sea wall, also exist. In some places, arable farming is an entirely new departure, the result of the post-war expansion of arable agriculture. In others, however, as on Romney, it represents a recent return to a form of land use common in medieval times, but abandoned in the late Middle Ages.

It is often assumed that the settlement and exploitation of coastal wetlands were always the consequence of deliberate draining and reclamation, in which areas of tidal marsh and saltings were surrounded with a bank or wall to exclude the inrush of the sea, drainage dykes dug within the embanked area, and some form of sluice installed to allow water to drain through the sea wall at low tide. But marshes did not always develop like this, and when land was taken in in this way it was sometimes at a relatively late date in the settlement process. The earliest phases of settlement often occurred as farmers moved onto small areas of land naturally made moderately dry by relative changes in land/sea levels, and which required only the deepening and maintenance of the existing natural drainage channels. Frequently this came about following the build-up of sediments behind a natural spit of gravel or shingle which had formed across an estuary, combined with changes in relative land/sea levels. Thus the area of open water and mud-flats now occupied by the Halvergate marshes silted up in later Saxon times as a result of the build-up of a substantial spit of sand now occupied by the town of Great Yarmouth (Norfolk). Similarly, the Romney marshes formed behind a complex of sand and shingle spits, including that at Dungeness. Once settlement had occurred in this way, it was often followed at a later date by the construction of embankments or 'walls', as the relative levels of land and sea changed again, so that land needed to be protected from the sea. In addition, once settlement had occurred in a naturally dry marsh, adjacent ground would usually be reclaimed by 'inning' neighbouring areas of low-lying saltings. Once reclaimed, and drained, such land often experienced a degree of shrinkage, leading to a further differential between the levels of land and water. Indeed, as a result both of shrinkage following draining, and of continued alterations in land sea/levels, most areas of marsh now lie below the level of average high tide, and would be inundated were it not for the embankments, sluices, and drains. In some areas, notably the north Kent marshes, the contrast in height between the protected 'inned' marsh and the salt marsh beyond has been accentuated by the growth of the latter through the accumulation of mud deposited by the high tide. Some marshes display a series of areas lying at slightly different heights above sea level, which mark the progressive inning and reclamation of land, with the oldest areas of reclamation lying at the lowest level. In other ways, too, the landscapes of the marshes carry evidence of their own history. Marshes are littered with the traces of lost watercourses, often surviving as 'creek ridges'—raised banks formed from the sand and gravel which originally lay on old river beds—or by 'levées'—similar banks formed on either side of old tidal creeks. As the land surface dried and shrank following drainage, these deposits experienced less contraction and thus stand slightly higher than the surrounding land.

Marshes generally remained dynamic environments throughout history: land was lost as well as gained in this uncertain world, and some districts have particularly complex patterns of reclamation and loss. In north Kent, for example, hundreds of acres of marsh lying to the north of Upchurch and Lower Halstow have disappeared since Elizabeth I's reign; yet large areas were first reclaimed only in the eighteenth century (near Yantlet Creek between Grain and Hoo, for example) and land near Egypt Bay was inned as late as the 1950s.

In most marshes, natural watercourses have been altered many times, both by the dramatic interventions of nature and by human interference—in order to improve drainage or navigation. On Romney Marsh an elaborate system of relic creeks is related to the rivers Brede, Rother, and Tillingham; the major outflow of this river system has relocated and changed course three times in the historic period, creating and abandoning a succession of flourishing channel ports: the Cinque Ports of Hythe, Romney, Winchelsea, and Rye. Erosion and deposition constantly worked to alter patterns of trade and communication in this way. Two thousand years ago the 'Isle' of Thanet was separated from the Kentish mainland by a river nearly 2 miles wide, the Wantsum. By the eighth century, according to Bede, this channel had been reduced to a width of 3 furlongs (*c*.600 metres). A fifteenth-century map of Thanet shows a ferry boat crossing the river at Sarre, but by 1485 the channel had narrowed sufficiently for the first bridge dating from that year to be built here by the Abbey of Minster, owners of the ferry since ancient times. Today the Wantsum is little more than a ditch.

Marshes were thus shifting, fragile, uncertain grounds. With hard work and careful organization men could effect radical alterations to the environment, and could make a good living. But the land was always vulnerable to the vagaries of climate, to natural changes in land/sea levels, and to the dynamics of coastal erosion and deposition.

Early Settlement and Reclamation

Many coastal wetlands were colonized on a large scale in Romano-British times. Unfortunately, the extent of such settlement is often unclear because the immediate post-Roman period saw further encroachments of the sea which in many areas blanketed the old land surface with layers of silt and clay. The technique of 'field-walking'—the systematic examination of the surface of ploughed fields in order to locate spreads of debris which indicate the position of settlements—cannot therefore be used to reveal the location or existence of Roman farms and hamlets. The Norfolk Marshland is, fortunately, not covered by later deposits in this way, due to the subsequent erosion of the land surface. Here it is clear that colonization on some scale was occurring from the second century AD, as farms were established on the higher ground beside tidal creeks, surrounded by small enclosures of arable land, and with extensive areas of grazing on the lower ground beyond. A similar pattern is currently emerging through a programme of extensive fieldwork in the Somerset Levels, although more land is under pasture here and field-walking therefore more difficult. Recent research on Romney Marsh has also begun to recover stray Roman artefacts, partly through field-walking and partly through the activities of metal-detectorists, and these likewise seem to indicate the presence of Roman sites buried beneath later deposits, in an area once thought to be empty of settlement before the early Middle Ages. On the low-lying Wentlooge Levels, on the northern shore of the Bristol Channel, it seems that the Romans attempted more systematic drainage schemes. Here the pattern of long narrow fields defined by parallel dykes, which is so characteristic of the area, has been shown by excavation to be of Roman origin. The north Kent marshes were similarly settled in Roman times, the land then lying some 5 feet above mean high tide level rather than, as today, some 10 feet below it. Traces of the extensive Roman road system are still visible to us, linking the Hoo peninsula and the Isle of Sheppey with Watling Street. Not all our coastal marshes were settled in this period—the Halvergate marshes in the Norfolk Broads, for example, still comprised a great open estuary at this time. But in most, settlement and land use appear to have been extensive.

Post-Roman changes in relative land/sea levels seem to have led to the general abandonment of coastal wetlands, however, and only as relative water levels fell once more in Saxon times did settlement recommence in particular districts. In

Fairfield Church, Walland Marsh, Kent 1904

This scene, including shepherd and flock, shows one of the many charming medieval churches that can be found scattered across the marshland landscape of the Romney region.

most, especially the coastal silts of Lincolnshire, the Norfolk Marshland, the north Kent marshes, and Romney Marsh, colonization was well under way by the eighth century. Elsewhere, as on Halvergate, it began slightly later, in the ninth or tenth centuries. As in Roman times, the earliest settlements seem to have appeared on land naturally rendered dry by silting and changes in sea levels. In north Kent, and on Romney, charters dating from the seventh to the tenth centuries refer to fishing, saltmaking, sheep-keeping, meadow, and arable land on the marshes, but not to walls or embankments.

From the tenth or eleventh centuries water levels were rising again. Coastal marshes became more vulnerable to flooding and were increasingly protected by embankments. In the same period, marshland communities began to embank new areas of salt marsh, adjacent to existing areas of colonization. Thus the area of Walland Marsh, to the south-west of Romney, was progressively embanked and settled, while communities in the marshes beside the Wash progressively 'inned' the low-lying silt marshes to the south and west. The embanking of new areas and the protection of old meant that from late Saxon times more and more protective banks or 'walls', equipped with sluices, began to appear. Today many areas of marsh, if still under pasture, are criss-crossed by a complex network of relict banks and walls, many of which are used as roads and farm tracks.

Marshes were not only occupied by communities of peasant farmers. During the Saxon period monastic communities were established in many areas. These locations provided a measure of isolation and seclusion, and an endowment of land which could be expanded by embanking and draining. Even where actual communities were not established, monastic houses often acquired extensive tracts of marsh. Thus Canterbury Cathedral Priory had extensive holdings on Romney; Battle Abbey on Pevensey (Sussex); while St Benet-at-Holme and Norwich Cathedral Priory held large areas of Halvergate (Norfolk). Monastic houses were thus often actively involved in embanking, although—so far as the evidence goes—the majority of reclamations were made by peasant farmers (who were often given land at advantageous rents or conditions) or lay lords.

Land Use and Settlement in the Middle Ages

The extent to which the marshes were settled, and the ways in which they were exploited, varied greatly in the Middle Ages. In some districts, as on Romney or on the silts beside the Wash in Lincolnshire and Norfolk, settlement was intensive—indeed, houses were sometimes more widely dispersed across the landscape than they are today. In both these areas field-walking has revealed a dense spread of farms and cottages, often located on the slightly higher ground provided by 'roddons', or creek ridges. Settlers preferred these slightly higher and drier locations for the security they offered against periodic flood. Both on Romney, and in Marshland, whole new communities came into existence in early medieval times, which in time developed into parishes: on Romney, parish churches are spaced at intervals of 2–3 miles. Indeed, benefiting from its proximity to the Continent, two significant ports soon developed on the edge of the marsh—Romney and Hythe. Large areas of land were under the plough in both districts. The marsh soils were naturally rich and fertile, and taxation records reveal how both rose steadily in wealth value in the course of the twelfth and thirteenth centuries.

In the Marshland of Norfolk, and in the adjacent areas of Lincolnshire, much of the land lay in a form of open fields with the individual strips—called 'darlands', 'darlings' or 'dielings'—bounded and drained by dikes. The earliest-settled land, usually near the parish churches, was laid out in small, irregularly shaped fields whose boundaries had partly developed from the earlier pattern of natural drainage channels. In areas subsequently reclaimed, in the course of the eleventh, twelfth, and thirteenth centuries, much larger fields were generally created, with strips up to one mile in length. On Romney Marsh, where again large areas appear to have been under the plough by the thirteenth century, rather different patterns of fields were laid out. Here there were no long, dyked strips. Instead, aerial photographs reveal the earthwork traces of hundreds of tiny fields, many as small as one or two acres, laid out in a distinctive sub-regular, rectilinear pattern, indicating the direction of colonization towards the contracting floodplain of the former Limen watercourse.

The extent of arable land use and settlement found in Romney, and around the Wash, was not shared by all areas of coastal marsh in medieval England. Many marshlands were largely or exclusively used for grazing. On the Halvergate marshes, on the marshes of the Suffolk and Essex coasts, and—at least by later Saxon times—on the north Kent marshes, vast numbers of sheep were grazed. Saxon charters reveal how, between the eighth and eleventh centuries, St Paul's Cathedral deliberately expanded its estates along the river estuaries and coastal marshes in Essex in order to increase its flocks. These supplied mutton, wool, skins, and cheese to a ready market in London, which was easily accessible by river and sea. By the time of Domesday over 18,000 demesne sheep were being grazed here, and ewes' milk cheese continued to be an important product for another 500 years. In the case of Halvergate, similarly, Domesday records large flocks on the demesnes of manors around the marsh: later documents provide more detail. Thus the Abbot of St Benet's had over 1,500 sheep grazing on the marshes here in 1343. The Earl of Norfolk had 1,800 sheep on his demesne marshes in Acle, Halvergate, and South Walsham in 1278: the animals were penned up in folds at night and their dung carefully collected and taken by boat to the neighbouring uplands, where it could be used as manure on the arable land.

Areas of marsh like these, used mainly for grazing, did not contain villages or separate parishes. Sometimes they simply formed extensions to parishes located on the adjacent 'upland', as on the Wantsum and Stour marshes in north Kent. But sometimes more complex patterns of territorial organization evolved. The Halvergate Marshes are a particularly interesting case. Here, some portions of marsh formed part of 'upland' parishes: some comprised detached portions of such parishes; but some constituted detached portions of parishes which did not lie on the immediate marsh edge, but several miles away. In Essex, similarly, the marsh pastures often lay in detached portions of manors lying at some distance from the main estate. Such arrangements clearly indicate the considerable value of these grazing grounds in the early medieval economy.

Although they lacked large villages, most grazing marshes contained a scatter of isolated marsh farms. Indeed, these were essential if detached properties were to be successfully managed. In north Kent and Essex these were often known as 'wicks'. Most have now disappeared but the distinctive pattern of dykes associated with them can still be spotted on the map. On Halvergate, in contrast, many isolated farms exist to this day, some with houses dating back to the seventeenth century. These are usually located on former levées, the natural silt banks beside rivers or former creeks. Documents refer to them back to late medieval times: a lease of 1550 for the Norwich Cathedral Dean and Chapter estate in Fowlholme and Skeetholme thus refers to the tenants' obligation to 'Maynteyn repare and kepe the reparacion off the howses which be buylded in and upon the premyses in good and sufficient reparacion'. Field archaeology can take their origins back further: where such farms have been abandoned in post-medieval times, and the areas in question subsequently put to the plough, field-walking reveals extensive scatters of late Saxon and medieval pottery, animal bone, and oysters.

Unlike areas of more intensive settlement and land use, these grazing marshes had a dyke pattern which mainly developed organically from the natural systems of salt-marsh drainage, either because this pattern had been preserved and adapted when 'inning' occurred, or because it had been fossilized when dry land had originally been formed by natural changes in land/sea levels. Such complex, curvilinear patterns were, however, often altered over time, by many piecemeal additions and alterations, mainly in the form of ruler-straight cuts to improve the flow of water. The number of new, straight dykes added was greatest in areas where—as we shall see—the post-medieval period saw some extension of arable land use, a particular feature of some Essex marshes.

Post-Medieval Changes

The late medieval period saw a number of significant changes in the landscape and economy of the coastal marshes. Storms and abnormal tides caused widespread flooding and damage in the thirteenth century and recurred throughout the later medieval period. Some areas, such as Romney and Pevensey, were particularly badly hit. Much of the newly drained land on Walland Marsh was lost, Romney town was devastated (descending the steps of the west door into the church of St Nicholas one reaches the original ground level) and its harbour blocked with silt and shingle, while Old Winchelsea and much of Broomhill were swept away, permanently changing the shape of the coastline. This period saw the construction of the great Dymchurch Sea Wall on the east of the marsh, and also the making of the Rhee, a man-made channel constructed to bring fresh water from Appledore to Romney in an attempt to keep the harbour clear of silt. It survives as a great bank along which the main road from Appledore to New Romney now runs.

Other areas of coastal marsh suffered similar damage. But few seem to have been permanently inundated, and the principal changes in the life of the marshlands from late medieval times are probably related to wider developments in the national economy—initially a marked decline in population levels, following the Black Death and associated disasters; and subsequently the growth of larger farms, the development of a more market-oriented national economy, and an increasing measure of regional agrarian specialization.

The first and most important change was a marked reduction in the extent of arable land on marshes, and a corresponding expansion of the area under grass. Initially stimulated by flooding, demographic decline and a fall in cereal prices in the fourteenth and fifteenth centuries, this change in land use was generally maintained even when population growth picked up again in the sixteenth century. The change was especially noticeable on Romney Marsh. In the course of the fifteenth century monastic investment was withdrawn; churches became redundant (the ruins of some can still be seen—Eastbridge, Hope All Saints, Midley); and many peasant farmsteads were abandoned. Their land was turned over to pasture and bought up and amalgamated into larger holdings by absentee landowners living on the Kentish 'mainland'. From this date onwards the Marsh became famous for sheep-keeping and large tracts of the medieval landscape were inadvertently preserved in relict form, sealed under a carpet of grass. Here it lay undisturbed until arable land use took off again in the course of the twentieth century.

Sheep were important in some other marshland areas in the sixteenth and seventeenth centuries—Norden in 1593 thus recorded that 'Neere untoe the Thames mouthe, there are certaine islandes converted to the feeding of ewes, which men milke and thereof make cheese, and of the curdes and the wheye make butter'. But in many places cattle replaced sheep as the most important stock. By 1594 the vast flocks of sheep on the Essex Marshes were giving way to cattle, and Norden described the Rochford Hundred, which lay between the Crouch and the Thames, as yielding 'milke, butter and cheese in admirable abundance'. This dairying industry had declined by the second half of the seventeenth century but sheep did not return: fattening bullocks for the London market now became the mainstay of the marsh economy. The same trend from sheep to cattle can be seen in other districts, such as the Lincolnshire silt fens, the Somerset Levels, Pevensey, or the Halvergate Marshes. In all these areas the local economy was, by the seventeenth century, based on the fattening of young bullocks. Those grazed on the great pastures of Pevensey came from the Weald: but in Lincolnshire and Norfolk they were brought by drovers from Scotland, Wales, or Ireland. Sold in local fairs,

A Looker's Hut, Romney Marsh

Sheep keeping has been important on the Romney Marshes for over a thousand years. The huts were originally numbered in their hundreds. They stood alone on the open windswept marshes providing an operational base for routine shepherding work. Modern agricultural methods have resulted in their abandonment and destruction.

they were grazed for one or two summers on the marshes (and normally fed in yards over the winter) before being sold off to the butchers—often after a further (and final) long walk, to Smithfield Market in London. In 1722 Daniel Defoe described the Halvergate Marshes in glowing terms: 'In this vast tract of meadows are fed a prodigious number of black cattle, which are said to be fed up for the fattest beef, though not the largest in England.' Grazing was thus of key importance in the post-medieval economies of marshland areas and landlords were keen to maintain the value of the rich pastures. In some areas they made strenuous attempts to limit, or even prohibit, the taking of hay crops, although such attempts were often abandoned where the market for hay was particularly good. Thus hay from the marshes of the Wantsum and Stour Levels not only provided winter fodder for the local farms but was also sold to the innkeepers of Margate and Ramsgate (Kent), as these watering resorts developed in popularity.

In spite of the general post-medieval preference for grazing, arable land did not entirely disappear from the marshes. The silts beside the Wash still had areas in cultivation, as did many on the Thames estuary, largely because of their proximity to London and the ease of waterborne transport. Foulness Island thus had a considerable amount of arable land in the fifteenth century on which wheat, beans, and oats were grown. London's demand for grain tripled in the seventeenth century, stimulating further coastal reclamation and enclosure, as at Iwade in north Kent where large areas of marshland were ploughed up for grain. Nevertheless, in the sixteenth and seventeenth centuries pasture dominated the marshlands of England, and it was only towards the end of the eighteenth century that growing population, and subsequently the Napoleonic blockade, encouraged more widespread conversion to tilth. William Cobbet described how, travelling near Old Romney in 1823, he rode for much of the way with

> corn fields on one side of me and grass land on the other. I asked what the
> amount of the crop of wheat would be. They told me better than five quarters
> to the acre … They reap the wheat here nearly two feet from the ground; and
> even then they cut it three feet long! I never saw corn like this before.

On the Halvergate Marshes in Norfolk there are signs of increased ploughing in the decades either side of 1800. In 1803 a marshman reported that marshes near Yarmouth had been so much damaged by salt-water flooding in the late 1790s that 'they have not since been ploughed': while other pieces of marsh here 'were also cultivated prior to that great flood which was occasioned by overflowing of the sea at Horsey'. Landlords generally attempted to limit the extent of arable, fearing that ploughing would cause permanent damage to the land. A valuation of the Norwich Dean and Chapter lands on Fowlholme and Skeetholme on Halvergate in 1790 thus noted:

> These marshes lye in thirty nine pieces every one of which I viewed
> separately, some of them I found sown & [...?] can Lessee break them up
> without consent of the Lessors they ought to be Covenanted against it under
> a penalty of five Pounds an Acre for every Acre so converted.

Nevertheless, the amounts of marsh in tilth seem to have increased gradually in many marshes in the course of the nineteenth century, and by 1881 ploughland accounted for nearly 20 per cent of the area of Romney Marsh. With the onset of agricultural depression from the late 1870s, however, the extent of ploughland declined. On Romney the area in tilth had fallen as low as 7 per cent by the 1930s: of Halvergate, it was said in the 1930s that 'Up to about 50 years ago quite a number of the marshes grew crops of mangolds, wheat and oats, but, with the exception of a few areas in the immediate neighbourhood of Yarmouth and Horsey, there is no arable land now.'

With the onset of the Second World War large-scale ploughing recommenced in a number of districts, in an attempt to counteract food shortages. On Romney this trend became permanent following an Agricultural Land Commission Enquiry in 1948 which recommended that 50 per cent of the marsh should be kept in cultivation. Today that figure has reached nearly 80 per cent. The Essex and north Kent marshes were similarly affected, and here many areas have also been lost to industrial development. Indeed, two-thirds of the north Kent grazing marshes have disappeared since the 1930s, much of it following the improvement of sea defences made in the wake of the 1953 floods. On Halvergate, and in the other Broadland marshes, ploughing came later, in the 1970s and 1980s. In all areas it was accompanied by the installation of under drains, and by the deepening of dykes—with disastrous effects on wildlife. In the case of Halvergate, fortunately, public outcry—coupled with the area's importance as a centre for tourism—ensured that the government took action: the area became the country's first Environmentally Sensitive Area (ESA), in which farmers are compensated for following 'traditional' forms of farming. Here the trend towards ploughland has been halted and in some places even reversed. On the Pevensey Levels, similarly, designation of extensive areas as Sites of Special Scientific Interest (SSSIs) has ensured that this remains one of the least spoilt areas of marsh in England. In most marshes in the south and east of England, however, old pastureland often now covers a small proportion of the land surface, usually under management agreements with English Nature and the Royal Society for the Protection of Birds. Such areas, teaming with wildlife and birdsong in the summer, are in sharp contrast to the dreary, lifeless landscape of the adjoining ploughlands.

The post-medieval period thus saw many changes in the agricultural economy of marshes. It also witnessed a continuation of the process of inning and embanking. The area of fresh marsh in Essex was thus greatly extended with several thousand acres reclaimed. Indeed, since the late sixteenth century the eastern margins of the Dengie peninsula have advanced seaward by more than a mile and in the south five successive walls have been built during this period. Active reclamation continued into the nineteenth century, creating landscapes with a distinctive geometric dyke pattern, as for example on Wallasea Island. Along the Somerset coast, similarly, drainage continued through the seventeenth and eighteenth centuries with the construction of embankments followed by the gradual conversion of the salt marsh behind to pasture. Not all attempts were a success. In the 1630s a group of Dorset gentlemen attempted to drain the Fleet, the area of marshland between the Chesil Beach and the mainland. Vast sums of money were spent before the scheme was abandoned: one of the principal investors, Sir George Horsey, described how while the area

> was putt into soe good a way of Drayneinge as that a man with boards fastened
> to his Feet have gone thereupon … neither the same nor any considerable part
> thereof was made soe Dry or Firme that any profitt might be made thereof, the
> same for the most parte being in Stormy Weather covered over with Salt Water

Living with Uncertainty

Marsh dwellers adapted to their environment and maintained a successful economy over the centuries. But their lands were often in danger of inundation, and great care and attention had to be paid to the maintenance of banks, drains, and sluices. Sea defence and land drainage were regulated by local custom in the early medieval period. On Romney, customs regulating the maintenance of sea walls and drains—the *consuetudo et lex marisci*—were administered by special officials elected by the proprietors of the Marsh. The first written reference to these rules dates from the twelfth century, and by 1251 they were being referred to as the 'antient and approved customs'. They embodied the idea that each man should contribute to the upkeep of defences from which he benefited, in proportion to the land which he farmed. These early customary practices eventually became more regulated and enshrined in law. They were subsequently adopted from late medieval times as a model for the Courts of Sewers, commissioned by the Crown to drain and manage areas of marshland all over England. Justices were empowered to oversee and enforce essential maintenance of walls and ditches, although as time passed increasing emphasis was laid on the powers of the Commissioners to carry out work themselves, rather than simply seeking to compel those responsible to do it. Many drainage initiatives continued to come from local landowners, however, especially in those areas in which Commissions were never fully established, such as Halvergate.

The drainage of marshland was steadily refined and improved during the post-medieval centuries. Sea walls were extended and rebuilt, like the great northern wall on the Kent marshes between Reculver and Minnis Bay, which was finally completed on its present alignment in 1808. In some districts gravity drainage was supplemented by pumps. Some drainage windmills were erected on Walland and in north Kent but it was in the Broadland Marshes, and especially on Halvergate, that they began to appear in large numbers in the course of the eighteenth century. Mill technology steadily improved in the course of the nineteenth century: patent sails replaced the old canvas or 'common' sails, and scoop wheels were replaced by turbine pumps. Steam drainage was employed here from the early nineteenth century but never replaced wind drainage: drainage windmills were still being built and rebuilt up until the time of the First World War. The last ceased to operate in the 1950s, when the spread of the National Grid allowed the widespread installation of electric pumps.

Although assiduous efforts were often made to improve drainage, the problem of flooding could never be entirely removed. Floods could result from increased volumes of water flowing off the adjacent 'upland' after heavy rain or a rapid thaw; from sea 'surges' and high tides; or from some combination of these events. Many areas of marsh have particular points of vulnerability, where the sea has broken through on a number of occasions over the centuries—like the sand dunes between Horsey and Winterton in the Broadland marshes. Salt water flooding is always more serious in its consequences than fresh water floods, for the salt does great damage to the quality of the grass and can kill crops growing in the fields. But all floods can cause serious damage and loss of life.

Few areas of marsh have escaped flooding. The coastal marshes of Essex and north Kent have all been flooded by salt water at one time or another during the last 500 years. December 1663 saw 'the greatest tide remembered in England' according to the diarist Pepys, and the Rushbourne Wall guarding the northern entrance to the old channel near Reculver was in danger of collapse. More recently, the Great Tide of 1953 flooded 49,000 acres of Essex, most of it farmland. In the Wantsum Marshes the great northern wall between Reculver and Minnis Bay was breached, the low ground behind inundated, and Thanet made, for a while, an island once again. In Essex alone more than a hundred people lost their lives and 21,000 were made homeless. The floods of '53 also caused phenomenal damage in Halvergate and the other Broadland marshes: the sea broke through the dunes at Horsey, poured down the river Thurne, and flowed freely throughout the district.

Floods were not the only difficulty to be faced by the marsh dwellers. Paradoxically, in this watery world, in some areas the shortage of fresh water was a problem. Although in most areas of reclaimed marsh the water within the drainage dykes is fresh, on Foulness and other Essex islands there are no streams flowing off an adjacent upland and the dykes are therefore filled only by rainwater. From medieval times a water boat was employed bringing supplies from the mainland, and in exceptionally dry summers livestock had to be removed to the mainland.

Above all, until recently the marsh dwellers were plagued by ill-health. Mosquitoes bred in the dykes and in pools of stagnant water and 'ague' or marsh fever—that is, malaria—was endemic. Indeed, the disease could still be found in

View across Halvergate marshes, Norfolk

An early nineteenth-century drainage windmill is shown.

some areas up until the time of the First World War. Hasted in 1798, writing of Murston on the north Kent marshes, typically described the parish as

> Most unpleasant as well as unhealthy, even in the highest grounds of it, but the greatest part lying so exceedingly low and watery, enveloped by creeks, marshes and salts, the air is very gross, and much subject to fogs, which smell very offensive, and in winter it is scarce ever free from them, and when most so, they remain hovering over the lands for three or four feet or more in height, which, with the badness of the water, occasions severe agues, which the inhabitants are very rarely without, whose complexions from those distempers become of a dingy yellow colour, and if they survive, are generally afflicted with them till summer, and often for several years, so that it is not unusual to see a poor man, his wife and whole family of five or six children, hovering over their fire in their hovel, shaking with an ague all at the same time.

Making a Living

The inhabitants of the marshes did not only involve themselves in agriculture. A number of specialized activities were pursued in these areas, exploiting their particular natural resources. From the earliest times saltmaking was an important industry. Evidence of late Iron Age and Roman salt works has been found in a number of places in the Norfolk Marshland, along the Essex Marshes, and on the Romney marshes (including Dymchurch, Lydd, and Snave). It continued to be an important industry in Saxon times: on Romney, for example, a charter of 732 granted land beside the Limen at Sandtun, West Hythe, to serve as a saltern and provided for 120 cartloads of wood annually for boiling the brine. In the same area Domesday Book records salt pans belonging to Eastbridge, Langport, and Rye. Large numbers of pans are similarly listed from the Halvergate Marshes and Marshland; in north Kent no less than forty-seven were recorded on the Chislet Marshes alone; while over a hundred were noted in the area around the Pevensey Levels. Early salt-working led to the build-up of mounds of debris, although these have often been levelled by subsequent agricultural activity. Salt-making largely died out on the marshes in the post-medieval period, although it still continues at Maldon in Essex.

Given that most marshes lie close to the sea, it is hardly surprising that fishing was an important occupation for marsh dwellers in many areas. Kiddels—fish nets on wooden frames—were set on the tidal mud flats in Essex and Kent and pits for storing fish constructed against the seawalls. Freshwater fish were also caught and, in particular, wildfowl were trapped and shot. As late as 1903 W. A. Dutt described how the marshmen in Broadland harvested the rich natural resources of the region:

> When the eels are 'running', he sleeps during the day, and is to be found by the riverside or at some dyke-mouth at night, busy with a sluice-net or sett. In winter, when he drives his cart to the nearest market-town, it is as likely to contain mallard as to be laden with pigs and poultry.

Some of the methods used to exploit such natural resources have left traces in the modern landscape. Elaborate complexes of ponds for overwintering oysters in

Elmley Marshes on the Isle of Sheppey

Mounds in the distance were formed by remains from the medieval salt-making industry.

the saltings beside reclaimed marsh are known in a number of places in Essex. More strikingly, the practice of 'decoying' wildfowl was introduced from the Low Countries in the early seventeenth century and remained an important aspect of the marshland economy until the nineteenth. Decoys consisted of a number of curving pipes—tapering channels covered by netting supported on a framework of wooden or iron hoops—leading off from an open area of water, about 3 feet deep, surrounded by trees and shrubs. Along one side of each pipe a series of reed screens was erected, arranged *en echelon*. Wildfowl would be attracted to the pool by food, and by the presence of tame decoy ducks. They would then be persuaded to swim up the pipe by the example of the latter, and also by the activities of a dog called a 'piper', who would run around each screen in turn, looking to the assembled wildfowl like a mysteriously appearing and disappearing dog, or fox. For reasons which have never been satisfactorily explained, they found this display irresistible. Once the birds were well inside the pipe, the decoyman would appear, startling them into flight down into the detachable purse net at the end.

The earliest decoys were probably built in the Broadland marshes—as early as 1620 it was reported that Sir William Wodehouse had constructed a 'device for catching DUCKS, known by the foreign name of a koye' at Wacton. During the next two centuries no less than fifteen decoys are known to have been constructed in and around the Broadland marshes; more than a dozen along the Suffolk coast; and no less than twenty-nine in or around the Essex marshes. A few were also constructed in the north Kent marshes, like those which can still be seen on Sheppey and near Iwade and High Halstow. On the other side of the country, more than forty were established on the Somerset marshlands and along the coast. The numbers of birds which could be trapped were phenomenal—the marshes teamed with bird life. The decoy at Steeple in the Dengie peninsula in Essex, for example, took more than 50,000 birds between 1714 and 1726.

One unusual nineteenth-century activity, which has left no trace in the landscape, was the ice industry. Samuel Hewett, a Barking fish merchant, organized Essex farmers for miles around to flood their low-lying lands in winter and harvest the ice crop. Between 2,000 and 3,000 men, women, and children were employed cutting, carting, and packing the ice to the Hewetts' store, which could hold 10,000 tons. From here the ice was sent out to the fishing smacks to preserve the catch. Hewett eventually built carriers capable of taking 18 tons of ice to meet his fleet at Yarmouth and transport it in fresh condition to Billingsgate market in London.

Although we often think of marshes as lonely, fog-shrouded, and largely empty expanses, some became important centres of industry in the eighteenth and nineteenth centuries, in part because of particular natural resources which they offered, in part because of the ease of communication for bulk cargoes provided by their waterways. Cement works were first established on the north Kent marshes in the 1790s: by 1851 there were twenty-one in the region, and millions of tons of clay had been extracted from the salt marshes. On the Halvergate Marshes, similarly, cement- and brick-making flourished. As early as the 1820s Thomas Trench Berney's Reedham Cement Works was flourishing in a remote spot in the centre of the marshes, beside the river Yare. Chalk, brought downriver from Whittlingham near Norwich, was mixed with mud dredged from the river, baked in kilns, and ground by a tall mill (which also doubled as a drainage pump). By the end of the century there was a sizeable settlement here, with cottages, pub, chapel, and railway station, although today only the tall mill, and the pub, still remain. Copperas, a stone containing iron pyrites from which dye was produced, was found in abundance along the shores of the Isle of Sheppey, and gave rise to an important industry which employed large numbers of women and children. A Fleming, Matthias Falconer, established the first copperas works at Queenborough in Elizabeth I's reign, and production continued until the early years of the nineteenth century. The modern sprawl of petro-chemical refineries, paper works, and the like along the Thames estuary thus stand at the end of a long tradition.

Defence has also left its mark upon the landscape of the marshes, especially those in the south and east of England, which face out across the narrow English Channel towards France, or across the wild North Sea to north Germany and Scandinavia. These were inviting, invaded shores. The Viking army first overwintered on Sheppey in 857; Ebbsfleet on the Kent marshes is the traditional landing-place of the Saxon warriors Hengist and Horsa. Because most marshes occupy areas of estuary which have silted up since Roman times, several boast examples of Saxon Shore forts—large military enclosures dating from the late third and fourth centuries, with walls of flint and tile and distinctive circular bastions. Thus the gaunt ruins of Burgh Castle dominate the Halvergate Marshes; Pevensey Castle overlooks the broad Pevensey Levels; Reculver and Richborough stare out across the Wantsum and Stour Levels; Lympne stands at the edge of the Romney Marshes. William I landed at Pevensey in 1066 and erected a castle within the old Roman fort, subsequently provided with a stone keep. Much later the construction of Martello towers all along the southern and eastern coasts as a defence against Napoleonic invasion added another distinctive element to many marshland landscapes. The marshes either side of the Thames, flanking the main naval approach to the capital, were particularly important in defence terms. The great naval dockyard at Chatham in Kent was established at the end of the sixteenth century; those at Deptford, Sheerness, and Woolwich during the seventeenth.

During the Napoleonic Wars batteries were constructed at Gravesend, Shorn-mead, and Lower Hope Point in north Kent, complementing Tilbury fort and the Coalhouse Battery on the Essex side of the Thames. In the 1860s a series of forts (which became known as 'Palmerston's Follies') was constructed all along the Thames estuary. Many were renovated and reused during the First and Second World Wars, when numerous new defences were also constructed, like the pill boxes which are so noticeable a feature in these flat landscapes.

Conclusion

The marshes of the imagination are a timeless, lonely, fog-enshrouded land of prison hulks and smugglers. In Dickens's *Great Expectations* Pip describes the moment he became aware that

> The dark flat wilderness beyond the churchyard, intersected with dikes and mounds and gates, with scattered cattle feeding on it, was the marshes; and that the low leaden line beyond was the river; and that the distant savage lair from which the wind was rushing was the sea.

Chetney Marshes, near Iwade

A typical view of the remote and windswept North Kent marshes beside the Thames Estuary. This area of reclaimed saltmarsh is still a haven for wildlife despite the threat from agricultural improvement and encroaching industry.

But marshes were also places teaming with life and industry, whose inhabitants often made a good living from the richness of the soil and the abundance of wildlife. There have always been many different kinds of marsh, and far from being timeless, static worlds most have undergone many changes—in their appearance, in their economy, and in the lifestyles of their inhabitants—during their long histories. Marshland landscapes, carefully created or maintained by man, contain numerous traces of their former history: but many, as we have seen, have suffered great loss and damage over the last four decades, as arable has steadily replaced pasture. They deserve more attention, and more respect.

7 Fenlands

Christopher Taylor

Introduction

The important distinction between fenland and marshland has been explained in Chapter 6. True fenlands exist widely over England. They range from small peat-filled depressions in north Shropshire and Cheshire, through the extensive mosses of Lancashire and Cumbria to the expanses of the Eastern Fenlands of south Lincolnshire and Cambridgeshire. Also included are the valleys of Norfolk Broadland, the Somerset Levels, and the land around the Isle of Axholme in south-east Yorkshire. Although these fenlands are different in detail, they all have three principal characteristics. First they are flat, although the Levels, Axholme, and the Cambridgeshire Fens have 'islands' of older rocks within them. This lack of elevation means that fenlands are difficult to understand for they seem to be covered by countless drains and watercourses criss-crossing each other in an incomprehensible way. Only from the air, or on large-scale maps, do patterns emerge and even then they are hard to interpret.

The second characteristic is that, unlike other parts of England, there are no permanent settlements earlier than the sixteenth century, except on the 'islands' and fen edges. What habitations there are are recent farmsteads or small hamlets. The third characteristic is the particular administration of the fens. Throughout their entire existence fenlands could only be exploited by cooperation between the landowners and users on a scale unknown elsewhere. Outside the fens necessary cooperation could be achieved usually at parish or township level. But in the fens collaboration had to extend over much larger areas. The moment any individual or community attempted to improve or to drain land, cut channels, dig peat, or build banks or dams, there was a knock-on effect. The precarious balance of water management could so easily be upset that from the very beginning of man's exploitation of the fens there had to be hierarchies of cooperative organizations. An additional complication in the larger fenlands was that most of the rivers were also used by waterborne traffic. This not only caused conflict with other fenland users but also eventually led to the creation of separate administrations for navigation.

Fenlands, showing the main drainage channels

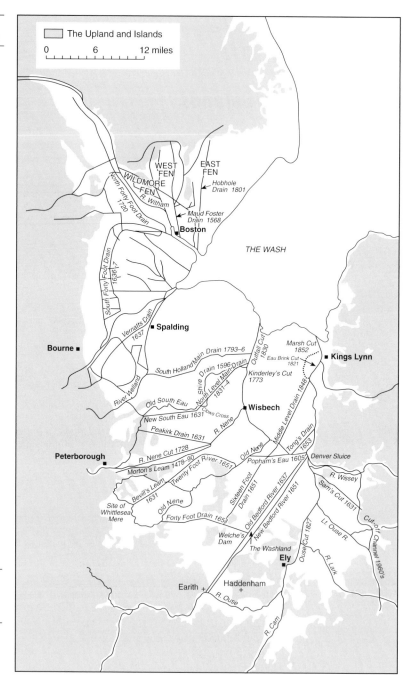

The triumph of arable farmers, Burnt Fen, Littleport, Cambridgeshire

The broad, meandering soil-mark is the bed of a channel of the medieval River Little Ouse. The rectangular fields on the left are Adventurers' Lands laid out in 1652, those to the right are of the eighteenth century.

Another conflict that has been inherent for most of its recorded history, and certainly until some fifty years ago, has been between two entirely different methods of agriculture. Until recently most fenlanders have always had, and have always desired, a largely pastoral economy supported by fishing and wild-fowling. Incomers and those with capital usually wanted an arable regime. Much of the large-scale drainage work in the fenlands was directed towards arable farming and thus, inevitably, impinged on those to whom pastoralism was the primary way of life. In this century in the Eastern Fens the battle has finally been won by the arable farmers. There grazing animals are a rare sight. On the Somerset Levels, however, cattle remain important.

Fenlands are apparently empty, boundless, and frequently windswept and yet have inspired few writers or poets. There is a long tradition of local doggerel and unlikely legends in fenland areas. The anonymous 'Powte's Complaint' of 1630 is an interesting document expressing opposition to drainage but its verse is crude in the extreme. Fenland farmers had little time or inclination to respond in verse or prose to their surroundings or way of life. Most of what has been written on fenlands has come from outsiders who have seen them in a very different light from their inhabitants. Even then most of it is hardly great literature. The appalling 'Ode on Whittlesey Mere' of 1676 by Dean Duport of Peterborough is best not quoted. What has been written of fenlands usually tells more about the writers than the landscape. Two twentieth-century examples illustrate this. Dorothy L. Sayers's *The Nine Tailors* (1934) has a fenland background but the story could have been set anywhere. And this is despite the fact that Sayers spent the first nineteen years of her life at Bluntisham, on the edge of the Huntingdonshire fens. Arthur Ransome, too, set two of his great children's stories, *Coot Club* (1934) and *The Big Six* (1940), on the Broads. Yet neither captured the essence of the landscape there nor the ancient way of life that was then ending. In any case, most writers ignore the fenland skies which make the landscape distinctive. With almost invisible horizons, the sky dominates everything and exhibits an ever-changing variety of clouds and light which is unique in England.

The Physical Background

True fenland is land which has been created in the recent past by the growth of decayed vegetable matter (peat) in areas of impeded drainage. But such land is not universal in fenland. In many places, notably in Axholme, on the Levels, and in the Eastern Fens, the areas of peat grade into, are interleaved with, or are covered by clays and silts, laid down by either rivers or incursions of the sea. It is thus difficult to separate silt fen from peat fen although it is the latter that principally is being discussed here.

All peat fens have been created within the last 12,000 years. The fen meres and mosses of north-west England developed within hollows left by the retreating ice of the last Ice Age. These filled with water in which decaying vegetation gradually became peat. The development of natural drainage was prevented by the nature of these hollows and the marshy environment was thus reinforced. The more extensive fenlands developed in a similar way but on a larger scale, within low-lying basins many kilometres across, close to the sea. Here the rivers flowing across these basins were held back by silts and clays deposited by incursions of the sea during periods of fluctuating sea levels. The result was that, behind low-lying coastal areas of clays and silts, the peat fens developed, first alongside the sluggish rivers and then over almost the whole landscape. Sometimes subsequent marine incursions covered the peat with further silt.

All these fenlands developed over long periods of time and at different rates. Chat Moss near Manchester was established 7,000 years ago. Yet the mosses of the Rivers Alt and Douglas, north of Liverpool, are only 4,000 years old. In the Eastern Fens the oldest peat was formed 6,000–7,000 years ago. But other peat did not develop until Saxon times. Locally the fenland landscape varied considerably. In the larger areas slow-moving rivers flowed between natural banks of silt. Elsewhere there were areas of open water which ranged from small pools to huge lakes. Soham Mere in south-east Cambridgeshire originally covered over 2,000 acres. In other places the vegetation varied from reed swamp and sedge fen, in almost perennial water, through boggy 'carr' with willow and alder, to dry land with grass or trees. Although changes continued to take place, fenlands wherever they lay were fully developed by about AD 500.

Early History

Unlike most other landscapes, fenland was still developing in prehistoric and Roman times. Thus the earliest prehistoric people lived on land which had not then become fenland. In contrast to elsewhere in England where prehistoric people settled in increasing numbers, the fenlands became progressively more empty as the areas of peat expanded. After about 5000 BC settlements of all types appeared around the fen edges and on the 'islands'. These settlements ranged from the well-known Iron Age 'Lake' villages of the Somerset Levels, actually only fen-edge and seasonally used groups of huts, to the permanently occupied fortress on the 'island' of Stonea in the Cambridgeshire Fens. The people who lived in such settlements took full advantage of what the fens offered in terms of peat for fuel, reeds for thatch, and fish and fowl for food. Land which was dry, either permanently or seasonally, was used for grazing. The discovery of dug-out canoes shows that the waterways were used, but much more impressive are the timber causeways constructed by prehistoric people across the developing fens. The oldest known is the Sweet Track in the Somerset Levels, about a mile long and dating from around 3800 BC. But there are many others and some thirty are known from the Glastonbury area alone. Similar ones existed in Cambridgeshire. Detailed analysis of some of the Somerset ones has shown that cattle were driven along them, emphasizing the early use of the fens and fen edges for grazing. Such immense causeways can only have been constructed by the co-operation of fenland communities, a factor which recurs continually.

With the incorporation of Britain into the Roman Empire it was inevitable that the technical and organizational ability of the Romans would affect fenland. But the Romans were aided by a lowering of sea levels and climatic improvements which resulted in an increase in the amount of drier land. One result was an explosion of settlement around most fenland. Hundreds of settlements, fields, and trackways are known from that time along the fen edges. Yet the peat fens themselves remained much as before, still providing a rich natural harvest for those who lived close by.

One feature which characterizes the Romans all over England is their engineering. Usually this is manifested as roads or aqueducts. But in the fens it takes the form of artificial water channels. The greatest of these is the Car Dyke which runs from Peterborough, along the western edge of the Lincolnshire fens. Its exact purpose is not known. It was either a canal for the transshipment of goods north from the rich lands of the fen edges, or a catchwater drain to prevent upland water entering the fens and so prevent flooding, or perhaps both. Either way it is an immense work 80 feet wide and 40 miles long. If it was a catchwater drain it implies the existence of reclaimed land on the adjacent fen. There are numerous other watercourses of Roman date that cross the fens linking uplands to rivers. One running east from Bourne (Lincolnshire) seems to be connected with an area of turbaries which suggests that the Romans were digging peat there

Cutting reeds in the fens

A nineteenth-century engraving of the traditional work that produced material for thatching. The reed beds were often deliberately managed in the ideal conditions that followed peat-digging. The reeds were used in the fenlands themselves and also far beyond.

on a large scale. Other Roman turbaries have been found near Upwell on the Cambridgeshire/Norfolk border. These comprise long parallel banks and depressions which are the silted-up remains of peat-cutting beds.

The end of the Roman period in the fenlands is not well understood. It was formerly assumed that settlements around the fens were totally abandoned in the fifth century AD. Now the picture is less clear. Certainly many Roman settlements became deserted, but how much this was due to wider economic problems and how much to natural causes is not known. There is a long tradition in the fens of blaming flooding for abandoned land, when outside events have caused the difficulties. Certainly there is evidence of flooding at the end of Roman times in the Somerset Levels and this led to the giving up of land and associated fen-edge settlements. But this does not appear always to have been the case in the Eastern Fens. Recent work there has indicated that the first Saxon people moved into, or perhaps more likely merely continued to live in, the Roman settlements.

The Saxon Fenlands

Fen slodgers

Wild-fowlers on the undrained Lincolnshire fens in the mid-nineteenth century, armed with long hooked poles for retrieving birds. 'Slodger' is an admirable dialect term for those who spent their lives in a watery fenland world.

From about AD 700 the first written records relating to the fens appear. Yet these sources can be misleading. The earliest documents were all written by outsiders and stress the inaccessibility and worthless nature of the fens. St Guthlac, who founded Crowland Abbey in Lincolnshire in 716, described the land there as 'a hideous fen of a huge bigness'. The Isle of Athelney in the Somerset Levels was described as late as the twelfth century as 'not an island of the sea [but] ... so inaccessible on account of bogs and inundations that it cannot be approached

but by boat'. Even the place names, mostly given to the fens by the Saxons, seem to indicate a similar landscape. Ely (Cambridgeshire) is 'eel island', Meare (Somerset) is 'lake'.

Yet for the people who dwelt around the fens life was probably easier than for those living elsewhere. For as well as crops from the dry grounds, the fens provided fish, fowl, hay, peat, rushes, timber, and grazing. The measure of their bountiful nature may be seen in the great monastic houses that were situated in and around them. These included Peterborough (founded 655), Ely (673), Crowland (716), and Ramsey (969), all in the Eastern Fens, Glastonbury (by the seventh century), Athelney (c.888), and Mulchelney (939) on the Somerset Levels, and St Benet's (?refounded 1019/20), Broadland. Yet remote though some of these were, they became extremely wealthy, at least in part because of the resources of the fenlands they possessed. By 1086 Glastonbury was the richest monastic house in England, Ely was second, and Ramsey was in the top ten. Other monastic houses far from the fens also gained great wealth from fenland. Cockersand Abbey in Lancashire, which acquired Pilling Moss on the northern edge of the Fylde, and Thornton Abbey, Lincolnshire, which owned large tracts of Axholme, are examples.

Domesday Book records much of the wealth of the fens in 1086. Around Bourne in Lincolnshire there were some thirty fisheries, while Soham Mere in Cambridgeshire was home to eight fishermen. Soham Mere also produced 4,000 eels a year but this was completely overshadowed by the 24,000 at Stuntney and over 27,000 at Doddington, both in Cambridgeshire. Rich though these fenlands were, it is clear that the wealth in Saxon times came largely from their natural resources. And while these resources continued to be garnered, from the tenth century onwards the scale of human intervention and exploitation increased.

The Medieval Expansion

From at least the early twelfth century increasingly favourable economic conditions, combined with a rise in population, provided the incentive and capital for both lowly peasants and large landowners to exploit the riches of the fens in a different way. Fishing, fowling, and reed-cutting went on as it always had done, but on an increasing scale. It was improvements to, and the extension of, pasture that produced the major changes to the fenland landscapes. These were aimed at both increasing the head of stock grazed and expanding the quality and quantity of meadowland to allow more animals to be overwintered. Sometimes, where conditions were suitable, even new arable land was established.

These changes involved improving the drainage to remove water and then preventing the flooding of the newly drained land. Water was removed by digging ditches around new fields, both individual ones and blocks of fields. Flooding was prevented by the construction of banks around the edges of the drained lands and by the cutting of channels to remove water. A consequence was that wooden sluices or dams had to be built at points in the banks to control the ingress and egress of water. These methods were used throughout medieval times to reclaim very large areas of fenland. The work was carried out by people at all levels of society, but its main characteristic was its piecemeal nature. Most schemes were limited in extent and were only gradually extended, often over centuries. It is likely that the greater part of the reclamation was actually carried out by peasants working together but their efforts were not recorded. Only the work which was undertaken by the larger and mainly institutional landowners such as the monasteries is known, for it is usually only their records that have survived. Nevertheless, they detail the kind of undertakings that were probably widespread.

In 1251, for example, on the Bishop of Ely's manor of Littleport (Cambridgeshire), sixty new tenants were holding almost 500 acres of newly reclaimed land. On another nearby manor, Doddington, 111 new tenants and many acres of new land are also recorded. In the Somerset Levels in 1234 750 acres of meadow in blocks ranging from a half to 25 acres in extent are recorded as having been reclaimed around the 'island' of Sowy. Occasionally documents give an insight into the work of lesser people. In 1281 Walter de Knolton was allowed to retain 30 acres of illegally reclaimed meadow on Southlake Moor. In return he paid the Abbot of Glastonbury 7s. 6d. rent, and gave up his rights of common in the rest of the moor which the Abbot was then draining. Improvements to already reclaimed land were also carried out. In the early fourteenth century three small landowners reclaimed some meadowland on the fen edge near Ely. The Sacrist of Ely, Adam de Walsingham, better known as the probable designer of the Octagon at Ely Cathedral, acquired this land for the Priory by a mixture of bribery, influence, and manipulation. He removed the existing 'hedges, ditches, deep drains, moats and trees' there and turned the land over to arable. The resulting income was specifically reserved for the monks' recreational allowances.

This example is characteristic of most medieval fenland reclamation. It was usually confined to the fen edges, around the 'islands' or along the banks of rivers, and little reclamation took place in the deep fen. The resulting fields are still clearly recognizable today. For example the 1251 survey of the Bishop of Ely's manor of Downham (Cambridgeshire) lists thirteen new tenants holding 70 acres of reclaimed land at a place called Apesholt. Its modern successor is Apes Hall, north of Littleport on a small island abutting a former course of the River Ouse. All around the island and against the old river banks are small straight-sided fields arranged in an overall irregular pattern. These are the medieval intakes and they stand out in sharp contrast to the later highly regular pattern of fields in the fenland beyond.

Even where there is no direct documentation, such fields can be recognized. At Burwell in south-east Cambridgeshire the fen edge is characterized by long narrow ditched fields extending into the fens. Beyond are much larger fields aligned at different angles. The fen-edge fields, which were the only ones subject to tithes, were almost certainly created by Ramsey Abbey which held Burwell throughout the medieval period. Medieval reclamation also took place on the mosses of north-west England. In 1268–84 John, Lord of Stalmire, abandoned his rights of pasture on Pilling Moor 'within the dykes' of the Abbot of Cockersand and then gave the Priory permission to dig dykes elsewhere. Although the exact whereabouts of these 'dykes' is unknown, eighteenth-century maps show the familiar pattern of fen-edge fields.

Inevitably this piecemeal reclamation produced conflicts. Because no overall plans lay behind the works, the removal of water through new drains into old watercourses in this haphazard fashion often led to more flooding, sometimes on the land of other people. This gave rise to disputes between communities, individuals, and institutions. In 1315 Glastonbury Abbey and the Bishop of Wells

(Somerset) clashed over the Bishop's swine pastures on Godney Moor which the Abbot's men had destroyed and over sluices which the Bishop claimed had 'submerged' his corn. Such conflicts led to other developments. Landowners were forced to go into the fens beyond the reclaimed areas to carry out work which would both protect the new fields and ease the passage of floodwater. These works included the heightening of the natural banks along the rivers. Such 'walls' are recorded alongside the River Axe in Somerset as early as 1129, and by the end of the medieval period almost all fenland rivers, as well as the outer perimeters of the reclaimed areas, had counter banks.

Another widespread work of improvement was the diversion or canalization of stretches of river both to prevent flooding and to remove old watercourses from reclaimed land. Southlake Moor in the Somerset Levels was reclaimed by Glastonbury Abbey in the thirteenth century, but was continually inundated by the River Cary which crossed it. In order to prevent this the Cary was diverted into a mile long new cut on the north side of the moor with a counter bank on its south side. In the Eastern Fens local people carried out similar works. These included a new 4-mile long channel from Littleport to Brandon Creek, which effectively switched the entire drainage of the south-east Cambridgeshire fens from its outfall to the sea near Wisbech to that at King's Lynn. Neither the date nor the instigator of this are known, but presumably the work was done in the thirteenth century and for the Bishop of Ely, to assist the contemporary reclamation of the land adjacent to the abandoned river north of Littleport. Another diversion, just over 4 miles long, was of the River Welland from the fen edge at Peakirk to Crowland, Lincolnshire. This is also undocumented but, as in 1428 the Abbot of Crowland was said to have repaired the banks there 'time out of mind', it is likely that the Abbey was responsible.

One final development aimed at solving the disputes which resulted from all these works was the emergence of regional administrative organizations to oversee drainage. The most important of these by late medieval times were the Courts or Commissions of Sewers set up by the Crown to deal with disputes. Such bodies were empowered to order remedial work, to oversee it, and, through taxation, produce the necessary finance. The system, although theoretically good, had the same failings as did similar contemporary institutions. The commissioners were local landlords, corruption and tax evasion were widespread, and there were few powers of enforcement. Nevertheless, these Commissions often did improve matters, if only through the self-interest of their members. More importantly they developed the principles of regulation and taxation of drainage which were to remain until the twentieth century.

A major activity in the medieval fenland landscape was peat-cutting. Although beginning as a cottage industry supplying fuel for fen-edge dwellers, in some places it had developed into major commercial undertakings by the late twelfth century. The largest of these was in the valleys of the Norfolk Broads. The Broads are not natural lakes, but merely flooded peat pits. Most of the evidence for their working comes from the records of St Benet's Abbey, which lay in the centre of what was to become Broadland. The earliest references are from 1153–68, although the industry must have begun earlier. By the mid-thirteenth century the quantities of peat being dug were phenomenal. In 1268–9 for example 250,000 turves were sold from South Walsham alone. It has been estimated that over 25 million cubic yards of peat had been removed from the Broads by the

A nineteenth-century drawing of women stacking peat

The peat was dug from long narrow trenches, always by men. It was cut out in small blocks which were stacked for drying, after which it was barrowed away, often being loaded into barges for transport. The mast of a barge is visible on the left.

fourteenth century. The Broads still show in their landscape how this peat was dug. The ragged edges of many Broads, together with the occasional lines of small islands, mark the sides of the long parallel trenches from which the peat came. In the late thirteenth century flooding reduced the output but increases in labour costs, and the arrival of cheaper sources of fuel such as sea coal, were probably more important.

Elsewhere the evidence for massive peat extraction is different. On the Axholme fens there are few physical remains of peat-digging. The original medieval land surface is now covered by silt from later warping. But the townships of fen-edge villages such as Swinefleet and Ousefleet are long narrow strips extending far into the fens, served by extended droveways. These droveways were access roads to turbaries which are well documented. Thus, in the late thirteenth century, Roger of Hook gave St Peter's Hospital, York, a turbary in Swinefleet which was only 360 feet wide but which extended south across the fens for about 4 miles. Another grant added 7 feet on one side of the strip for a drain and 7 feet on the other for an access road.

Similar records suggest that by the early fourteenth century the whole landscape here was full of peat-diggers working their way across the fens. Most of these workings were owned by monastic houses, although individuals also had small turbaries. Around 1300 John de Yortheburg had one measuring only 7 feet by 20 feet at Swinefleet. The sale of the peat produced great wealth. In the early fourteenth century Thornton Abbey (Lincolnshire) was delivering 16,000 turves a year just as rent on turbaries it held from the Earl of Lincoln. Peat was also dug on the Eastern Fens and on the Somerset Levels, although not apparently in such quantities as elsewhere. While there is both documentary and ground evidence in the

Stalham Staithe, Norfolk Broads, 1933

At almost the end of centuries of commercial trade on the Broads and fenland rivers, a sailing wherry unloads its cargo at the tiny harbour south-west of the village.

Levels, nothing survives in the Eastern Fens, where later wastage of the peat has removed all trace. Nevertheless it is well documented there. The labour services at Billingborough, Lincolnshire, in the late thirteenth century included peat digging.

One final activity that affected the medieval fenlands was navigation. Most of the main rivers in the principal fenlands were navigable, often as a result of improvements. Vast quantities of goods were moved along these rivers to and from the numerous developing towns on the fen edges. In the Eastern Fens places such as Cambridge, St Ives, Peterborough, Bourne, and Lincoln were all supplied

by water. Indeed the annual Stourbridge Fair, held on the banks of the River Cam outside Cambridge, was recognized as one of the greatest fairs of medieval Europe. At Ely in the twelfth century the Bishop actually diverted the River Ouse from its natural course so that it flowed alongside the newly established wharves there. On the Somerset Levels, the towns of Bridgwater, Langport, and Glastonbury were all on navigable channels. In the Broads a whole series of short 'cuts' have been identified, mostly apparently medieval in date and all made to improve the navigation. Even small inland ports were created and connected to rivers. One was the fen-edge village of Reach in Cambridgeshire, created in the twelfth century as a collaborative venture by Ely and Ramsey Abbeys, each of which had land there. A hythe and a market place were laid out and an old Roman canal recut to join Reach to the River Cam. Locally too, most fenland movement was by water. Almost every fen-edge village had its hythe or staithe, often some distance from the village itself, to which the riches of the fens were brought by boat.

Great Schemes: The Sixteenth and Seventeenth Centuries

From the early fourteenth century, economic decline, the Black Death, and the resulting changes in population, decreased the demand for goods, altered social conditions, and slowed down the exploitation of the fenlands. Records of the fourteenth century are full of complaints about poor crops and abandoned lands, all the result of floods. But, although these complaints were in part valid, the underlying reason was that the marginal fenlands were now less profitable for cereal growing. Small farmers living on the fen edges continued to graze their stock and harvest the natural resources, largely unaffected by such problems. As a result, when outsiders resumed reclamation and drainage in the sixteenth century, the outcry from the dispossessed locals was considerable.

The beginnings of the renewed work can be detected in the late fifteenth century. Landowners started to see the fens as valueless but with potential. Much discussion ensued, but one of the first visible results, and an indication of what was to come, was the cutting in the 1480s of Morton's Leam, a 10-mile long artificial watercourse from Peterborough to Guyhirn in Cambridgeshire. This was intended to shorten and limit the River Nene, then flowing in a series of distribution channels, and to confine it to a single cut. It was assumed that this would increase the flow, prevent silting, and reduce flooding to a large area of fenland.

Initially, changes to the fenlands were slow. An event that was both a setback and a force for future change was the dissolution of the monasteries. This removed the institutions, which for centuries had developed and maintained some of the drainage systems over large areas. Yet conversely the conservative monastic landlords were replaced by new landowners, often successful merchants and lawyers, who sought to use science, influence, and the law to exploit the fenlands. Small farmers and landless fishermen, all of whom lived well on the resources of the natural fenlands, were no match for these new lords. They fought them legally and illegally, engaging in open riots and clandestine terror. But ultimately they lost. Indeed, on the Somerset Levels, many were actually forcibly removed by the new owners.

At first the new drainage work remained locally based. At Kirkham on the Lancashire Fylde, half of the moss was largely pastureland. In 1553 enclosure by new lords began. Despite objections and disturbances by the commoners the enclosures proceeded slowly all through the seventeenth century and beyond. The same process went on all over the smaller fenland mosses of north-west England, gradually creating the typical pattern of regular fields, grouped in rectangular blocks, which mark the stages of reclamation.

Even in the more extensive fenlands reclamation remained small-scale. But significantly one major organizational device was introduced. In 1605 Lord Chief Justice Sir John Popham, a Cambridgeshire landowner, with a group of 'undertakers' agreed to drain an area of the Eastern Fens in return for grants of 12,500 acres of common fenland. The scheme failed after three years, although Popham's Eau, an artificial cut 5 miles long east of March, remains to mark the attempt. The originality of the plan was that by granting common land to the drainers the expense of drainage work would be covered. The practice was quickly taken up in the Somerset Levels, where large areas of undrained peat fen had come to the Crown from Glastonbury Abbey. The impecunious James I was persuaded to organize the reclamation of King's Sedgemoor, the largest undrained area in the Levels. The necessary capital was found by granting large blocks of land to investors in the drainage. The work, the first of a number of similar royal plans, began in 1618 but it did not reckon with the fierce opposition from other landowners and commoners. Neither this nor the other schemes were immediately successful, partly because of poor engineering but also because of the battles that followed with the local people.

These were fierce and prolonged in the Axholme fens where, under the direction of Cornelius Vermuyden, a Dutch engineer, a great drainage enterprise was initiated. Vermuyden came to England in the 1620s to advise on various drainage plans and in 1626 was employed by Charles I to drain Axholme, much of which was royal property. Charles offered some 25,000 acres of land to thirty-seven 'drainers', wealthy landowners and investors, a number of whom were Dutch. The King was to keep another 25,000 acres and the remaining 25,000 were to be left for the local people to use in the traditional way.

Accustomed to the silt-lands of Holland, Vermuyden assumed that the new drainage required only two types of work. First, new artificial embanked channels were dug, to divert or shorten the existing rivers, to increase their flow, and to prevent them flooding the land they crossed, and second, new internal drains were made to remove water from the fens themselves. In little more than two years Vermuyden carried out all the work that he believed necessary. The River Don, which had formerly crossed the fens in a number of separate channels, was confined to a single embanked fen-edge cut flowing north to the River Aire and thence to the Ouse. The old medieval, and artificial, Bykerdyke which took the waters of the River Idle and its tributaries across the fens to the River Trent was recut and a whole series of new internal drains was constructed. The work was completed in 1629 and a year later all the new fields were laid out and granted to the drainers. These rectangular fields, arranged in blocks, and edged by drains, can still be identified on the ground.

By 1634 some of the newly reclaimed land had been settled by incomers, who erected the first farmsteads there. These included 200 Protestant families coming as religious refugees from the Low Countries. Their lives were made very difficult by the fenlanders, who disputed their claims to the land and some removed to the Bedford Level when invited by the Earl of Bedford. Further, the drainage work itself was not a success. Flooding continued, albeit in different places, partly because Vermuyden had underestimated the size of the channels required. The newly reclaimed pasturelands gradually lost their fertility, for Vermuyden had also failed to grasp the importance of natural warping. The embanking of rivers meant that the floods no longer deposited the silt which had produced the rich grasslands. In 1632 Vermuyden returned and, in order to reduce the flooding, cut the New or Dutch River which enabled the Don to flow directly into the Ouse at Goole. This was only partly successful and it was to be many years before the Axholme fens became valuable farmland. Yet the drainers had destroyed the ancient economy by enclosing the old commons. The 12,500 acres of the commons of Epworth township, for example, were reduced to under 6,200 acres. What remained was overstocked and thus poorer. Disturbances by the dispossessed had begun in the 1620s and continued into the eighteenth century. None of these changed the situation, although eventually the wet fens were drained.

Long before this Vermuyden had moved to the Eastern Fens and an even larger drainage scheme. In 1630 a group of major landowners, led by Francis, fourth Earl of Bedford, and later known as Adventurers, set up an organization to drain the southern part of the Eastern Fens, to be called the Bedford Level. Vermuyden was employed as engineer. The plan was that in return for 'adventuring' their capital the landowners would be granted some 45,000 acres of reclaimed land, the King 12,500, and 42,500 acres were to be retained for a new Corporation to finance the maintenance of the drainage.

Vermuyden began in 1631 and in some six years transformed the landscape of the southern fens. He worked to the same principles as in Axholme, but on a larger scale. The river improvements included the recutting of Morton's Leam, Sandal's Cut, near Ely, some 2 miles long, and the New South Eau near Crowland, almost 8 miles long. But his major work, completed in 1637, was the [Old] Bedford River, a channel over 60 feet wide and nearly 21 miles long, which carried the waters of the River Ouse from Earith on the fen edge to Denver in Norfolk. This was probably the greatest engineering achievement in England since Roman times, yet it is poorly documented. Internal drains included the 10-mile long Bevill's Leam and the Peakirk Drain of similar length. All the work was completed in 1637 and the fens declared 'drained'. At once complaints followed that land was still being flooded regularly, and riots and other disorders broke out as commoners objected to the loss of grazing, peat-digging, and fishing rights. But before action could be taken the Civil War started and all work stopped.

With peace Vermuyden returned to the fens and improved matters. Some new internal drains were constructed, including the Forty Foot Drain, east of Ramsey (Cambridgeshire), and the Sixteen Foot Drain near Chatteris. Old drains, including Popham's Eau, were recut and a Wash or embanked floodwater reservoir was added to Morton's Leam. But again the largest undertaking was the cutting of the New Bedford River, parallel to the Old Bedford River, leaving between them the Washes, a 21-mile long and 0.3-mile wide strip of land which was to hold and still holds the floodwaters of the River Ouse. In addition new roads were laid down, bridges built, and navigation rights protected. All was completed by 1652 when the fens were again declared 'drained'. Soon afterwards the blocks of land allotted to the Adventurers were laid out. As elsewhere, these can still be recognized. A few are called Adventurers' Lands or Fens but all have their boundary ditches, central drains, and rectangular form which stand out from the pattern of later fields that surround them.

In the Lincolnshire fens and on the Broads similar drainage work took place, but on a much more limited scale, without the overall planning and only when the apparent success of the Cambridgeshire draining was seen. Thus in 1666 an Act of Parliament gave the Earl of Manchester and other landowners powers to drain Deeping Fen, north-west of Crowland.

Disasters and Successes, 1660–1870

At first the newly drained lands prospered. New crops, including the especially important rapeseed, appeared where the plough had never before reached and, in particular, rich pastures extended into old marshes. But this situation did not last and until the mid-eighteenth century there was little to show for all the work of the seventeenth-century drainers. For most of the fenlands of England the late seventeenth and early eighteenth centuries were at best a time of stagnation, at worst disastrous. Attempts at reclamation continued but with limited success. In the Somerset Levels further work on draining the extensive peat areas such as King's Sedgemoor failed partly because of the power of the commoners who wielded more influence there than elsewhere and partly because of the lack of co-operation between the larger landowners. The only major works were concerned with trying to increase the flow of the main rivers and to prevent incoming high tides causing flooding. Such schemes, for example, the cutting-off of the Viking Pill meander of the River Parret in 1677–8, were few and were the work of single landowners, who were the principle beneficiaries. The same situation pertained in Axholme, the Broads, and the Mosses of north-west England. All saw some works of reclamation, but all also suffered continual flooding. Nowhere was there any overall planning.

In contrast, in Cambridgeshire, overall control of the drainage was exercised through the Bedford Level Corporation, set up in 1663. The Corporation, whose members were elected by the landowners, had powers to levy taxes, and had permanent staff to organize and maintain the drainage works. Yet, paradoxically, it was these same southern fens that suffered the greatest disasters between 1660 and the 1750s. The reason for this was a phenomenon which no one, least of all Vermuyden with his experience of the silt-lands of Holland, had anticipated. This was the lowering of the peat surface following drainage. As water was removed from the fens the peat dried out and shrank. It also disintegrated and was blown away or disappeared following bacterial action. The more water that was removed the faster the peat sank. As the surface of the newly reclaimed land fell, and yet the main rivers and channels remained at their original levels, it became impossible to remove the internal fen water. Further, the embanked rivers and watercourses, now perched above the fens, were much more likely to burst their banks in time of flood. It was this lowering of the peat that led to many of the disasters of the late seventeenth and early eighteenth centuries, particularly in the Eastern Fens. And disasters there certainly were for much of the newly drained land was said to have reverted to fen within thirty years. The ancient Soham Mere in south-east Cambridgeshire was emptied in 1664, a central drain installed, and the lake-bed divided into ditched fields. But by the early eighteenth century it was being flooded regularly and was reported as 'of very little value'.

Dugdale, who visited Cambridgeshire in 1657, saw what must have been experimental crops such as onions, peas, and hemp being grown at Willingham and fruit trees and vegetables near Whittlesea. Elsewhere much rich grazing land existed. New farmsteads were built on the reclaimed lands, usually close to rivers or on their former beds and thus on the stable silt rather than on the unstable peat. Tubbs Farm which stands on the north-west side of the Old Bedford River in Sutton parish is recorded as 'Tubbes his howse' for the first time in 1665. Yet when Celia Fiennes went to Sutton in 1695 she described the fen there as 'mostly under water' and by 1700 large parts of the Cambridgeshire fens were in a similar condition.

The Eastern Fens, and indeed to a lesser extent other fenlands, were saved by the development and then the widespread use of windmills. Unspecified 'engines' for lifting water are mentioned from the late sixteenth century. Windmills, their sails driving large slatted waterwheels or scoop wheels, are first recorded in the early seventeenth century and by its end were common in many fenlands. Their efficiency was limited. They worked only when there was enough wind and they could scoop water by little more than a metre at best from one drain to another. But by the middle of the eighteenth century hundreds of windmills were in operation in the Eastern Fenlands and many others on the Broads and in Axholme. Curiously they were very rare in the Levels for reasons which are not entirely clear.

Without these windmills the fenlands ultimately would have reverted to their natural condition. With them farmers, now increasingly the descendants of the commoners who had fought so hard against the drainage, made the fens prosper. The establishment of windmills allowed the reclamation of abandoned land and

**Drainage mill,
Conington Fen,
Cambridgeshire, 1853**

A typical fenland mill, its sails turn the timber-clad scoop wheel on one side. The wheel was set in a narrow trough, its flat paddles lifting water from the inlet drain in the foreground to a higher drain beyond.

enabled further reclamation to take place. These developments were assisted by the introduction of a new system of administration, Drainage Commissions, later called Internal Drainage Boards. These comprised groups of landowners, usually from a number of adjacent parishes, who obtained an Act of Parliament which allowed them to organize the drainage of those areas which lay outside the remit of the Bedford Level Corporation in Cambridgeshire or the still-existing Commissions of Sewers elsewhere. The Acts allowed the Drainage Commissions to tax the local landowners and to use the revenue to construct and maintain all the principal drains and to erect windmills to lift water from the Drainage Com- mission's area into rivers or high-level drains. Individual farmers then built other windmills to lift water from their land into the main internal drains.

The first Drainage Commission was set up at Haddenham (Cambridgeshire) in 1727 and over the next 150 years similar Commissions or Boards were set up in almost all the fenlands. This system of administration was combined with normal Enclosure Acts, used elsewhere to enclose open-field arable and wastes, which enabled vast areas of fenland to be enclosed and drained. Much of this work covered the fenlands of two or three parishes although sometimes very large areas were drained, such as the Witham Fen in Lincolnshire in 1762.

In some places this form of reclamation and drainage was not completed until the late nineteenth century. Many of the Drainage Boards of the Somerset Levels were not established before the 1860s. Elsewhere, especially on the mosses of north-west England, the work of piecemeal reclamation never stopped, although new and wealthier landlords increased its speed. For example, on the Rawcliffe and Stalmine mosses, Wilson Ffrancis obtained an Enclosure Act in 1830 and promptly settled all the existing common rights, deepened all the old drains, and laid out a new road system and fields on 2,000 acres of land within a year. He then inserted under-drainage, ploughed, marled, and planted the land with a rotation of potatoes, turnips, and wheat within another year.

By 1900 very little fenland was left except for a few small patches and the open waters of the Broads. Everywhere the ancient fenlands had been divided into regular blocks of rectangular fields and crossed by wide droveways and engine drains. And these enclosure-type landscapes had also acquired the same characteristic isolated farmsteads as other enclosure landscapes. These appeared on the fenlands in their thousands during the nineteenth century. Few were, or are, of any architectural merit, usually being built as cheaply as possible and of local brick.

One locally important landscape change was the development of warping, especially in the Axholme fens. Natural warping, the deposition of silt by river- flooding, had always occurred in the fenlands but as reclamation and drainage went on this process ceased. Artificial warping, by letting water onto the reclaimed land through small sluices, thus improving the land, was widely prac- tised from the medieval period onwards. But in the late eighteenth and nine- teenth centuries it was much developed on the Axholme fens where the Rivers Ouse and Trent especially carried enormous loads of silt in suspension. Huge warping drains, to carry the Trent water well inland to the enclosed fenlands, were constructed. These, together with complex arrangements of sluices, are still a major feature of the present Axholme landscape.

Elsewhere the massive expansion of reclamation and drainage led to an increased need for more and improved main drainage channels both to remove internal water and to carry external water safely across the fenlands. As a result, throughout the eighteenth and nineteenth centuries, numerous major water-courses were constructed in most of the fenlands. The present 12-mile channel of the River Nene from Peterborough to Guyhirn was cut in 1728. The meandering course of the River Ouse north of Ely was diverted into a straight 74-mile cut in 1827 while the Middle Level Main Drain, 10 miles long, was made in 1834. All these were in the Cambridgeshire Fens, but similar work went on elsewhere. In the Axholme fens a series of Acts of Parliament from 1783 to 1831 authorized the construction of many internal drains including the 4-mile long South Level Engine Drain. On the Somerset Levels too the final drainage of King's Sedge-moor between 1791 and 1796 led to the diversion of the River Cary along a new 12-mile channel into which all the local drains flowed. The bare recital of the names and dates of such channels means little on the printed page. But on the ground they can be seen as great engineering achievements made before modern means of excavation existed.

An equally important result of the late eighteenth- and nineteenth-century expansion of reclamation was the continuing fall in the level of the peat fens as more and more water was removed. This affected most of the fenlands, but was worst in the Eastern Fens, where it is still clearly visible. For example, at Coveney in Cambridgeshire the waters of the New Bedford River flow at about 13 feet above sea level. Yet the land just outside the river bank is actually now 7 feet *below* sea level. Likewise, since the great Whittlesey Mere, south of Peterborough, was drained in 1852 and its bed cultivated, the adjacent peat has shrunk by over 10 feet. Such developing differences between the internal fenland and the outfall channels meant that increasingly the windmills became ineffective. One solution was to set up 'double' or even 'triple lifts', with windmills set in a row, each lifting a little water. But this was never satisfactory and again disaster was only prevented by the somewhat belated introduction of a new technology, steam. Despite their earlier use in mines, there was much discussion before steam engines were employed in the fenlands and the first one was not built until just before 1820, in Cambridgeshire. They then spread rapidly. These engines were not very efficient, at first merely driving similar but larger scoop wheels than those driven by wind-mills. But they operated twenty-four hours a day and could lift up to 50 tons of water a minute. One steam engine could take the place of numerous windmills and thus they came to dominate the fenland landscapes. However, individual farmers retained windmills to drain their own fields, and in the Broads even the Drainage Commissions retained windmills until the twentieth century. Steam engines even inspired their own poetry. An inscription on the existing engine house built in 1830 beside the New Bedford River in Cambridgeshire records:

> These Fens have oft been by water droun'd.
>   Science a remedy in Water found.
> The power of Steam she said shall be employ'd
>   And the Destroyer by Itself destroy'd.

None of these steam engines is now working, although many of the buildings that housed them, tripartite structures for the boilers, engine, and scoop wheel, survive. They continued to drain the fens until well into the twentieth century, often with the addition of a more efficient centrifugal pump. They were then gradually replaced by much more efficient diesel engines which have been superseded in turn by automatic electric pumps. These modern developments in technology have left their mark on the fens where one of the most characteristic features is crude and often dilapidated diesel-engine sheds with attached slatted wooden cooling towers set next to neat brick huts holding the electric pumps.

The Modern Fenlands, 1870–2000

Long before this new technology arrived, the agricultural prosperity, which had encouraged and supported the expansion of reclamation and enclosure and had led to the cultivation for the first time of large parts of the fenlands, collapsed with the onset of the great Depression of the 1870s. This depression, with a brief respite during the First World War, was to last until 1940. On the whole fenland farmers survived it better than those elsewhere. This was partly because their farms were small and tended to be more manageable in the circumstances, and partly because they specialized increasingly in profitable vegetables and flowers. Even so, times were hard. Drainage taxes went unpaid, drainage work was restricted or abandoned, and owners and tenants refused to co-operate. Retrenchment and isolation were the watchwords, and a spiral of limited finance, poor drainage, floods, and the abandonment of land developed. More and more land was given up to the water and even such measures as the 1930 Land Drainage Act, which tried to simplify and improve the administration of fenland drainage, failed. Navigation on the fenland rivers came to an end, as the railways and, later, road transport took over the movement of goods.

By the end of the 1930s probably more land was liable to flooding in the fens than since the seventeenth century. Yet the fens were again saved, this time by the onset of the Second World War and the demands of a beleaguered nation for food. From 1940 to 1946 the Ministry of Agriculture, through its local War Agriculture Committees, poured money into the fens. Cost was irrelevant as drains were recut, banks rebuilt, new pumping machinery installed, and then thousands of acres of land brought back into cultivation.

Perhaps the greatest change the war brought, certainly for the people living and working on the fens, was the construction of a new road system. Up to 1940, most fenland roads were unsurfaced droveways which cut up so badly in winter that they often had to be ploughed flat. During the war hundreds of miles of concrete strips were laid down on the droves to give access to even the most inaccessible corners of the fens. These roads not only changed the landscape and aided agriculture as was intended, but also revolutionized the way of life of the inhabitants.

Unlike 1918 the end of the Second World War did not bring renewed recession. The political and economic conditions of the 1950s and 1960s meant that the fenlands continued to be intensively farmed and thus provided with all the required drainage infrastructure. Many millions of pounds have continued to be invested in the fens. In the Somerset Levels during the 1960s a major flood relief scheme was constructed to reduce flooding along the River Parret. Even larger was the Cut-off Channel excavated in the 1950s and 1960s for almost 25 miles along the south-eastern margin of the Eastern Fens from Barton Mills to Denver. This intercepts in flood time all the waters of the Rivers Lark, Little Ouse, and Wissey on the fen edge and carries them north to Denver and thence to King's Lynn along the 11-mile Flood Relief Channel, cut at the same time.

Street Heath, Somerset Levels

The dog-leg channel extending across the picture is the South Drain, cut in 1804 to remove water from the myriad fields reclaimed between medieval times and the eighteenth century. The most obvious features, however, are the water-filled trenches of modern commercial peat-diggings that scar the landscape and destroy its history.

 186

Today then the fens, perhaps for the first time in their history, are fully exploited. The Eastern Fens and those around the Isle of Axholme are amongst the richest arable land in England. The Somerset Levels remain more pastoral but are still highly valued. The Broads, as well as fenland rivers elsewhere, are important tourist attractions. And still, albeit on a large-scale commercial basis now, peat continues to be extracted, especially in the Levels and on Axholme. The fens are still yielding their natural wealth. Yet the result of the centuries of work of drainage and reclamation has been to produce a completely man-made land-scape which requires constant human intervention. And it will always need this intervention. Without it, it would revert to true fen within a few years.

8 Moorlands

David Hey

The word 'moor' has long been used to describe two very different types of landscape. When most people think of moors their thoughts turn to the barren uplands of the Pennines and the North York Moors, or to the south-western uplands of Exmoor, Dartmoor, and Bodmin Moor, the areas that we shall be concerned with here. An equally ancient usage, however, refers to low-lying, wet districts such as Sedgemoor, Inclesmoor, or the Weald Moors of Shropshire. Perhaps the word originally denoted boggy, barren land, but if so, its meaning must have changed over the centuries, for eventually 'moor' was also applied to dry heathland. Perhaps, then, the original sense of the word was simply 'waste', before any of these vast stretches of land acquired an economic value? We can only guess at this, for the word is an old one, used by both the Anglo-Saxons and the Vikings and preserved in the first element of settlement names such as Morton, Morley, or Moorby, or as a second element attached to Old English or Old Norse personal names.

The vegetation of the upland moors is far from uniform. Huge stretches of the uncultivated hills of northern and south-western England are largely covered in peat, ranging in depth from an inch or two to many feet and supporting only heather or ling, bilberries, crowberries, cranberries, and whortleberries. Other parts are characterized by tough, almost inedible grass, and increasingly by bracken, or by patches of wet, boggy ground marked by reeds, mosses, bog-myrtle, and white cotton-grass. For much of the year these moors are dark and devoid of colour, but in late summer they are transformed gloriously by the flowering of the purple ling. Apart from the modern conifer plantations of the Forestry Commission—particularly evident in Northumberland—upland moors are now largely devoid of trees. Pollen analysis has demonstrated, however, that in prehistoric times oak, with alder and willow in wetter places, and some hazel, birch, and rowan covered much of the present moorland landscape. Woodland clearances began as far back as the Bronze Age, and though the pace of change varied considerably from one part of England to another, by the Roman period the transformation of the moors was well under way.

Moors acted as formidable barriers that separated one distinctive race from another. Westmorland, for instance, was 'the land of the people west of the moors'. The most forbidding moors, such as the aptly named Bleaklow in the northern Peak District, bear little evidence of past human activity. They consist of blanket peat, many feet deep, dissected by meandering gullies known as groughs. The rambler who stands in the gluey ooze at the bottom of a grough is presented at eye level with a section cut through the millenia of accumulated remains of dead plants which have been pressed down until they form peat. Other moors, however, such as the smoothly green Exmoor (Devon), present little difficulty to the walker. Eyam Moor (Derbyshire) is characteristic of accessible moors where peat is found only in shallow patches, and where grass is as plentiful as ling. Here the surviving evidence of human activity ranges in time from the Bronze Age (a stone circle, barrows, and numerous cairns) to the nineteenth century (grouse-shooting butts, enclosure walls, and small quarries for building stone). The plentiful evidence of past human endeavour indicates that the history of Eyam Moor is very different from that of nearby Bleaklow.

The upland moors of today were formerly much more extensive. In 1774 Arthur Young noted that enclosure had changed vast tracks of the Derbyshire Peak District 'from a black region of ling to fertile fields covered with cattle'. He reflected admiringly on 'the improvement of moors in the northern counties where Enclosures alone have made those counties smile with culture which before were as dreary as night'. The district now known as the White Peak was totally converted by miles of limestone walls into pastures and hill meadows. In the early eighteenth century, two or three generations before parliamentary enclosure changed this landscape, Daniel Defoe had ridden over 'a large plain called Brassington Moor, which reached full twelve miles in length'; his party had 'eight miles smooth green riding to Buxton bath'. The names of many of these former moors are preserved in local speech and are marked on the maps of the Ordnance Survey. They also survive in built-up areas on the edges of the Pennines. Sheffield (Yorkshire) now has a shopping centre called The Moor and suburbs named Crookesmoor and Shalesmoor, which were commons until they were divided up in the later eighteenth century. Here, as in many other parts of the country, the terms 'moor' and 'common' were interchangeable. Defoe continued his journeys north by travelling from Tankersley to Barnsley and Wakefield across a 'vast moore' and thought that 'Black Barnsley' had acquired its nickname not from the smoke of industry and domestic fires but from the dark character of the surrounding moorlands.

Early travellers were united in their condemnation of these bleak landscapes. On reaching Exmoor, Defoe reflected that 'Campden calls it a filthy, barren ground, and, indeed, so it is'. When he rode from Chesterfield (Derbyshire) to Chatsworth he had to cross 'a vast extended moor or waste, which, for fifteen or sixteen miles together due north, presents you with neither hedge, house or tree, but a waste and houling wilderness, over which when strangers travel, they are obliged to take guides, or it would be next to impossible not to lose their way'. He complained afterwards about 'this difficult desart country' and the 'comfortless, barren … endless moor'. His horror increased when he reached the High Peak, which, he thought, was 'the most desolate, wild, and abandoned country in all

England'. Further north, where the Pennines divided north-west Yorkshire from Lancashire, he wrote: 'Nor were these hills high and formidable only, but they had a kind of an inhospitable terror in them … all barren and wild, of no use or advantage either to man or beast.' Westmorland, he thought, was 'a country eminent for being the wildest, most barren and frightful of any that I have passed over in England, or even in Wales it self'.

Such descriptions are a far cry from those of the Romantic writers of the later eighteenth century, who first began to perceive beauty, innocence, and spiritual calm in wildness. Earlier taste was for a tamed, cultivated landscape. The moors were dismissed tersely as 'waste' or as 'barren ground' of no economic value and with no aesthetic appeal. When seventeenth- and early eighteenth-century writers searched for adjectives to sum up their feelings about moorland landscapes they came up with words like frightful, awful, horrible, hideous, dismal, gloomy, unpleasing, and inhospitable.

Prehistoric Times

These feelings of revulsion, frequently expressed, must not mislead us into thinking that the moors were avoided at all cost. Human beings have occupied many of these moors since prehistoric times and have left their mark upon them. The scree on Pike o' Stickle at the top end of Langdale (Westmorland) contains massive boulders on which thousands of stone axes were hammered into shape in Neolithic times, then polished and distributed far and wide. Some moorland landscapes—notably Bodmin Moor (Cornwall), Dartmoor (Devon), the East Moors and Stanton Moor of the Peak District, and parts of the North York Moors—are exceptionally rich in archaeological remains, for they have been extensively occupied in the remote past. This may seem surprising when we view such unwelcoming terrain in harsh winter weather, but the climate was warmer in prehistoric times and the landscape had a different appearance, being lightly covered with trees. The buried remains of oaks, birches, and pines, and the systematic analysis of pollen preserved in the peat prove that in Mesolithic times the English moors were wooded as far as the highest summits. Trees stretched up the mountains of the Lake District (Westmorland) to about 2,000 feet. Most of the summer settlers would probably have retreated to more sheltered sites upon the onset of winter, but evidence of their activities off the moors is more difficult to find. The rich archaeological record on the present moors survives because the land there was less intensively occupied in later times, whereas most evidence of prehistoric settlement in districts beyond the moors has been obliterated by subsequent farming, industry, and housing.

The astonishing scale of prehistoric settlement is revealed by a recent systematic survey of Bodmin Moor by the Royal Commission on Historical Monuments (England), which has identified four long cairns, 354 round cairns, sixteen stone circles, eight stone rows, sixteen standing cairns, over 300 miles of ancient boundaries, and 211 settlements, including 1,600 round houses and seventeen defensive earthworks, together with many more medieval sites. Of course, only a proportion of these were in use at any one time, but the sheer number of finds is very impressive. Pollen analysis has shown that although Bodmin Moor had some woodland in prehistoric times, it was dominated by grassland and heath. Here

and on other moors, the increased activities of Bronze Age people gradually produced the open moorland with which we are familiar today. The removal of the tree cover meant that the peat grew in depth and the grassland became more acid.

Dartmoor, too, is exceptionally rich in archaeological remains. Mesolithic hunters found good hunting among the trees, Neolithic people constructed 'chamber tombs' of large stone slabs, encased in cairns, and long, stone ceremonial rows, of which about seventy survive, and in the Bronze Age upland pastures and small cereal fields were created, and the moor became a major ceremonial area, with stone circles and large cairns placed prominently on ridges and summits to mark the final resting places of the leading inhabitants. About 1700–1600 BC parallel lines of low, broad stone walls, known as reaves, were laid out over the hills and across the valleys to mark the boundaries and the subdivisions of the grazing land of the various communities which used the moor. The Bronze Age population was clearly far larger than was once believed. Reaves survive in surprising numbers and have been systematically mapped since the 1970s. The discovery of this extensive network of boundary walls and associated enclosures and field-systems has played an important part in the modern recognition of the huge scale of the prehistoric contribution to the development of the English rural landscape.

Aerial view of a reave

Walkhampton Common Reave is a good example of the long lines of stone walls built by Bronze Age people on Dartmoor. These 'reaves' acted as boundaries for the grazing lands of different communities.

Aerial photography, which has been so useful in surveying the archaeological features on the moors of south-western England, is less helpful in identifying similar features on the northern moorlands, which are often covered with ling. In the Peak District, for example, less than 50 per cent of the archaeological remains are visible from the air. Careful field-walking and measurement has, however, led to the discovery of 130 small, circular platforms of 13–33 feet in diameter, which show that Bronze Age houses were scattered throughout the field systems of the East Moors. These moors, together with the former moors of the White Peak, to the west of the river Derwent, form another of England's major Bronze Age landscapes. Some Mesolithic finds and a few Neolithic axe-heads suggest small groups of wandering hunters and summer settlers, but the considerable Bronze Age material that survives from *c.*1700 BC includes extensive field-systems, clearance cairns, and settlement earthworks, indicating a large settled population. Two sites at Swine Sty and Gardom's Edge have been excavated and selected areas of moorland have been thoroughly surveyed; the results of these inquiries point to mixed farming, especially in the flatter, better-drained areas, with an emphasis on livestock. Ceremonial sites marked by stone circles and hundreds of burial cairns stand close to the cultivated fields. A few miles to the south, the prominent burial mounds and the 'Nine Ladies' stone circle on the high, exposed plateau of Stanton Moor have inspired the apt description of this landscape as 'practically an open-air museum of the Bronze Age'.

On the North York Moors a small group of Neolithic long barrows survive, but here too most prehistoric remains date from the Bronze Age. The most common landscape features from this period are the burial mounds, known as barrows or howes, which serve as striking landmarks on the skyline, though large numbers of small cairns and a few stone circles can also be found. It has been suggested that as the moorland barrows are mostly of similar size they were the memorial tombs of local leaders of equal status. Here and elsewhere, such tombs are the raised heaps of stone and earth that were opened by the amateur archaeologists of the nineteenth and early twentieth centuries in search of cremation urns, pottery, daggers, and other weapons. Many of these barrows had been raided even earlier in the hope of finding buried treasure. Much valuable archaeological evidence has thus been destroyed.

A major problem in interpreting Bronze Age features of the moorland landscape is that they are difficult to place in a wider context, for evidence from nearby cultivated districts has been largely obliterated in succeeding centuries. A study of Swaledale has suggested that most of the Bronze Age people in the Yorkshire Dales lived where they do now, on the sides of the dale, rather than on the upper moors. This was probably the common situation in other parts of England. We need to remember that the impressive evidence of Bronze Age activities on the various English moors was unaffected by later farming activities and to remind ourselves that it was accumulated over several centuries.

Another puzzle is that very little evidence survives on the various moors that can be dated to the Iron Age or to the Roman period. The likely answer is that the local farmers grew their grain on lower land nearby and used the uplands mostly for pasturing their livestock, as in later times. The climate seems to have changed for the worse after the Bronze Age and the upland soils, now cleared of their trees, became podzolized: shallow, acid peat started to spread over large areas of upland and the moorland landscape of today began to be formed. It was a process that took centuries to complete and which varied considerably from one district to another.

The Middle Ages

In his *Survey of Cornwall* (1602) Richard Carew wrote: 'the middle part of the shire lieth waste and open, sheweth a blackish colour, beareth heath and spring grass, and serveth in a manner only to summer cattle.' An intensive field survey of Bodmin Moor has shown that after the Bronze Age the upland parts were indeed left as extensive tracts of rough grazing. Traces of herdsmen's huts, built as temporary summer accommodation, provide the sole physical evidence of renewed settlement on the moor before the end of the twelfth century. Place-name elements suggesting transhumance, in both English and Cornish, are found on most parts of Bodmin Moor. The principal use of this moor seems to have been for grazing cattle and sheep by people who lived off the moor or along its edges. The same picture can be drawn for all the other English moors. Communities more than 20 miles away used the moors for common grazing throughout the Middle Ages. In the north the Old Norwegian word *skali*, which has been turned into the minor place-name element 'scales' or 'scholes', identifies the sites of some of the summer huts of these herdsmen. Names ending in '-sett' or '-seat' often have a similar derivation. At first, the fells were mainly grazed by cattle, but by the end of the Middle Ages the farmers had turned instead to sheep.

Under the Normans, huge stretches of moorland were turned into royal or seigneurial forests, such as the Pennine forests of Macclesfield, Peak, Sowerbyshire, or Bowland, the Cumberland Forest of Inglewood, and the Bishop of Durham's Forest of Weardale. The whole of Dartmoor was a forest under the jurisdiction of the Duchy of Cornwall. These were not forests in the popular sense of 'dense wood', but were special jurisdictions which included settlements and cultivated areas, as well as large tracts of moorland devoted to hunting and grazing. In the Middle Ages much of the North York Moors—or Blackamoor, to give the district its medieval name—belonged to the royal Forest of Pickering; the rest was divided between a small group of lay and ecclesiastical magnates. Their centralized demesne farms were worked by villeins from the surrounding villages and hamlets, while land newly cleared from the moors was divided into holdings of between 10 and 15 acres and occupied by bondmen who paid heavy rents and rendered services to their lords. At the same time, a few freeholders were allowed to create their own farms on the North York Moors and some monasteries cultivated their outlying properties from isolated farmsteads known as granges. For ecclesiastical and local government purposes, the moor was divided up into long, narrow parishes and townships, which formed a distinctive pattern stretching up the dales from the villages on the moor edge.

The growth of the national population during the reigns of the Norman kings brought renewed pressure on the moors. The demand for grazing even on the roughest ground led people into boundary disputes and into moves to enclose small blocks of pasture which they could improve, while their need for more cereals encouraged families to clear small, irregular shaped fields known as 'intakes', 'assarts', or 'royds' in which oats could be sown. Between the late eleventh and the early fourteenth centuries the more hospitable moorland edges were increasingly colonized by small groups of settlers, who created new hamlets and farmed their arable land communally in strips, like their contemporaries in the settlements off the moors. They dug deep ditches to separate the newly won land from the surrounding moor.

The formation of one hamlet, that of Dunnabridge on the edge of Dartmoor, has been dated precisely, for a manorial account of 1306 records a 7s. 6d. rent for 96 acres there, 'in the waste of the king', whereas no such rent was recorded in a similar account two years previously. The rent was paid by five men, each of whom held 19.2 acres. One of the men, Hamlin de Shirwell, probably came from Sherril, in the parish of Widecombe, 2 miles away. They each pledged that the following year they would lime the land to the value of 15s. The strip pattern of the new arable fields at Dunnabridge survived until the seventeenth century. The five 'tenements' of the hamlet were amongst nineteen that were created on Dartmoor in the first 45 years of the fourteenth century. Most of these small farms were grouped in hamlets, in a way suggesting that the clearance itself had been a joint venture by energetic young men, but a few stood alone, until they too developed into hamlets when the sons of the first farmers built houses for themselves.

The new farmhouses of such hamlets were built in the form of longhouses levelled into the hill slopes. The walls consisted of two roughly-coursed faces with a core of smaller stones and some larger, flatter stones used for binding.

Hound Tor

The medieval hamlet standing at about 1,000 feet on the east side of Dartmoor has been excavated to reveal a cluster of rectangular longhouses within an abandoned field-system. The settlement is thought to have been abandoned after the Black Death, when land was no longer scarce.

They were generally about 3 feet wide and on Dartmoor they rose to a height of 6 feet; the roof truss was supported directly on the side walls. The earliest excavated examples on the south-western moors have been dated to the later twelfth century, but the style goes back at least a couple of centuries earlier in settlements off the moors. The smaller, detached buildings that have been found were probably animal sheds or barns, or they may have been corn-drying barns with an oven; a few with hearths may have housed labourers.

During the Middle Ages clearances by groups of people with the permission and probably the active support of the lord's officials, were more common than assarts made by individual peasants. The North York Moors provide documentary support for this assertion. They appear to have been largely empty of settlement in the eleventh and early twelfth centuries. Then in sudden spurts in the thirteenth century, when the demand for land was growing rapidly, large-scale assarting was carried on by the lord's agents in Goathland and Farndale, and the holdings were let to tenants. Although no obvious regularities survive nowadays in the settlement pattern, the drive to create more farms on the moors appears to have been planned. On Skelton Moors and in Bilsdale the clearance process was much more varied. Here, assarting resulted in hamlets surrounded by a small block of 'townfields' (moorland that was ploughed to make arable land, divided into strips), which were farmed by bond tenants. Meanwhile, a number of large separate farms with irregularly shaped fields on the moorland fringe belonged entirely to a substantial class of freeholders who probably were responsible for administering the manor for a lord who never came near.

The farmsteads and hamlets that were created by the initiative of peasants or lords often became a source of surnames for the families who lived in them. Areas of scattered settlements have a much higher proportion of such 'locative' surnames than elsewhere in England, rising in some districts to 50 per cent of the whole. Thus, on the western fringes of the Pennines, arose such distinctive surnames as Akroyd, Holroyd, and Murgatroyd, Barraclough, Gledhill, and Hey, names that still have a marked regional character to this day. Likewise, the Lake District has its Braithwaites, Fells, Gaitskells, and Mossops.

The great lay and ecclesiastical landowners whose estates included enormous stretches of moorland were pleased to increase their revenues by encouraging assarts and charging whatever rents the market would bear. They also established 'vaccaries' (cattle farms whose tenants produced butter and cheese and reared young cattle, which could later be fattened on lusher pastures off the moors). Thus, in 1307 the *inquisition post mortem* of Thomas de Audley recorded vaccaries at Fawfield and Quarnford and numerous 'stalls' in other parts of the Staffordshire Moorlands. Further north, in the upper Calder valley (Yorkshire), that part of the huge manor of Wakefield which was designated the Forest of Sowerbyshire was developed in the same way. In the thirteenth century, six vaccaries were recorded at Upper Saltonstall, Fernside, Withens, Crattonstall, Nettletonstall, and Hathersfield. In 1309 Upper Saltonstall, Fernside, and Crattonstall were each keeping a bull, 30 cows, and 20 calves, which were fed through winter on hay and holly and pastured on the moors in summer. The presence of other vaccaries in upper Calderdale is suggested by the place names Shackletonstall, Heptonstall, Rawtonstall, and Wittonstall. Across the Pennines in the Forest of Rossendale a similar system of vaccaries was well established. In other districts an Old Danish word which has passed into northern speech as 'booth' was used instead of 'tonstall'. Several of the farms in Edale, in the northern part of the Peak District, for example, incorporate 'booth' in names such as Barber Booth, Grindsbrook Booth, Ollerbrook Booth, and Upper and Nether Booth.

Meanwhile, the great Cistercian abbeys and other monasteries were not only transforming the landscape of their own immediate vicinities, but were building granges or lodges on their outlying estates in order to organize their sheep runs and to develop other agricultural and industrial activities. In some parts of England, notably the Yorkshire Dales, where a quarter of the land was eventually owned by religious houses, the monastic contribution to the shaping of the medieval landscape was enormous. Fountains Abbey, which held extensive tracts in the Craven region, produced up to 30,000 pounds of wool each year for sale to Italian merchants. Their vast sheepwalks were surrounded by miles of boundary walls, such as that which survives on Fountains Fell, Malham. The flocks of Fountains' daughter foundation, Newminster, were largely responsible for the creation of the bare Northumberland fells between the river Coquet and the Scottish border. In some moorland areas the monks relied on lay brothers to graze cattle as well as sheep and to smelt lead or iron on windy 'bole hills', commemorated now by minor place names and half-buried deposits of slag. Upon the dissolution of the monasteries, former grange-keepers and lessees of Fountains were often able to acquire the freehold and to establish yeoman dynasties that lasted for several centuries.

The abbeys were responsible for many of the curious stones that marked estate boundaries in the centuries before accurate, large-scale maps were first made. The 'crosses' on the North York Moors are not usually made in the form of a cross, but are shafts set in a solid stone socket. About thirty medieval examples survive, mostly on the high moors to the south of Eskdale. As they are frequently sited alongside tracks, they may have had the additional purpose of waymarker. Some have acquired nicknames such as Fat Betty and Stump Cross, but the associations that led to crosses acquiring the names Anna, Job, John, Percy, Ralph, and Redman are lost to us. Such crosses are notoriously difficult to date.

The steady growth of the medieval population came to an end with the Black Death of 1348–50 and further epidemics in the late fourteenth century. Some of the hamlets on the edges of the moors, such as Hound Tor (Dartmoor), were gradually deserted or reduced to single farms. But the story was not one of general retreat, for great lords and peasant farmers used the opportunity to drive many more cattle over long distances to graze the moors in summer. On Dartmoor turves were dug to be made into charcoal by 'colliers' whose numbers grew remarkably in the late fourteenth century, so that large tracts of the moor were completely stripped of deep peat, both then and in the post-medieval period. An increasing number of landless tinners moved on to the moor as the total output of the Devon and Cornish stannaries rose threefold in the century after 1370. New assarts were needed to grow cereals for these tinners, and for millstone hewers and stone getters.

Other English moors were also exploited for their industrial resources in the later Middle Ages. The Peak District and the Northern Dales were famous not only for their millstones, but for their lead, which was mined in the carboniferous limestone districts but often smelted on the edges of the gritstone moors. Meanwhile, the occupants of farmsteads, hamlets, and villages in the huge parish of Halifax (Yorkshire) were turning to cloth-making and earning a reputation for their cheap woollen cloths. The dual occupation of the craftsman-farmer or the miner-farmer offered a sufficient living in these upland districts. It was not 'marginal' land in the late Middle Ages. Rather, it offered alternative ways of getting a livelihood.

In the changed conditions of the later Middle Ages the lords of great estates, lacking a ready supply of labour, abandoned pastoral farming on their own demesnes. Thus, in what had been the Forest of Sowerbyshire in the huge manor of Wakefield Erringden Park was 'disparked' in 1449–51 and divided between eight families. A century later the former park was farmed by fifty householders. The religious and political upheavals of the sixteenth century considerably modified the old pattern of lordship, though many great estates remained intact despite frequent changes of ownership. Their tenants were given relatively short leases and had to pay full economic rents, a matter which was often argued fiercely before the courts. On the whole, however, families were able to earn a sufficient living. A survey in 1570 of the Earl of Northumberland's estates in Cumberland observed that, 'Albeit the country consist most in waste grounde and is very cold hard and barren for the winter, yet it is very populous and breedeth tall men and hard of nature, whose habitations are most in the valleys and dales where every man hath a small portion of ground; which, albeit the soil be hard of nature, yet by continual travail is made fertile to their great relief and comfort.'

The changing pattern of landownership at the end of the Middle Ages is nowhere more evident than in the lower and middle reaches of the northern Yorkshire Dales. Teesdale, Swaledale, and Wensleydale each contained nucleated villages surrounded by extensive open fields, while the upper reaches and some of the remoter side valleys were occupied by monastic granges and the vaccaries of lay estates. From the late thirteenth century onwards, and especially after the Black Death, these large farms were gradually broken up and leased out, often as small units. By the sixteenth century the present pattern of scattered farms had been established throughout the remoter uplands. After the dissolution of the monasteries, many of the new landlords sold off their farms to the sitting tenants, so that by 1640 large tracts of the Dales were occupied by small freehold farmers, who began the enclosure of the common fields. They made butter for a market that was eager to buy and the women knitted stockings which brought welcome extra income in these Northern Dales during the seventeenth century. It is no coincidence that the first stone vernacular buildings date from this period.

The Early Modern Period

As the population began to grow from the second quarter of the sixteenth century onwards, so the pressure on the moorland edges increased. Although no new moorland villages were created (except in those parts of the Peak District and the Northern Dales where large numbers of lead miners came to live near the mines), hamlets and farmsteads reappeared in the landscape. Thus, on the Staffordshire Moorlands the older farms of over 100 acres, occupying the best land in the valleys and on the hillsides, can readily be distinguished from the small farms of under 20 acres, which were created on the higher ground from Elizabethan times onwards.

Cruck-frame barn being demolished

This two-bay cruck barn on the northern edge of the Peak District was photographed after the later stone walls and roof had been removed. The cruck blades have been raised on stone footings and fastened by a tie-beam. A ridge-beam and purlins help the blades to support the roof.

The inhabitants of the moorland edges framed their houses with crucks and roofed them with thatch. Dendrochronological analysis has dated the timber frames of a lot of the surviving cottages and barns to the sixteenth and even the seventeenth century. A good example is Spout House, Bilsdale, on the North York Moors, originally a single-storeyed building with a thatch roof, thick, rubbly walls, and internal timber partitions. Its only concessions to style are chamfered and decorated window and door surrounds. Smoke blackening of the roof timbers indicates that the inhabitants used an open hearth for heating and cooking before a fireplace and a loft were inserted. Single-storeyed buildings such as this were usual in moorland districts until well into the eighteenth century, long after the abandonment of timber frames. The hearth tax returns of the 1660s and 1670s show that most houses on the edges of the moors were still small structures, usually with a single hearth. Normally, a house or a cottage had a through passage connecting the front and rear entrances, running behind the hearth, and separating the living and sleeping accommodation from the service end. This hearth-passage type of vernacular housing seems to have been a natural development of the medieval longhouse, where the livestock had been moved elsewhere and the lower end had been converted into a kitchen and perhaps a buttery.

The upper Calder valley is exceptional amongst moorland districts in the quality of its housing, paid for out of the wealth generated by the cloth trade. The humble cruck-framed building is largely absent here, and large numbers of houses built by yeomen and lesser gentry between the late sixteenth and eighteenth centuries survive, sometimes in villages, but more often strung out along the southward-facing hillsides, close to the moors. They display considerable variety of architectural detail, but nevertheless give a strong impression of visual unity with their use of millstone grit walls, low pitch roofs of thin sandstone slabs, and rhythmic patterns of mullioned windows.

The small size of many of the farms was the result not only of renewed assarting but of inheritance customs. In the upper Calder valley the general practice was to divide freehold and copyhold lands between the surviving sons and to provide daughters with a lump sum or an annual payment. Many of the smallest farms were rented by subtenants. Such families had to supplement their income by another occupation and by claiming their common rights to pasture animals on the wastes. Some moorland farmers held their land by attractive tenures, however. The seventeenth-century disafforestation of royal forests allowed many former tenants to purchase the freehold of their properties. On the Duchy of Lancaster estates in Derbyshire, land in the manors of Castleton and High Peak was mostly freehold, and in Wirksworth manor the copyholders of inheritance had the right to sub-let land for lives or years and to pay entry fines that were fixed at one year's customary rent.

The decline of the great medieval lordships and ecclesiastical estates, culminating in the dissolution of the monasteries, brought other changes to the farming systems of the moorland edges. The vaccaries became indistinguishable from neighbouring farms, and farmers enclosed the communal strips of the townfields privately by numerous separate agreements and converted them into permanent or temporary pastures. Often this was a long-drawn out process. For example, the South Field at Heptonstall was enclosed piecemeal and by agreement in the sixteenth century, but North Field remained open until 1783. In many cases, the strips were not rearranged in new blocks but were fossilized in the landscape by the building of drystone walls around their boundaries. The long, narrow, slightly curving fields at the edge of the moorland villages often contrast sharply with the rectangular fields created a little further out by the (later) enclosure of the commons and wastes by private Act of Parliament.

Another change in farming practice on the moorlands saw the abandonment of the use of holly as a winter feed, as new fodder crops became more widely available in the early eighteenth century, and as deer were removed from the remaining forests and chases. Abraham de la Pryme had noted in his diary in 1697 that 'In the South West of Yorkshire, at and about Bradfield and in Darbishire, they feed all their sheep in winter with holly leaves and bark … To every farm there are so many holly trees, and the more there is the farm is dearer, but care is taken to plant great number thereabouts.' The evidence for the cultivation of 'haggs' of 'hollins' goes back well into the Middle Ages, but peters out in the 1720s. Only a few minor place names now commemorate this former widespread practice.

The integration of England's markets and improvements in transport allowed the moorland farmers to buy all the cereals that they needed in exchange for their meat and dairy produce and the products of their industry. Arable farming was therefore increasingly abandoned in favour of a concentration on livestock, often combined with an industrial occupation. The early modern period saw a rapid development of lead fields, millstone and other stone quarries, lime kilns, and (in some places, such as the North York Moors) the mining of coal and iron. Describing Wirksworth Hundred in his *History of Derbyshire* (1712), William Woolley wrote, 'It is an indifferent country to live in and if it were not for the profit of the mines it would be barren of inhabitants.' The extensive wastes were devoted to the 'breeding of great numbers of sheep'.

In his *The Natural History of Staffordshire* (1686) Robert Plot observed that 'Both Moorelands and Woodlands have goodly Cattle, large and fair spread, as Lancashire itself, and such as the Graziers say will feed better ... The warm Limestone hills of the very Moorelands ... though in an open cold country ... [produce] fine sweet grass.' The commons and wastes of the High Peak were graded into the best, middle, and worst sorts. The best were set aside for summer grazing by cattle; the rest were left to the sheep. Farmers enclosed many of these commons by agreement during the seventeenth and early eighteenth centuries, long before the era of parliamentary enclosure, often as a consequence of the disafforestation of Peak Forest. Thus, Hayfield commons were divided equally between the King and his tenants in 1640, Hope and other parts of the High Peak in 1675, Castleton in 1691, and Chapel-en-le-Frith between 1640 and 1714. Some of the moorland pastures of Macclesfield Forest were enclosed upon the Restoration, by 1665 nearly half of the 900 acres of Fairfield, near Buxton, had been divided up by stone walls, and in 1687 about 200 acres of Morridge common were brought into private ownership by agreement. A unanimous resolution to enclose 153 acres of pasture at Eyam in 1702 made the freeholders 'very well pleased contented and satisfied'. The grazing of the remaining commons had to be restricted by the manorial courts. For example, in 1706 an agreement

Walker's peat carting

George Walker's water-colour of Langstroth Dale was included in his Costume of Yorkshire *(1814) in the belief that the peat cart was 'the only one now remaining of this original construction'. The wheels were designed to pass easily through the peat, which Walker noted was 'the general fuel used in the mountainous and moorland districts'.*

was made to stint the commons of Chapel-en-le-Frith: 'Suppose 5 Sheep to a whole Beast. A Beast and halfe to a Horse so in proportion to all younger cattel and horses. And 2 Sheep gates to every Pound according to their accustomed Rents or value of their respective Lands Farms.'

The moorland wastes were also a valuable supply of fuel for those families who possessed common rights. A holloway that descends into Edale from Grindslow Knoll is still remembered as the old Peat Way, by which commoners brought down their peat on sleds, pulled by horses or men. A number of moorland holloways that apparently lead nowhere probably served a similar purpose until all the peat was stripped from the moor. The deepest holloways, however, were deliberately cut to facilitate the passage of millstones across the moors from the quarries to the network of roads that headed for the nearest inland ports.

In inclement weather the difficulties encountered in travelling across the moors were immense. A memorandum in the burial register of the Pennine Chapelry of Bradfield for 1718 notes 'a coffin put in the earth with Bones of a Person found upon the high Moors, thought to be Richard Steade, August 25.' Other registers also record the discovery of bodies, especially of 'Strangers lost in the snow'. Most journeys across the moors were sensibly restricted to the summer months. Nevertheless, many moors are criss-crossed with old tracks, marked by guide stoops, causeys (paths constructed of flagstones), and packhorse bridges that were erected mostly in the first half of the eighteenth century, but sometimes a little earlier, in response to the increasing volume of traffic. The East Moors of the Peak District, in particular, provide rich evidence of the attempts to improve the routes by which salt, lead, and millstones were brought from the west and coal, corn, and malt were taken as back carriage. Further north, the West Riding textile district has a similar legacy of guide stoops, bridges, and causeys from the era immediately before the more substantial improvements made by turnpike trusts and enclosure commissioners. This was the same period that saw a significant shift in mental attitudes, as the first tourists came in search of the Wonders of the Peak. Soon, they were venturing further: to Snowdonia, the Lake District, and the Scottish Highlands. Moorland travel was difficult, but in the summer at least it was not impossible. Huge loads could be taken across the moors, albeit with difficulty. The Revd J. C. Atkinson recorded his astonishment upon his arrival on the North York Moors in the 1840s at seeing how men with a team of twenty horses and oxen managed to draw a large block of stone, weighing 5 tons, up the 'hill-side road, which rises like a house-roof on the eastern side of Stonegate Gill'.

The Modern Period

The creators of the turnpike roads of the middle and later years of the eighteenth century improved the major highways that crossed the moors, but largely preserved the original lines. Some of these routes were already centuries old, perhaps even prehistoric in origin. Thus, the old track by which men with packhorses brought salt out of Cheshire up the Longdendale valley crossed the Yorkshire border at Saltersbrook. The Cheshire portion of the road was turnpiked in 1734, the Yorkshire section, heading for the Don Navigation at Rotherham and Doncaster, was improved seven years later. Drainage ditches were dug, the worst slopes were made more level, bridges were built wide enough for wheeled traffic, and milestones were erected to mark the way. The old route over Saltersbrook remained intact until diversions were made in 1828 to form the present Woodhead road.

The turnpike roads attracted most of the moorland traffic and so the minor routes were allowed gradually to decay. Further changes came in the late-eighteenth and the first half of the nineteenth century, when men decided to enclose their local moors. The enclosure commissioners created new roads and lanes and straightened and walled old routes to standard widths. They even straightened and walled the new turnpike roads where they crossed the former commons and wastes. The distinctive long, straight lanes that were laid out at the time of parliamentary enclosure were a more important contribution to the improvement of the local transport network than were the moorland turnpikes. They were unmetalled until well into the twentieth century, and thus presented a different appearance from that now provided by a strip of asphalt, surrounded by wide grass verges. Some have remained long green lanes, accessible only to walkers and the occasional vehicle. The ones that were old tracks which were improved at the time of enclosure are often marked as such on earlier maps and sometimes have old guide stoops alongside them. Modern travellers can soon sense that an older-settled landscape has been left behind as they enter into a regular pattern of straight roads and rectangular fields that were clearly drawn up by a surveyor sitting at his desk, before they were planted in the countryside.

The extent of the commons and wastes just before parliamentary enclosure can be judged from the county maps that were sponsored by the Royal Society of Arts on the scale of one inch to one mile. Enclosure effected a huge transformation of the moorland landscape, with an enormous investment of capital and labour. In Cheshire, the enclosure of 8,750 acres of Delamere Forest was completed in 1819, after twenty-three years of surveying and considering the complex, competing interests in grazing, pannage, and timber rights. In Northumberland an Act of 1800 permitted the enclosure of the estimated 42,000 acres of Hexham and Allendale Commons. The last Northumberland moor to be divided up was the 12,000 acres of Knarsdale Common in 1859.

On Bodmin Moor a large number of people built new farmsteads so as to be near the rectangular, straight-sided fields that they had created on those parts of the moor that had been allotted to them by the local parliamentary enclosure awards. The new holdings were gradually won from the moors by sheer hard labour, often in remote and exposed locations. Many are now deserted, the struggle having been abandoned as hopeless. The same pattern can be observed on all the other English moors. The Royal Forest of Exmoor was enclosed in 1817 but in adjacent parishes the process continued until 1872. When the Crown decided to sell its allotment on Exmoor, the highest bidder was John Knight, a Worcestershire ironmaster, who offered £5 per acre for a moorland landscape that contained but a single farmhouse. Knight was a champion of agricultural improvement who thought that he could transform Exmoor by sheep-corn husbandry using unmarried labourers and shepherds, but twenty years of intense activity and huge capital investment (including the conversion of 2,500 acres of rough moorland into enclosed pastures) ended in failure. His son, Frederic Winn Knight, who took over in 1841, tried a different approach, encouraging families to settle as tenants of farms and cottages at low rents. Attempts to grow cereals other than oats were abandoned and traditional livestock husbandry was reintroduced.

By the 1880s some 7,000–8,000 acres of the Exmoor landscape had been changed for ever.

The war with France (1793–1815) and the following two decades formed the peak period for the parliamentary enclosure of the English moors. Some of the former wastes were converted (though usually not for long) into arable land, because cereal prices were high. Narrow ridge and furrow patterns, following a straight course and confined by the new field boundaries, still identify these attempts to grow more cereals. They can easily be distinguished from the broader, slightly curving, ridge and furrow patterns of older open-fields. The Board of Agriculture's *General View* of the West Riding in 1794 noted that a fifth of new enclosures were arable, chiefly oats, and that the rest was pasture, mostly for dairying and the breeding of stock. In 1801 the curate of Bolsterstone and Midhope (south-west Yorkshire) responded to a government enquiry about crops by writing 'I can assure you that forty years ago there was not one tenth part of the wheat grown in this place, that is grown at present; nor half of the oats or barley. NB all the Inhabitants here live wholly on Oat Bread and use no wheat, but for Pies and Puddings.' His neighbouring clergyman insisted that 'The Chapelry of Bradfield is more proper for breeding sheep and other cattle than growing corn', and the vicar of nearby Penistone noted that 'there is little land upon the Plow … the Farms in general being grazing and Stock Farms.' Here, and elsewhere, the enclosed pastures and remaining moors were grazed by new breeds of sheep, especially Cheviots and Blackfaces, but also local breeds such as the Penistone which adapted well to a moorland environment.

Local labourers, using stones that had been cleared from the new fields or dug from shallow quarries nearby, built the walls of the new enclosures. The new farmhouses, too, were constructed from stone that was obtained as near to hand as possible. Some were given jocular names such as North America (the furthest point west), but the reality of the hardship involved in earning a living from such poor land has led to many post-enclosure farms being abandoned. In the White Peak, however, the former moorland on the carboniferous limestone has been successfully converted into pastures and hill meadows.

On the southern Pennines the characteristic farmstead that was built upon enclosure is the laithe house, a type of building that can be traced back to 1650 and which proved suitable for small, moorland farms. The farmhouse and barn were built in one range, with gritstone walls and stone slab roofs, but unlike the medieval longhouse the building contained no common entrance for humans and livestock, and no internal connecting door. Other barns were built in isolation far away from the farmhouse. They are a particular feature of the landscape of the Yorkshire Dales and the White Peak, where they are dotted around the fields that were taken in from the upper pastures, near to the 'dew ponds', puddled with clay (and, in modern times, concrete), which were constructed to retain rainwater for the animals.

On the North York Moors the traditional longhouse plan was replaced during the late eighteenth and early nineteenth centuries by double-pile farmhouses with a central front door and rooms to each side. At the same time the standards of cottage accommodation improved, though many single-storeyed cottages remained in use. Writing in 1891, the Revd J. C. Atkinson remarked that, 'The replacement of these ancient, incommodious, comfortless, hovel-like dwellings by new and substantial and decently arranged houses, or at least the substitution of such for them, did not begin, certainly was not in full progress, much, if at all, before the last quarter of last century.' The same remarks could have been made of moorland settlements in other parts of England.

Many of the new cottages on the moorland edges were built for industrial workers. A great burst of activity on Bodmin Moor produced mining and quarrying villages such as Minion, Henwood, and Upton Cross. The slate quarries at Coniston and Honister were 'the most considerable' in the land. In the Peak District the mining of lead, coal, ironstone, and ganister clay, and the quarrying of building stone and millstones were major sources of employment for much of the nineteenth century. Many settlers came to Alston Moor, high on the northern Pennines, after 1735, when the moor was given to Greenwich Hospital and the mines, which had not been worked for many years, were leased in lots, with large portions going to the London Lead Company. In 1857 the chief agent of that company reckoned that nine-tenths of the population of Teesdale was connected with mining. He observed that the farming and mining population were so mixed up that it was 'scarcely possible to go into a family occupying half an acre in Alston Moor or Teesdale, without finding that one or more members of the family are workmen employed in the mines.' The subsequent decay of most of these occupations, particularly lead mining, meant that many of these settlements were short-lived. The quiet countryside of today was created when industrial workers left in droves, often for a new life overseas.

Meanwhile, the Romantic movement and improvements in transport had encouraged some wealthy families to move for at least part of the year to 'lodges' erected amidst moorland scenery. The outstanding example is Cragside, built between 1869 and 1884 by the river Coquet near Rothbury, Northumberland. Other lodges were built to house guests during the grouse-shooting season. Game had been shot on the wing since the second half of the seventeenth century, but not in such a hugely organized fashion. Improved gun design encouraged the sport,

Walker's moor guide

George Walker's portrait of a guide on Stainmoor (North Yorkshire) in 1814 depicts a lead miner in his autumnal employment directing grouse-shooters and relieving 'their shoulders from the irksome load of the game bag and ammunition'.

and parliamentary enclosure made possible the creation of private grouse-shooting moors extending over thousands of acres of land. Because the moors were otherwise of such little value, great landowners received enormous acreages as compensation for their loss of common rights. They then quickly consolidated a moorland estate through the purchase of adjoining portions. Thus, when the Hathersage, Totley, Holmesfield, and Baslow commons and wastes in the Peak were enclosed, the Duke of Rutland was able to create a grouse moor only a few miles from one of his seats at Haddon. By about 1830 he had constructed Longshaw Lodge, Fox House Inn, and gamekeepers' lodges, all in mock Jacobean style. The surrounding landscape has changed little since then. Many of the neighbouring moors have rows of butts, dating from the mid-and late-Victorian period, when the practice of employing local men and boys to drive grouse over the horizon towards the guns began. The annual slaughter, beginning on 12 August, then reached incredible figures. The record 'bag' in the Peak District was on the Broomhead Moors in August 1913, when 2,843 birds were shot by nine guns in a single day.

The present landscape of most of the English moors has been frozen in time by this obsession with shooting grouse. Not only were the public denied access, the moors were carefully managed for the rearing of game. On Stanage Moor (Yorkshire) in the early years of the twentieth century, the Wilsons of Sheffield went so far as to employ masons to cut channels and holes into the rocks to collect water for young grouse in the breeding season, so that they would not fly off and be shot on someone else's moor. The 108 troughs are arranged in three sequences along the boundaries of the moor. Controlled burning of sharp-edged sections of the moors over a cycle of years now became a regular practice, so that young heather shoots could be produced every ten or twelve years, for heather is the staple food of grouse. Meanwhile, sheep and cattle were driven off the moors and rabbits were encouraged so as to vary the 'sport' in the shooting season.

What, we may ask, would have happened to the English moors after enclosure if grouse-shooting had not become a summer passion for the wealthiest members of society? The likelihood is that they would have been clothed with conifers, like the poor soils of much of continental Europe. This most distinctive of British landscapes could have been lost had the rearing of birds for slaughter not become the main object of moorland estate management.

During the late nineteenth and twentieth centuries the traditional English moors have been threatened by afforestation, drainage schemes, the increased scale of extractive industries, reservoir construction, new roads, and recreation. Thus, on the North York Moors the Forestry Commission began planting conifers in 1921; now 14 per cent of the area is covered with trees. In Northumberland the Forestry Commission began work in 1926, and six years later acquired most of the Duke of Northumberland's upland estate; they now own over 152,000 acres of plantations, which stretch in an almost continuous belt from Hadrian's Wall to the Scottish border. Between 1800 and 1946 Bodmin Moor shrank by nearly 50 per cent and since then has shrunk still further, so that now it covers only 41 per cent of the area that it contained in 1800.

From the mid-nineteenth century onwards, the growing industrial towns needed a supply of water for their inhabitants and industries on a scale that had not been imagined before. The deep valleys of the upland moors, where rainfall was the highest in England, were a natural choice for the construction of large impounding dams whose reservoirs could provide a regular supply of piped water, even to distant places. For example, Derby, Nottingham, and Leicester are supplied by reservoirs in the upper Derwent valley in the remote moorland district now known as the Dark Peak, and Manchester gets most of its water from reservoirs in the Lake District. A series of private Acts of Parliament authorized the flooding of valleys and the destruction of small villages, hamlets, and scattered farmsteads to make way for the new reservoirs. The quarries that provided the stone and the railways that were constructed to take stones to the dams are now interesting aspects of industrial archaeology, but the temporary villages that were quickly erected to house the workers in 'huts' with corrugated iron outside walls have disappeared entirely and are remembered only from old photographs and through oral history. Thus, Birchinlee, which was laid out alongside the river Derwent in three streets in 1901–3, with a school, recreation hall, two hospitals, a bathhouse, public house, and shops, and which accommodated 967 people in August 1909, was demolished in 1914–15, upon the completion of the reservoirs. The site is now marked only by a memorial cross.

The Army took over large stretches of moorland during the Second World War and has remained in some of them to practise tank manœuvres and erect rifle ranges. These moors are inaccessible to the general public. The battle for access to other moors has been fiercely fought since the beginning of the twentieth century, nowhere more so than in the Peak District, where ramblers and climbers from the surrounding industrial towns sought weekend recreation in an environment far removed in character, though not in distance, from that of their homes and places of work. The battle for the right to roam has been won on many moors, but not on all. Meanwhile, the Council for the Preservation of Rural England campaigned successfully for the prohibition of housing developments that would have threatened the traditional heather moorland landscape. They have been less successful in the fight against large-scale extractive industries.

The creation of National Parks, beginning with the Peak Park in 1951, has been the most important factor in the conservation of the English moors. The National Trust, too, has become a major landowner and has actively sought to restore the flagging health of its property. For example, when the Trust took over the Kinder Estate in Derbyshire (once the scene of the fiercest access battles) in 1982, it acquired one of the worst sites of upland erosion in the country. Overgrazing by sheep, fires caused by careless walkers, and atmospheric pollution had eroded the peat, destroyed much of the heather, and caused the spread of bracken and tough, inedible grass. In earlier times, bracken was trodden down by cattle and often gathered for bedding. The spread of bracken on all the English moors is now a serious problem, made worse by improved drainage which allows bracken to colonize areas that it could not tolerate before. On Kinder the Trust has provided firm surfaces for walkers on the Pennine Way, introduced effective fire-fighting methods, rebuilt derelict walls, constructed boundary fences, rounded up stray sheep that should not have been there, and applied light dressings of fertilizer and limestone dust to prompt indigenous seed to germinate. The results have been dramatic.

The human contribution to the making of a distinctive British landscape that at first sight seems to be 'untouched' and 'natural' has been enormous. In the late twentieth century a consensus has been reached that the moors must be carefully managed to ensure their preservation. For those who know them well, they are, in Emily Brontë's words, 'In winter nothing more dreary, in summer nothing more divine'.

Howden reservoir dam under construction

In 1907 a large crowd was attracted to the public ceremony of placing an inscribed stone in the huge retaining wall of one of the spectacular reservoirs which were constructed in the Derwent Valley on the Yorkshire–Derbyshire border. This one was begun in 1902 and completed in 1916.

9 Common Land

Alan Everitt

The Survival of Common Land

Probably few of us today realize how much common land there once was in England or how widely distributed it used to be. It underlies many of our industrial areas; it moulded the development of many of our suburbs; it gave birth to a world of local crafts and occupations; and until the early nineteenth century it still underpinned the economy of innumerable rural parishes. During the 1950s, rather more than a million acres of common were recorded in England by the Royal Commission on Common Land, or about 3 per cent of the country (see note at the end of the chapter regarding these figures). In the 1690s the equivalent figure was probably between 8 and 9 million acres, or 25–30 per cent. Estimating the extent of common land at any time is a task that raises many problems, and I shall return to it in more detail later. But the loss of probably more than 7 million acres over the last three centuries alone is plainly a matter of fundamental importance in the history and economy of the English countryside.

At the present day, the most extensive commons are to be found, as we should expect, among the fells and moorlands of the Highland Zone: on the Pennines, in the Lake District, and in the West Country. But although that was generally true in earlier centuries too, the further back we trace the history of common land, the more complex and unexpected the ancient pattern becomes. Until the early nineteenth century, such areas were far more numerous and extensive in many parts of lowland England, as well as in the north and west, than might be supposed. Nothing has more surprised the present writer than the discovery of so many heaths and wastes, so much half-wild country, in all the woodland counties of southern and south-eastern England. In his *Hints to Gentlemen of Landed Property* in 1775, Nathaniel Kent remarked 'that within thirty miles of the capital there is not less than 200,000 acres of waste land'. A generation later, nearly 97,000 acres of common and heath survived in Surrey alone, or one-fifth of the entire county, of which Bagshot Heath accounted for 31,500 acres.

The extent to which London itself has developed over former commons, and the way in which it consequently remained embedded in a world of the past until the days of Queen Victoria, is a little-known yet fascinating theme in English history. It is worth discussing here because it sheds light on the way in which past and present were similarly interwoven in many parts of the provinces too. The last vestiges of that world are still recorded in the names of London's numerous parks and open spaces: Hyde Park, Blackheath, Wimbledon Common, Oxleas Wood, Hackney Downs, Hampstead Heath, Wormwood Scrubs, Shooter's Hill, Brook Green, Bethnal Green, Peckham Rye, and many other places are all remnants of ancient common land. Their varied names and suffixes have much to tell us—something slightly different in each case—about their origins in a profoundly rural countryside.

Soon after the South-Eastern Railway through Croydon, Reigate, and Tonbridge was extended to Tunbridge Wells in 1845, the *Railway Chronicle* issued one of its 'Travelling Charts' for 'perusal on the journey'. These charts were designed to encourage people to make use of the railway to explore the surrounding country on foot. The first 13 or 14 miles of the route through Surrey and Kent, twenty-two years before the present line through Sevenoaks was built, are described in the following quotation:

> Sydenham hills and woods offer many charming walks. The well-known
> Beulah (i.e. beautiful) Spa, about a mile from Forest Hill station, is a famous
> place for 'pic-nics'… . The walk from Sydenham station to Beckenham,
> thence to Bromley, on to Chislehurst and Sundridge [in Bromley parish] can
> be highly recommended for a day's excursion… . Of late years the hills [about
> Croydon] have been inclosed, and the pleasant walks over the purple heather
> contracted … [but] the walk over the Addington Hills to Wickham Court is
> alone worth the journey from town. The old red brick Elizabethan house, and
> its ivied walls, furnish capital studies for the artist, and at the proper season
> the hills, especially Crohamhurst, are luxuriant with the lily of the valley… .
> The pedestrian should alight at the Croydon station, and stroll over and
> about Crohamhurst to Sanderstead, and then make a circuit back through
> Addington. The woods in the spring resound with the plaints and the joyous
> 'jug-jug' of the nightingale. It is a most delicious country.

How strange it sounds today! The whole of this area now lies in Greater London. Yet here, in the 1840s, were charming walks about Sydenham and Bromley, hills luxuriant with heather and the lily of the valley, and woods resounding with the song of nightingales: and much of that 'delicious country' was only 'of late years … inclosed'. True, what chiefly interested the anonymous author of the Chart was its picturesque character: its homely woodland churches, its timber-framed farms and manor houses, then often in a state of some decay, its mingled scenery of woods, heaths, parks, commons, 'bottoms', sunken lanes, winding shaws (strips of woodland round a field), irregular little fields, and its wealth of wild flowers. But other sources confirm his picture and tell us that within this area, or just beyond it, numerous commons and woods still remained unenclosed at that time: at Bromley, Chislehurst, Keston, Hayes, West Wickham, Sydenham, and Forest Hill in Kent, for example, and at Addington, Shirley, Norwood, Penge, and Croydon in Surrey. They covered some thousands of acres on either side of the boundary, reaching nearly to the Thames in Clapham, Camberwell, Peckham,

Deptford, Greenwich, and Plumstead, and extending eastwards for 14 miles in an almost continuous line to Dartford Heath. These areas were but vestiges, moreover, of an intercommonable tract of country which has left its mark in more than 100 place names specifically denoting commons down the borders of Kent and Surrey in the 35 miles between Lingfield and Deptford. At its northernmost end, the county boundary continued to fluctuate over wastes and commons in Deptford and Camberwell until the seventeenth century.

If it seems strange today that so much common land still surrounded London in Queen Victoria's reign, we have to remember that the city was embedded in a good deal of comparatively unrewarding or densely wooded country. Away from the rich alluvial parishes bordering the Thames, one quickly reaches the intractable soils of the London clay in Middlesex, the stony Blackheath and Woolwich beds in Kent, the notoriously infertile Bagshot Sands in Surrey, and the thin, flinty lands and desperate clays capping much of the chalk. Partly as a consequence of these conditions, and in striking contrast with the far larger medieval city of Paris, London early had to search distant parts of England for its food supplies. By the seventeenth century, it was already drawing on such counties as Wiltshire, Herefordshire, Cheshire, Yorkshire, and Norfolk, as well as those bordering the river upstream to Oxford. The great south-easterly drift of cattle and sheep from the Pennines, Wales, and Scotland, bred up among moorland commons themselves, and destined to be fattened on the lush pastures of the south Midlands or the marshlands of Essex, Sussex, and Kent, was already an almost legendary aspect of metropolitan tradition. By Queen Elizabeth's reign, 'sea-coal' from Newcastle was beginning to displace wood and charcoal from the Surrey hills and the Chilterns as the city's principal fuel.

Yet the fact that London had to seek far afield for its food and fuel was also one reason why so many commons remained within its orbit until the coming of the railways began to transform them. They remained not merely as picturesque relics of the past, moreover, but with a distinctive native life of their own. They had given birth over the centuries to an unsuspected range of local crafts and industries, drawing on the woods and heaths themselves for their resources. By providing Londoners with some of their more unusual raw materials, such as boxwood for printers' blockmakers, and with many familiar domestic necessities, such as brooms, besoms, mats, baskets, woodware, earthenware, bricks, tiles, garden produce, culinary and medicinal herbs, geese and poultry, they afforded them also a ready glimpse (if they would) into an unfamiliar world of cottage-farmers, woodlanders, and rural craftsmen.

Ten miles south of London, the ancient market town of Croydon itself owed much of its historic prosperity, like other woodland towns elsewhere in England, such as Frome (Somerset) and Berkhamsted (Hertfordshire), to this same kind of environment. By 1841, with 14,000 inhabitants, it was the largest urban centre in Surrey apart from Southwark. Yet its parish of nearly 10,000 acres was upwards of 36 miles in circumference and still comprised extensive commons in the neighbourhood of its numerous outlying hamlets. It had long been known for its 'broom-squires', dwelling among the woods of Shirley, and producing brushes and besoms for the London market. Until the eighteenth century it had been especially noted for its charcoal-burners, likewise drawing upon the surrounding woodland commons. Their strange lives and blackened faces had been the butt of Cockney jokes since the early medieval period, and after the Restoration a comedy was written about them, entitled *Grim, the Collier of Croydon*.

Mushroom Green

A rare survival of the squatters' settlements once typical of the Black Country. It developed on the Worcestershire–Staffordshire border between Dudley Wood and Cradley Heath. Early nineteenth-century cottages are scattered haphazardly on a bank above a stream, linked by lanes and small patches of open green. The first inhabitants worked nearby in the Earl of Dudley's colliery at Saltwells, but the local speciality of chain-making quickly developed and survived until recently. It epitomizes the humble common-land origins of many an industrial district.

Unique though London was on account of its size, some of the same themes were echoed in many parts of the provinces too. Most of our major cities, such as Manchester, Birmingham, Sheffield, and Leeds have developed from early market towns, or from ancient incorporated boroughs like Nottingham, Bristol, and Leicester. Yet they owed much in their historic evolution too to the commons, woods, or moors surrounding them. The complex moorland parish of Sheffield (South Yorkshire) covered nearly 23,000 acres, reaching far to the west in its outlying townships, and set in an area well endowed with mineral resources. The suburbs of Birmingham, though in quite different country, are still peppered with names in heath, green, wood, end, and common, which, like the London parks, have something to tell us about the city's origins in forested country. Though places like Leicester and Nottingham may seem at first sight straightforward examples of Midland open-field boroughs, they too owe something in their past to their common rights in the forests of Leicester and Sherwood respectively. In industrial regions such as the Black Country, the Staffordshire Potteries, and the Leicestershire–Derbyshire borders, early settlements were often embedded in tracts of heath, moor, or woodland, from which they drew their raw materials, and in which many new industrial villages developed in the nineteenth century, such as Lye Waste in Old Swinford (Worcestershire), and new towns on

parish boundaries like Coalville (Leicestershire) and Woodville (Derbyshire). In some cases, perhaps many, the staple industries of districts like these developed from the humble crafts of the commoners.

Both in the south-eastern counties and in some other parts of England, a tentative understanding of the life of the commons was beginning to emerge, as people travelled more widely, in the days of Celia Fiennes and Daniel Defoe. We find it again among some of the more perceptive antiquaries of that time too, such as Robert Plot in Staffordshire, William Woolley in Derbyshire, and John Morton in Northamptonshire. But it blossomed afresh, and with new insight, among a host of literary writers and explorers from the days of Gilbert White of Selborne in Hampshire in the late eighteenth century to those of Flora Thompson in Oxfordshire in the early twentieth. Their interests, themes, and viewpoints varied as widely as the kind of country and people they depicted. But for us today it was fortunate that something of the native economy of the commons survived until this period of rediscovery, and that they had the imagination to perceive it as a survival of the past. They are too numerous, and in many cases too little known, to discuss here. But in their very diverse ways, writers like William Cobbett, George Borrow, Mary Russell Mitford, J. C. Atkinson, W. H. Hudson, Richard Jefferies, George Sturt, James Thorne, Alfred Williams, Louis Jennings, and many others enable us to think ourselves back into an older world. They remind us of that strange intermingling of ancient and modern in English history which in their times was still a reality, and which so often haunted the Victorian imagination in its literature and its art.

Definition and Distribution

What exactly is common land? We are not here concerned with common fields or meadows, but with what in historic language is called 'common waste'. It was land that lay outside the arable fields and meadows, whether open or enclosed. In general it was most extensive in those parts of England where the classic common-field system, as it is usually envisaged, did not exist, or where, as in Hertfordshire and Staffordshire, it formed only one element in a complex landscape of scattered hamlets and farmsteads rather than single nucleated villages. Prominent among those areas where commons were especially notable were, first, the great fells and moorlands of the north and west, where much 'waste' remains today; second, the fens of eastern and parts of western England, where it has now almost all been drained and enclosed; third, the forest and woodland countrysides once to be found in many counties, and still extensive in the south-eastern shires and the Welsh Marches; and fourth, the numerous tracts of heathland also formerly familiar in many parts, though they are now chiefly characteristic of the south.

Commons were also at one time to be found in many of the common-field parishes of the so-called 'Midland Plain'. But here they were more closely integrated with the life of the fields and villages themselves, so that their history is often quite different. In many places they were gradually ploughed up and added to the village arable lands as population increased in the early medieval period. Where they survived into more recent times, they eventually succumbed as a rule to parliamentary enclosure in the eighteenth or nineteenth century. Very little common waste remains in these areas as a consequence today: only 353 acres in the whole of Leicestershire, for example, and none at all, apart from a few village greens, in Northamptonshire outside the Soke of Peterborough. Behind all these generalizations lies a pattern of local diversity in every county. They give us no more than a framework, and there is a sense in which the variety of the English landscape gives every common a history of its own.

Over the last 350 years numerous legal treatises have been written on the status and usage of common land. In this respect, too, local custom varied greatly, but a few general points must be borne in mind. The first is that, although all commons were in law the 'waste' of some particular manor or group of manors, not all 'waste' was necessarily 'common', and the term 'manorial waste' is a quasi-technical one. It does not mean what we mean by 'waste' today, something that is intrinsically useless or worthless. Probably little of England can be so described over the past six or seven hundred years, perhaps longer. It means land that was not intensively exploited for agricultural purposes, and in the main consisted of some kind of moorland, woodland, heathland, rough pasture, boggy land, fen, or scrub. Before adequate clay-pipe drainage in the nineteenth century, such areas were far more widespread than today, and in a peasant society even scrub had its uses. It is the same word as 'shrub', and it gave birth to some important occupations, such as basket-making. It appears in numerous English place names, moreover, such as Shrubland in Suffolk and Whitley Scrubs in Kent.

The second point to note is that common land is not, historically speaking, public property. Until relatively recent changes in the law, it was not land in which the public generally held any rights, and those changes related mostly to limited matters of access for purposes of recreation. Common land has thus never belonged to the 'people of England', as the Diggers under Gerrard Winstanley believed during the Interregnum, though the dawn of that idea is an interesting point in constitutional history. In the early phases of English settlement, and perhaps long before, certain areas such as Dartmoor and the Weald were 'intercommoned' by the people of a whole region or shire. But in historic times—since, say, the tenth or eleventh century—commons have always been the private property of manorial lords. They are property, however, of a somewhat special kind, in or over which all the local commoners, including the lord—but not the general public—traditionally possessed certain rights of usage or 'common'.

Such rights varied greatly in character and degree from place to place and from one period to another. They varied according to the extent of the waste, the diversity of its resources, the number of commoners, the manorial structure of the area, and the nature of local custom. In many places, particularly in the common-field parishes of the 'Midland Plain', they tended to be restricted to the inhabitants of a single village or township. But where commons were extensive or complex, especially in the outer or 'peripheral' counties of England, they were frequently shared—often over many centuries—by a number of surrounding settlements. The 'Great Wood' of Norwood in Surrey, referred to earlier, was 'intercommoned' in this way by the five parishes of Croydon, Battersea, Lambeth, Streatham, Camberwell, and their dependent hamlets. It adjoined the common land of Lewisham, Sydenham, and Forest Hill in Kent, moreover, which was likewise intercommoned by several neighbouring places. In Hertfordshire, the 2,000 acres of Berkhamsted Frith and its adjacent commons were shared by eleven nearby hamlets and villages on the Chilterns; and in Essex, Tiptree Heath, south-west of Colchester, by no fewer than sixteen places in the vicinity.

There were also many parishes even in the early nineteenth century to which common rights appertained in areas quite detached from them. In the part of Surrey just mentioned, Penge was the detached hill-pasture of Battersea, 7 miles away on the Thames. Its name, *pen + cēt* ('high or chief wood'), is a Celtic one and it is recorded as a swine-pasture before the Conquest. But it remained an uninhabited spot of some 700 acres until the late seventeenth century, and it was still marked as Penge Common on the early Ordnance maps. Many comparable instances might be cited, especially in the fell and moorland areas of the north, the fenlands of the eastern counties, and the old woodland districts of southern England and the Midlands. The further back we go, moreover, the more numerous and complex such detached pastoral areas and rights of intercommoning tend to become, until in many places they merge into the 'Great Commons' of the Old English period. The wooded border-country of Kent and Surrey, the Selwood Forest region dividing Wiltshire and Somerset, the Derbyshire Woodlands adjoining Nottinghamshire, and the Hertfordshire–Middlesex boundary are all notable examples of this aspect of the history of commons. They are riddled with place names, vestiges of ancient jurisdictions, and early rights of pasture bearing witness to its former existence.

Of the many uses of commons, those of pasture are the most familiar, together with the right to obtain fuel—wood, peat, coal, furze, 'tods', 'hassock', and so on—and also timber for building and farm-gear. But diversity of use is the keynote, and in many ways it is that which gives common land its greatest importance in English history, especially in recent centuries. For although it was usually the poorer, more inaccessible, or intractable land of the parish, it was often land that was rich in mineral resources such as stone, coal, glass-sand, iron-ore, gypsum, fuller's-earth, and brickearth, and rich too in a great diversity of trees, shrubs, and plant-life. In *A First Book of Wild Flowers*, published as recently as 1909, M. M. Rankin described no fewer than 110 species which might readily be discovered by 'any country child'. Most of these plants are now all but unknown, even to those of us with an interest in the subject. But they were not unknown to the people of the commons. Necessity bred in them a power of observation denied to most people today, and an ability to exploit even the humblest resources that lay to hand.

We need to think of such areas, therefore, not simply as poor outlying tracts of rough pasture or merely unrewarding land, but also as reservoirs of resources which gradually, over the centuries, gave rise to a surprising variety of local occupations, and in many places to entirely new settlements. Most of these occupations, like the resources themselves, were of a humble kind; but in some areas they developed into major local industries or gave birth to manufacturing districts. Many dyes, for example, were ultimately derived from plants or trees of the common waste, though some like saffron, which gave its name to Croydon ('wild saffron valley'), later came to be cultivated as commercial crops. Four distinct colours were obtained from the alder alone, a tree that likes boggy spots and sodden stream-banks. Until recently numerous familiar medicines also originated from wild plants, such as the foxglove (*digitalis*), St John's Wort (*hypericum*), and feverfew (*chrysanthemum parthenium*). Many were still included in the British *Pharmacopoeia* until the 1950s or 1960s, and some are now being rediscovered by enterprising pharmacists. In the sixteenth century both medicinal and culinary herbs were gathered by poor country women in places like Mile End Common in

Lubberland Common, on Catherton Clee Hill in south Shropshire

The common is pock-marked by remains of early bell-pit coalmines. Round the edges and on island sites in the middle are the encroachments of squatters on the extensive wastes, widely dispersed rather than closely packed as at Mushroom Green.

Middlesex, and sold in the London herb-market at Bucklersbury. Three centuries later they were still being gathered by poor people among the Surrey commons near Surbiton, and hawked from door to door in the suburbs or sold to London druggists.

The dual status of common land, as the property of manorial lords, yet land in which all local commoners had rights, inevitably occasioned conflicting interests between lords and peasants. They were often exacerbated by rivalry between neighbouring communities, especially in intercommonable areas, each claiming rights in the waste. Evidence for the history of commons often arises from lawsuits relating to such disputes, from cases recorded in court-rolls, and from Special Commissions and Depositions in the Exchequer. Such suits and commissions became especially numerous at times of expanding population, which occasioned pressure on the commons and their resources. The thirteenth and early fourteenth centuries, the sixteenth and early seventeenth, and the period from *c.*1750 to 1870 thus stand out as times when conflict was particularly intense. It did not die out, however, when population declined, as in the late medieval period, or when it was relatively stable, as in the late seventeenth and early eighteenth centuries. It remained an endemic problem, and it was complicated by many other matters, such as the local extent of the waste, the kind of countryside in the vicinity, the nature of its resources, the structure of settlement and society, and the rural economy of the area.

There was a continuous tendency, moreover, for landless people to drift towards the woods and wastes and establish a toehold for themselves if they could, especially on the boundaries of parishes, townships, or tythings, where jurisdictions might be ill-defined or uncertain. That not only aggravated pressure on commons, but often occasioned conflict with local inhabitants and magistrates. On the other hand, sheer necessity sometimes drove incomers to search for new resources in the vicinity, and so win a livelihood for themselves by setting up their own local occupations. Many of the greenside-hamlets and boundary-settlements in areas as diverse as Hertfordshire, west Leicestershire, the Black Country, the Forest of Dean, and Kent originated in that way: as obscure communities of squatters on the waste, without holdings of their own, or even as migrant or seasonal places—of matweavers, brickmakers, potters, charcoal-burners, and fishermen, for instance—which eventually became permanent villages, or occasionally developed into industrial towns like Coalville or small ports like Whitstable (Kent).

The origins of common-land hamlets of this kind need more investigation. But there can be no doubt that in the early modern period many developed from a scattered group of what were then called 'night houses', 'clod-houses', 'clay houses', or (in Wales) *ty-un-nos*. They were built under the widespread belief that any house erected on the waste overnight entitled the man who built it to possession, provided smoke was seen rising from it by the morning. The origin of this belief is unknown; it can only be called a folk-custom; it is not quite extinct even today. I came upon it three or four years ago in Leicestershire. In Cumberland, such 'clay-houses' were said to take only three or four hours to build, and were simply 'rude earth and timber shanties that would not readily burn'. Sometimes the judges at assizes, as in a Somerset case in the 1630s, ordered local landlords to allow squatters to retain 'quiet possession' of their hovels merely 'in regard of [their] misery'. Not always, however: when two Kentish labourers built cottages on Longbridge Leaze in the Stour valley, and told the manorial court of Willesborough that they had done so 'for that they were destitute of houses and had seen other cottages upon the same waste built by other poor men', they were fined by the court—itself only goaded into action by the Exchequer—and ordered to pull them down, and move on. Yet wherever manorial organization was weak or slack, and increasingly as it decayed further in the eighteenth and nineteenth centuries at a time of rapidly rising population, these humble, outlying, squatters' hamlets, or 'scattering houses' as William Woolley aptly called them in Derbyshire, tended to develop on the wastes. By the 1850s some of those in Hertfordshire had more than 500 inhabitants, and in Staffordshire still greater numbers.

Charcoal-burners outside their hut

This was a seasonal occupation, surviving in its traditional form in Kent and other woodland counties until after the Second World War. They often worked alone or in small elusive groups, recognized only by their slowly smouldering circular stacks in the woods. Some slept in huts of this kind; others sheltered under an old sail or tarpaulin draped over a convenient bough, with an open fire or brazier outside it.

Recent Extent

It is for these reasons predominantly, as well as others beyond the scope of this chapter, that the history of common land over the last thousand years has been a matter of continuous attrition. At the time of Domesday, the population of England was probably about 2 million, or nearly so; by the early fourteenth century it had almost trebled, to some 5½ million, before dropping back to perhaps 2½ million by the late fourteenth. Significant recovery does not seem to have begun until the early sixteenth century, but by the mid-seventeenth population had almost regained its medieval peak. It then declined slightly, or perhaps remained broadly stable for about three generations, though continuing to increase in London and the new industrial districts of the provinces. Rapid growth began once again after 1750, bringing the total to about 8.6 million in 1801 and almost 17 million in 1851. The earlier figures are subject to debate, and all conceal widespread regional and local variations. But it is against this demographic background that the extent and decline of common land needs to be considered.

Estimating that extent at any period is difficult. The sources are inevitably patchy and unsystematic, and there can never be indisputable figures. But the county maps of the eighteenth century, and the first edition of the Ordnance Survey in the early nineteenth, make it quite clear that commons were still much more numerous and extensive at that time than might be thought. Even though the surveyors were not primarily concerned with the legal or customary status of land, and the eighteenth-century maps vary widely in reliability, there can be no doubt on that point. Rocque's great survey of London and the area 10 miles round, at 2 inches to the mile, provides the most detailed early representation of commons (1741–5). Thomas Jefferys's map of Bedfordshire (1765) and Joseph Hodskinson's of Suffolk (1783) are typical examples of the better county surveys; but even rather poor maps, such as Andrews and Dury's of Kent (1769), may yield useful details to those with intimate local knowledge. Obviously, cartographic sources always need to be combined with other forms of evidence, particularly with that of place names and historic jurisdictions. Essentially the task of reconstruction is a matter of working back from the known to the unknown, and ultimately that means working from parish to parish and township to township as well as county to county.

The material collected by local authorities throughout England and Wales for the Royal Commission on Common Land in 1955–7 provides the most useful starting point. In England itself—that is, excluding Wales—some 4,000 commons were recorded, together with 1,354 'village greens' still thought to be subject to common right, the entire area amounting to 1,055,000 acres. In 1873, when the only previous 'census' of commons was compiled, a total of 1,700,000 acres was returned. Though the local reliability of these estimates varies and the figures for 1873 may be seriously incomplete, it is likely that at least 700,000 acres of common were lost between the two periods.

One of the main problems faced by local authorities in the 1950s was the legal status of much land where common rights may once have existed but which in recent generations had fallen into desuetude, or ceased to be regularly enforced. In Derbyshire, the total of only 174 acres is obviously difficult to credit for an area where there is still much upland waste—though such waste is not necessarily common—and where more than 21,000 acres were reported in 1873. In Kent, a number of commons known to the writer were omitted in the 1950s, and the true figure may have been nearer 5,000 acres, or even more, rather than the 3,500 acres officially listed. On the whole, however, the returns for most counties sound plausible. The sheer mass of local detail recorded is itself a telling point, emphasizing the diversity of custom, and some authorities made exceptionally thorough investigations. Hampshire commissioned a special survey at Southampton University and arrived at a figure of 67,000 acres; Hertfordshire also commissioned its own survey and reported about 5,500 acres; and County Durham, with its many Pennine moors, reported 64,500 acres, or some 9 per cent of the county. These three shires, together with Worcestershire (4,819 acres) and Huntingdonshire (1,238 acres) were alone in recording a greater area in the 1950s than the 1870s, no doubt because of more thorough inquiry.

The figure for Hampshire was the highest recorded in the Lowland Zone of England and of course included the New Forest; but the 27,000 acres recorded in Surrey equalled it in relation to the size of the county. In the Highland Zone figures were naturally greater: more than 100,000 acres were reported from Cumberland, Westmorland, Devon, and the West Riding, and more than 220,000 from the North Riding—the highest figure for any English county. Yet in highland parts, as in lowlands, there were also striking exceptions. In the very large county of Northumberland (1,249,299 acres), only 25,244 acres were returned, and in Somerset (1,007,000 acres) no more than 8,900. In the north of England vast areas of common have been lost since c.1830 to grouse-moors, just as in lowland shires many common woods were enclosed in the nineteenth century as pheasant-preserves, developments, not unnaturally, that often occasioned fierce social unrest at the time.

In almost every county unexpected, and often fascinating, details were brought to light in the 1950s, shedding light on the history and usage of common land, and on the diversity of local custom behind it. But perhaps the most interesting general point was that 'village greens' were still so often subject to rights of common. Of the thousands of places popularly so called today, some are genuine centres of ancient villages, though the great antiquity once attributed to them is now questioned by scholars. Others, especially in the north-east of England, were deliberately created for newly planned villages after the Conquest. The vast majority of 'village greens', however, are in reality centres of outlying hamlets in areas of dispersed settlement. They are especially numerous in old woodland counties such as Hertfordshire, where no fewer than 116 were reported to the Commission as subject to common right, the highest figure in one of the smallest English counties.

Detailed investigation by the writer in a dozen or so shires shows that places of this kind are usually remnants of former common land, quite often situated on or near parish-boundaries, and rarely recorded as settlements before the late-medieval period. Though they may well have existed earlier as open spaces among the woods, perhaps with some habitations or seasonal huts scattered round them, they are not ancient village-centres or church-settlements, and it is historically misleading to call them 'village greens'. They may now be quite populous places, and many have been engulfed by suburbia or by industrial development. But in origin they were humble common-land hamlets, and they still provide much vital evidence of former commons.

Historic Extent since 1690

If the survival until recently of more than a million acres of common land in a country as crowded as ours seems remarkable, far more so is the scale of its disappearance over the last 300 years. The earliest general estimates of land use we have are those to be derived from Gregory King's statistics in the 1690s. They have occasioned much debate and raise obvious problems of interpretation, since he was concerned with the usage of land rather than its legal or customary status. But they are well worth considering. For England and Wales together King suggested about 10 million acres of 'heaths, moors, mountains and barren land'; 3 million acres of 'forests, parks and commons'; a further 3 million acres of 'woods and coppices'; and 500,000 acres of 'roads, ways, and waste lands'. Though he overestimated the area of the two countries by nearly 2 million acres (39 m.: 37.3 m.), the fact that, without accurate maps, he came so close to the true figure tends to endorse his broad conclusions rather than discredit them. Excluding the 5 million acres of Wales, and making allowance for the many problems involved, his estimates possibly suggest a total of some 8–9 million acres of wastes and commons in England as a whole, including, that is to say, the greater part of his heaths, moors, mountains, barren land, and commons, together with some part of the forests, woods, coppices, and roadside wastes.

Is such a high figure seriously credible? The complexities behind any calculation lie outside our scope here. But intensive local investigation by the present writer and others suggests, I believe, that it may not be far from the mark. If so, at least one-quarter of the area of England (32.2 million acres) was still 'manorial waste' at the end of the seventeenth century as compared with 3 per cent at the present day. In other words, quite 85 per cent of the common land surviving in the 1690s—perhaps more—has subsequently been enclosed and turned over to other uses. Clearly so massive a transfer of property in the usage, if not the actual possession, of the soil and its resources, entailed revolutionary consequences for the entire economy of this country.

Between the 1690s and 1870s no comparable statistics survive for reconstructing the national acreage of common land. In the first decade of the nineteenth century, three different estimates were made; but they are difficult to interpret, and appear to vary from 4 million acres, or less, to more than 6 million. More useful in the present context are the figures recorded for individual counties during the 1790s, or thereabouts, and again in the 1830s. They are not systematic and do not exist for all areas. The earlier figures are mainly derived from the county *Reports* to the Board of Agriculture during the Napoleonic Wars. For some shires, such as Suffolk (100,000 acres), they are obviously rounded; for some, such as Kent, they are gross underestimates, based on the reporter's inadequate knowledge; and for some, especially where small commons were numerous, as in Hertfordshire, no total was attempted. Nevertheless, they are of much interest, and in some counties they have been checked against contemporary maps and other sources. In Northamptonshire, Professor J. M. Neeson has shown that the Board included no commons of less than 100 acres, and she has raised its total of 42,000 acres by a further 7,500. Her conclusions may well apply in other areas.

The estimates of the 1830s were also compiled locally, many of them for the county entries in Samuel Lewis's *Topographical Dictionary of England* (1833 and other editions). Sometimes they seem to be merely repeated from those of the

1790s, but in some shires serious efforts were apparently made to arrive at genuine totals. At least they suggest the order of magnitude to bear in mind, and in some counties, such as Surrey, they may be compared with other figures. For many particular commons, other local estimates survive and provide a further check: 6,000 acres, for example, at Mousehold Heath in Norfolk, 4,200–6,500 (at different dates) for Hounslow Heath in Middlesex, 2,020 acres for Finchley Common (Middlesex), 1,000 acres of 'common wold' at Cotgrave in Nottinghamshire, and so on.

Despite their limitations, all the figures we have, like the cartographic evidence mentioned earlier, point to a massive decline in common land since the late eighteenth century. Wherever we look, it is plain that scores if not hundreds of local commons, great and small, have disappeared. In the table, tentative acreages for eleven counties for which estimates exist are brought together.

County	County Acreage (1861)	Estimated Acreage of Common Waste			
		c.1800	1833	1873	1956/58
Surrey	479,000	96,000+	65,000	42,900	26,900
Sussex	937,000	110,000+	c.100,000	21,200	14,100
Berkshire	451,000	40,000	?24,500+	7,700	3,600
Suffolk	948,000	100,000	'still extensive'	7,500	7,400
Kent	1,039,000	?c.50,000+[a]	c.20,000+	8,200	3,500+
Herefordshire	535,000	—	?c.30,000	10,200	5,500
Lancashire	1,219,000	108,500	98,000	69,000	33,100
Staffordshire	728,000	140,000	100,000+	12,300	3,400
Northants	630,000	49,500[b]	—	2,900	208
Northumberland	1,249,000	450,000+	?c.450,000[c]	53,214	25,244
Norfolk	1,354,000	?c.200,000	'still of great extent'	16,510	8,386

[a] My own estimate.
[b] This figure excludes part of the county.
[c] This figure apparently repeats that of c.1800

To these figures a brief comment may be added on the enclosure of waste in Leicestershire. The county is often taken as a classic example of Midland open-field country, where commons had shrunk to almost negligible proportions by the seventeenth century. There were many parishes, such as Wigston Magna, where that was certainly true. But although, as already remarked, only 353 acres were reported in the 1950s, and less than 700 in 1873, more waste formerly survived than may be supposed. In the 1820s the 'rugged wilderness' of Charnwood was still one of the largest tracts of common in the Midlands, extending to some 18,000 acres when finally enclosed in 1830. Elsewhere in the western half of the shire, local investigation suggests that some thousands of acres of open heath and vestigial woodland survived until the preceding century or so. Much of it was enclosed piecemeal in relatively small parcels; but the great waste of Ashby Woulds, enclosed with some neighbouring patches of common under two Acts of Parliament in George III's reign, extended to 3,000 acres. Several other commons on this side of the county, such as Normanton Heath (enclosed in 1629), covered more than 500 acres. In the wooded parts of eastern Leicestershire, some areas of 200 or 300 acres survived into John Nichols's time (c.1800). Even in the heart of the common fields, in the south of the shire, the open tract of Kimcote Heath was still intercommoned by at least three neighbouring villages when it was enclosed under two Acts of Parliament in the 1770s.

The parliamentary enclosure movement of the eighteenth and nineteenth centuries gave rise, of course, to a massive corpus of official documentation. It is systematic evidence, moreover, and the literature devoted to it by generations of historians is well known. But in turning to it to estimate the extent of waste, two major problems arise. The first is the difficulty of disentangling enclosure of *commons* from enclosure of *field land*. Where the awards relate almost entirely to waste, as in several of the outermost or 'peripheral' English counties, that difficulty does not arise. But in most counties, especially those further inland, common fields and wastes were often enclosed together under the same Act, and detailed examination of all the local apportionment acreages is needed to reconstruct valid figures.

That herculean task has yet to be undertaken for England as a whole. But in a pilot survey based on a 10 per cent sample of the awards, interim totals were worked out a few years ago by Dr John Chapman. Though his techniques were criticized by some at the time, and indeed the problem bristles with difficulties, his broad conclusions were of great interest. Cutting through the technicalities, it seems likely that altogether some 4½ million acres of manorial waste, or thereabouts, were enclosed under Act of Parliament. If that figure is added to the 1.7 million acres said to have existed in 1873, and some allowance made for reputed underestimation at the latter date, it seems that there must have been quite 6½ million acres of common land when parliamentary enclosure began in the 1730s. That tentative figure brings us, however, to the second, and in practice far more serious, difficulty: how widespread was enclosure of waste by piecemeal, or 'concealed', methods—that is, apart from Acts of Parliament—after 1730?

Old Woodland Country

On this question opinion has varied greatly. Some historians have thought piecemeal or 'concealed' enclosure became relatively insignificant after the mid-eighteenth century. In classic common-field country in the Midlands, where more than 70 per cent of the area concerned was enclosed by parliamentary methods, that may not be far from the mark. Professor Neeson has shown that in many parishes, such as West Haddon in Northamptonshire, commoners were so numerous and their rights so deeply entrenched in local custom that enclosure by other means than Act of Parliament was normally impossible. Only the State—itself the instrument of landed magnates—had sufficient power to uproot such communities, and in doing so to enclose commons along with common fields. At least 50,000 acres of waste disappeared in Northamptonshire alone in that way, and almost none survives today.

But over much of lowland England some thousands of commons, heaths, woods, greens, and other stretches of waste still remain: at least 240 were recorded in Surrey in 1956–8, for example, and 186 in Hertfordshire. Such areas are rarely as extensive as those of the fells and moorlands in the north and west. With rare exceptions, such as Ashdown Forest's 6,400 acres, they are often mere vestiges of what once existed, much of it until the 1830s. But like the fragments of a mosaic, they enable us in the light of other evidence—place names, jurisdictions, maps, documents, and topography—to piece together something of the earlier pattern of common land in countryside of this kind. They point up for us, moreover, a landscape distinction of profound significance in English history.

It was personal experience that opened my own eyes to that distinction. I was brought up in a part of Kent where you could wander for miles unmolested, by a maze of footpaths, holloways, and sunken lanes, through commons, heaths, woods, and copses, embedded in crooked fields and winding shaws, scattered with ancient manor houses, farms, and green-hamlets, and rich in wild flowers, shrubs, and trees. Although only limited areas were still subject to common right, many woods remained unenclosed, and few landowners or farmers much minded in those days if you walked through them. In places, local customs such as the annual gathering of sweet chestnuts still lingered on as traditions, moreover, though no longer as formal rights. When I came to Leicestershire in 1960, after several years spent elsewhere—in Herefordshire, Yorkshire, Lancashire, west Devon, and west Surrey—a new and startling difference suddenly came home to me. There seemed to be nowhere in this part of the world where you could go for a walk in the kind of broken, wooded, piecemeal country I had known at home. I knew Leicestershire was old common-field land and was bound to be different. But only then did I realize that I had been brought up in what I now think of as 'old woodland', 'old coppice', or 'intermingled' country.

These terms are no more than a mental shorthand. The proportion of wooded land may be relatively small, rarely perhaps more than 25 per cent, and its antiquity is not the present point since much may have been replanted. Everywhere they require local qualification, moreover, and probably no county is quite devoid of such country. Even in east Leicestershire some patches of 'intermingled' land remain: as for example in the vicinity of Keythorpe, Tugby, Launde, and the lost priory of Bradley, where place names and topography clearly underline the evidence of a woodland past. Nevertheless, in tracing the history of common land, the distinction between 'old common-field country' and 'old woodland or intermingled country' is as crucial as the familiar dichotomy between Highland and Lowland Zones, and in many areas it cuts across that divide.

For in intermingled country gradual, piecemeal enclosure of the waste, rather than parliamentary enclosure, has always been a normal custom. It began far back in the early medieval era if not before the Conquest, and it remained a dominant mode throughout the parliamentary period too. On the whole it was only in some of the surviving royal forests, or in areas like Bagshot Heath (Surrey) where manorial waste was exceptionally extensive, or in places where numerous parties or several parishes were involved, as at Cox Heath in Kent, that landowners found it imperative to resort to an Act of Parliament if they wished to enclose in countryside of this kind. Though it will never be possible to assess the full scale of piecemeal enclosure, for so much is unrecorded, detailed investigation in a dozen or so widely separated counties suggests that it may quite possibly have amounted to 2 million acres, more or less, since the seventeenth century. If that is anywhere near the mark and is added to the acreage enclosed by Act of Parliament, the tentative estimate of 8–9 million acres of common in the 1690s, based on Gregory King's figures, does not sound incredible.

In tracing the history of commons in old woodland country, one usually finds that a more loosely organized form of society lies behind it than that of the clas-

Greenstead, one of the early out-parishes of Colchester

Colchester was a town still surrounded by extensive heaths in the eighteenth century. By 1066 Greenstead had its own church, but its population remained minute. Its boundary crossed Whitmore Heath and Wivenhoe Heath, and many other signs of former common land survive in such minor place names as Parson's Heath, Gallows Green, and Rovers Tye. 'Tye' or 'tey' is a frequent word in Essex for a small pastoral enclosure in waste land.

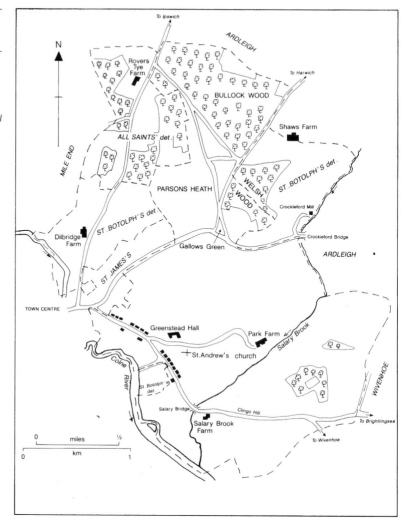

sic common-field village. The commons themselves were rarely if ever so closely integrated with the life of a single township or a single field-system. Where open fields existed, they formed only one element in the local economy; they were often grouped in small complexes around separate hamlets; and in many places there was no systematic township organization like that of the Midlands to enforce village customs. As a rule, hamlets of this kind were also interspersed with isolated farms often worked in severalty, and with outlying sub-manors which had themselves been reclaimed from the waste, and which sometimes remained embedded in woods and commons for centuries. In counties where extensive wooded tracts were placed under forest law after the Conquest, such as Staffordshire, Wiltshire, and Hampshire, there were obviously certain variations on these broad themes. These are the areas that have attracted most attention from historians, moreover, because royal forests had an official status and gave

rise to a corpus of formal records. Yet there were many predominantly old woodland counties, such as Surrey, where very little royal forest existed, save for a brief
phase in Henry II's reign; in some counties, such as Kent and Hertfordshire, there
was none at all; and even in more typical 'forest'-shires, like Staffordshire, much
old woodland or intermingled country lay outside forest law.

If it seems remarkable that relatively little attention has been paid to the old
woodland countrysides of England, unless they also happened to become royal
forests, the relative paucity of local documentation is no doubt often the main
reason. Yet their loosely organized and intermingled character is a crucial phenomenon in the history of common land. The further back we trace them, moreover, the more numerous and extensive they tend to become. They were certainly
not originally confined to the southern counties and the Welsh Marches, where
they are still plain to see. In many places in the north of England, and parts of the
Midland Plain itself, place names, jurisdictions, local sources, and topography
indicate common-field villages which are manifestly embedded in vestiges of
'old-coppice' country. Cranfield, Wootton, and the neighbouring places in Bedfordshire, for instance, are still surrounded by outlying farmsteads, moated sites,
green-hamlets, and minor medieval settlements bearing woodland names,
though little woodland remains today and much of Cranfield has been built over.

Characteristics of this kind are nevertheless more striking, as we should expect,
in the kind of intermingled country we find in areas such as Essex and Hertfordshire, Kent and Surrey, Cheshire and Staffordshire, the border-country of Hampshire and Berkshire, the Derbyshire Woodlands east of Derwent, and the *cœd mawr*
or 'great forest' of Selwood dividing Wiltshire and Somerset. Obviously these
regions are not identical, and ultimately they developed in different ways. Yet to a
discerning eye they still bear an underlying resemblance. It is no surprise to find
that in later centuries it was areas like these, and not the common-field parishes of
the Midland Plain, that became the classic home of new squatters' settlements, and
in many places the home of new crafts, new industries, and new ways of life.

Patterns of Piecemeal Enclosure

Some of the issues these regions raise are best illustrated by studying a single
county. Though none can be wholly typical, Kent exemplifies many of the features found in old woodland areas generally. By the seventeenth century it is clear
that the great tracts of common pasture characteristic of the Old English shire
had long given way, through persistent reclamation and enclosure, to a countryside of smaller and more scattered heaths, wastes, and woods. Yet it is equally
plain that the 6,500 acres or so enclosed by Act of Parliament formed but a fraction—perhaps, at a guess, no more than a fifteenth—of the common land surviving into the parliamentary era. Evidence of one kind or another for hundreds
of former commons may be found, though only now and then is there any direct
indication of their acreage. How was it, then, that these other wastelands of
Kent—on the Downs, on the Chart, and in the Weald—came to be enclosed?
There seem to have been eight dominant ways, apart from an Act, in the early
modern period and the nineteenth century, of which some had certainly been
customary in the medieval period too. None was peculiar to Kent, though their
relative importance varied widely from county to county.

First, in many places commons were enclosed through cases brought before Quarter Sessions, usually for the diversion of roads, but often also enabling local landowners to enlarge or create parks, as at Montreal in Sevenoaks in 1764. Second, much land was enclosed through manorial courts, probably far more than we know since court-rolls often fail to survive, though contemporaries sometimes commented on their decisions. Third, many commons disappeared by way of legal, or sometimes merely personal, agreement between neighbouring landowners, especially where they straddled parish boundaries or where commoners were few and in a weak position. Fourth, much piecemeal encroachment on the waste occurred, particularly though not solely by small owners and squatters. Sometimes—perhaps quite often—encroachments of this kind were sanctioned retrospectively by manorial courts on payment of a small fine or in some other way. Many squatters' cottages may be readily recognized in the field today, since they tend to have little or no land attached to them, and are often scattered about haphazardly in unenclosed areas, beside outlying greens, or on parish boundaries. Certain types of place name, moreover, such as those in 'row', 'lees', 'minnis', and 'common', usually refer to hamlets of this kind. Brabourne Lees, Old Wives Lees, Whitley Row, Stelling Minnis, Goathurst Common, and Hosey Common are typical examples. As population increased at an exceptional rate in Kent after 1750, the total acreage enclosed must eventually have been considerable, though individual encroachments were usually small and squatters' settlements rarely became as populous as those in Hertfordshire and Staffordshire.

Piecemeal enclosure for industrial purposes, as well as for the kind of suburban development mentioned earlier, frequently occasioned the extinction of common rights and was very extensive. Most of the dominant industries of Kent were heavily dependent on extractive processes: brick-making, ironworking, lime-burning, chalk-working, flint-mining, and digging for glass-sand and fuller's earth all had a long history behind them. Tile-making, potting, and quarrying went back in places at least to the eleventh century, and all experienced a major local revival in the nineteenth century. In many places, as in other parts of England, these industries originated on commons which later, whether by legal means or not, are found to be wholly enclosed. Some industries of this kind, and others such as charcoal-burning, also drew heavily on nearby woodland for fuel, and that occasioned further enclosure and conversion of common woods to managed coppice.

The erection of public buildings such as workhouses and hospitals, and the use of commons for military purposes, were both legally recognized reasons for enclosing land. Hertfordshire is particularly notable for the number of immense hospitals, asylums, and other institutions erected on the waste in the nineteenth century, usually by charities or authorities based in London. In the border country of Hampshire, Berkshire, and Surrey, the development of Aldershot, Camberley, Farnborough, Fleet, Sandhurst, Pirbright, and Bordon Camp entailed enclosure of the waste on a truly ferocious scale. There was no real equivalent to such extensive developments in Kent, but workhouses were often erected on commons, and during the Napoleonic Wars local barracks were built at many places, such as Gallows Common in Sevenoaks. A long tradition, moreover lies behind the use of areas like Blackheath, Coxheath, Pennenden Heath, and Barham Downs for military gatherings, reaching back at least to the sixteenth century, and in later times giving rise to permanent enclosure at places like Shorncliffe Camp near Hythe.

The heaths and wastes of the Surrey–Sussex–Hampshire borders covered many thousands of acres. They were the last of the 'great commons' to survive almost unaltered in southern England. 'As you reach the top of the hill which stands between Haslemere and Hindhead, a picture reveals itself which one would not be surprised to find five hundred miles away…. The sides of the hill are covered with heath and ferns … sheep are scattered over it here and there, a few foot-tracks wander up its side … and a small farm or cottage stands quite alone, almost at its foot.' (Louis Jennings, Field Paths and Green Lanes, *1878)*

One major reason for the attrition of commons arose from the creation of new farms, both in the early medieval period and again in the eighteenth and early nineteenth centuries. Along the Downs and the Chart in particular, scores of farms were thus newly taken in, either wholly or in part, from the upland waste. The marginal nature of those established after 1700 is often given away by their curious or amusing place names. Filchborough or Fitchborough, to the east of Wye, means 'polecats' borough'; Kettlebender in the same vicinity is a jocular word for a tinker; while names like Starvecrow, Starveall, Owls' Castle, Rats' Castle, Spiders' Castle, Birds' Kitchen, Monks-in-the-Hole, Heart's Delight, and the like, tell their own wry tale of local humour. Many farms of this kind are also significantly sited near parish boundaries, and often in surprisingly bleak or isolated spots reclaimed only by necessity. Little Betsoms on the Downs north of Westerham stands at 800 feet above sea level, exposed to every winter gale; one of several farms called Starvecrow lies buried among the woods and commons of Seal Chart, on the 650-foot contour; Kettlebender was likewise carved out of dense woods, near the border of Waltham ('forest *hām*') and Godmersham; across the same boundary, and close to the outlying lands of Chilham and Chartham—another significant early name, in *ceart* (wild common land)— Woodsdale was formed, after 1801, from Godmersham Common. Elsewhere in England there are frequent parallels to places of this kind, both in old woodland country as in Kent, and among the fells and moorlands of the northern counties. In the Lake District (Westmorland), for example, many farms in Langdale, and many beyond Loweswater, were created by the addition of 'intacks' from the fellside to earlier smallholdings or seasonal shielings in the medieval or early modern period.

A second broad conclusion to emerge from the Kentish evidence is that, by the late eighteenth century if not before, the *pattern* of surviving waste was quite different from that still found in many parts of England. There were no longer any vast semi-barren tracts like Hindhead and Bagshot Heath in west Surrey or Charnwood Forest in Leicestershire. Except perhaps among the Blean Woods north of Canterbury, there was nowhere like the great wooded heathlands of Ashdown Forest (Sussex), the New Forest (Hampshire), or Epping Forest (Essex), and nothing resembling the huge, bare 'blowing sands' which contemporaries noted in the Brecklands and Suffolk Sandlings. Save in the north-east corner of Kent and in Romney Marsh, commons were very numerous and notably diverse. But by 1750 few if any exceeded 1,000 acres and many covered less than 300—sometimes much less. Something of the same pattern of small-scale commons had often developed in old woodland country elsewhere: in Hertfordshire, for instance, in eastern Surrey, on the Chiltern Hills, among the Derbyshire Woodlands, and along the Hampshire–Berkshire borders. It was a regional characteristic that led observers like Arthur Young to underestimate the importance of commons in areas like these or even to ignore their existence. No one would guess from Young's *Report* to the Board of Agriculture on Hertfordshire that in the joint-parishes of Great and Little Munden, covering 5,556 acres together, there were no fewer than thirteen commons, six communal greens, and several other patches of waste at that time. John Boys of Betteshanger, near Deal, made much the same mistake in his *Report* on Kent.

The final conclusion to comment on here is that there were many places in Kent, as in similar woodland regions, where small commons of this kind tended to occur in chains or clusters, often linking up with one another to form more or less continuous though very irregular tracts of mingled heath and woodland. Occasionally these extended for several miles, but contemporaries rarely seem to have noticed them as a chain because they were usually split up between several parishes and were often known by unfamiliar local terms. In Kent, such words as 'chart', 'minnis', 'warren', 'hoath', 'leacon', 'lees', 'tye', 'forstal', 'scrubs', 'bushes', 'roughs', and 'roughetts' are remarkably frequent. Sometimes, too, old woodland terms like 'frith', 'shaw', 'weald', and 'hurst', and here and there 'moor', 'plain', and 'wold', were also used of common land. Words of this kind often expressed subtle differences in the origin, nature, or usage of the waste, and tended to develop their own local connotation. Many of them may be found elsewhere, of course, though not necessarily with the same sense, while other parts of England employed their own distinctive expressions.

Beyond the early modern period, the problem of assessing the extent of common land can only be solved by intensive local scrutiny. Important work has been done in this field by Jean Birrell in Staffordshire and by other scholars elsewhere, to whose publications the reader must be referred. At the time of Domesday it was estimated by the late Sir Clifford Darby that some 8½ million acres of land was arable, and that would imply that quite 70 per cent of the English countryside still lay beyond reach of the plough. Of that area, a small though vital proportion was meadow. But Domesday rarely records pasture as such, save in a handful of counties, and it is impossible to say how much of the remaining area was in any sense 'common', whether as woodland, moorland, fenland, marsh, or scrub.

By following up the kind of place names mentioned above in the light of other evidence, however, it is possible to glimpse an older and more extensive network of commons well beyond that of the early modern period. Ultimately, as for example on the borders of Surrey and Kent, of Hertfordshire and Middlesex, of Wiltshire and Somerset, or Derbyshire and Nottinghamshire, we can then sometimes see how the great woods and commons of the Old English period were gradually eaten away, under the multiple impact of rising population, the foundation of monastic houses, the enclosure of parks and chases, the establishment of new towns and the reclamation of new holdings, in the centuries following Domesday. But that long story of persistent erosion, of great importance in English history, is too detailed and complex to relate here.

People of the Commons

The common lands of England bring us into touch with a profoundly human world: with a world, in the main, of homely things and humble people: with a pastoral, woodland, or moorland economy of homespun crafts and small-scale farming: of husbandmen, cottagers, woodmen, quarrymen, potters, and squatters. In industrial areas there was often more to it than that, because the varied resources of the commons sometimes gave birth to specialized manufactures demanding substantial capital, such as the glass industry of the Stourbridge area and the stoneware of Woodville and Moira on the Derbyshire/Leicestershire border. Wherever there are resources to exploit, moreover, there will always be those who wish to take them for themselves and expropriate other people. That was often how communal quarries came to be enclosed, and exploited by private entrepreneurs in later centuries.

It is with the humbler occupations of the common waste, however, that we are specially concerned here. Nothing has more surprised me in studying the history of commons than the great range of local crafts to which they gave birth. Some of these, such as potting, tile-making, and quarrying have long been familiar, and have given rise to an extensive literature. But most of the activities in question are scantily recorded, forgotten perhaps even in the locality today, and discovered only through chance references. Some of them certainly go back to the medieval period, and gave rise to local surnames; but many do not seem to have developed until early modern times or the beginning of the nineteenth century, and all too often we cannot trace how or when they originated.

Quite a few of the crafts, such as the making of hones (whetstones for sharpening cutting tools) at High Halden in the Weald of Kent and Penselwood on the Wiltshire–Somerset border, were by their nature somewhat obscure, depending on very localized raw materials and employing small numbers of people. Sometimes they were migrant, seasonal, or part-time activities, such as brick-making, and the full numbers engaged in them cannot be precisely ascertained. Even in the most exhaustive Victorian directories it is rare to find references to such people as charcoal-burners, crook-makers, spindlers, spooners, truggers, cloggers, mat-weavers, hassock-makers, ash-burners, peg-makers, shovel-makers, hoopers, hurdlers, mop-makers, broom-squires, and besom-makers. Those that are usually mentioned, such as brick-makers, basket-makers, and potters, are often under-recorded because they were frequently combined with other occupations, particularly farming or innkeeping.

In the main, however, it is clear that the craftsmen of the commons were independent people rather than employees, usually operating on a small scale, and in some trades working together as families. It is equally clear that their products, though seemingly humble to us today, were invariably essential to contemporaries, sometimes on a large scale until late in the nineteenth century. The domestic demand for common red earthenware, for example, was enormous, and it was still met very largely from local potters rather than from Staffordshire. The necessity for some quite unexpected or forgotten products, such as wooden shovels, remained widespread both on farms and in towns and cities into the present century. Every harvester, moreover, still needed his own hone, as well as his scythe or sickle, and carried it around with him as he moved from farm to farm.

In some cases the inspiration behind industries of this kind stemmed from old market towns like Berkhamsted, Chesham, Croydon, Wymondham, Brandon, and Castle Donington. Such places as these, with extensive common wastes of their own or ready access to them near at hand, often found themselves in an advantageous position, especially if they were unincorporated places, untrammelled by overmuch control. In the eighteenth century the Suffolk river-port of

Brandon, for instance, became the principal centre of gun-flint manufacture, when the prehistoric flint-mines on Brandon Common, which had long been exploited for local building materials, came to be reworked for a new purpose. Wymondham in Norfolk and King's Cliffe in Northamptonshire were among several places once widely known for their turnery or 'treenware', again drawing on nearby commons and copses. In Hertfordshire, Berkhamsted, with its great open Frith on the Chilterns, was also known for its turners, and especially for its wooden shovel-makers. In Buckinghamshire, Chesham, in a heavily-wooded parish of 12,700 acres, and in Wiltshire, Aldbourne, the historic centre of Aldbourne Chase, were both well known for their chair-makers. One tiny firm of ten men at Aldbourne was still producing 15,000 chairs annually in the early years of the present century, and delivering them by carrier's cart throughout Wiltshire and Berkshire. In Leicestershire the small market town of Castle Donington became a notable centre of basket-makers, probably drawing on the parish's withy-beds by the Trent and perhaps on its coppice-woods as well.

In many places, however, the initiative behind such occupations stemmed directly from the people of the commons themselves. The cottagers and squatters who had somehow established a toehold among the woods, or on parish boundaries, outlying greens, and tiny patches of roadside waste, were driven by sheer necessity to search for some means of livelihood. The people bred among woods were often said to be more stubborn and uncivil than those of champion districts, but they were also at times perhaps more inventive. So we find them drawing on almost any homely resource that lay to hand, such as brickearth, potters' clay, fire-clay, glass-sand, wild herbs and shrubs, or mere hedge-loppings: things almost worthless themselves until someone noticed them, and recognized their potential use. At Coleshill, for example, a detached part of Hertfordshire divided between the two woodland parishes of Amersham and Beaconsfield in Buckinghamshire, they discovered potters' earth and—as in scores of other places in England—began to make domestic earthenware. In Kent a long line of potters' centres in the early nineteenth century extended eastwards from Deptford Common and Norwood as far as the Darent valley. On the Hampshire/Surrey border, George Sturt in his biography of his grandfather, *William Smith: Potter and Farmer*, gives us a fascinating picture of another hamlet of this kind at Cove Common, where at one time there were no fewer than thirteen potteries.

Probably most numerous of all were the commoners' settlements dependent on a wide range of woodland crafts. About 3 miles from Coleshill, the inhabitants of Seer Green, a detached hamlet of the parish of Farnham Royal on the boundary of Beaconsfield and Chalfont St Giles (Buckinghamshire), were engaged in hurdle-making. Elsewhere, as in the neighbourhood of Croydon (Surrey) and Bromley (Kent), household brooms were made with a simple machine from the wild broom-bush and the fast-growing saplings of ash-trees for their handles. Our modern word 'broom' is derived from the name of the bush, and it appears in many English place names like that of Bromley ('broom clearing'). In other places, scrubbing-brushes for butchers and housewives were made from the hard, spiky shoots of the butcher's broom, a characteristic shrub of south-country heathlands. Bobbin-spindles and butchers' skewers were turned in great numbers on a portable pole-lathe, utilizing the hard wood of the spindle-tree. The tough-springing shoots of the wild dogwood, the fast-growing shoots of hazel and alder, the rushes, reeds, and sedge of ponds and stream-banks, as well

Hazel-hoop making

Making hazel-hoops was a major coppice industry in Sussex until quite recently and often had common-land origins. The demand for hoops for binding empty or 'slack' barrels was enormous: cheese, cement, apples, sugar, crockery, and glass were among the many articles thus conveyed. Where hoops were bent in the woods while green, as here, the hoopers erected their own cone-shaped cabins of rods and turf during the season, and thatched their sheds with shavings.

A chair-bodger's hovel in the Chiltern woods

Note the pole for the portable lathe on which legs were turned before being stacked to season. This ancient craft survived into the present century in this area, and ultimately gave rise to the furniture industry of High Wycombe; but migrant chair-makers were also found in many other parts of England and in Scotland and Wales. Beech was commonly used for the legs of hoop-back or Windsor-type chairs, elmwood for the seats, and ash for those parts that required bending.

as many kinds of willow, were all put to use in weaving the countless specialized types of basket. In one village in Wiltshire, and probably others elsewhere in England, even the stray locks of wool left by grazing sheep on briars and thornbushes were not neglected. They were carefully gathered by cottage-women on the Downs, spun at home into a thick, coarse fibre, and then made into household mops for sale in the area, particularly to local dairy farms.

Even as the commons dwindled, moreover, their very attrition at times forced into action the native spirit of ingenuity. Perhaps many of us can still recall the gypsy pegmakers of the 1930s, who manufactured clothes-pegs in tens of thousands, year after year, from the hedge-loppings of the roadside waste in old woodland areas like Hertfordshire and west Kent. Every individual peg was rapidly, deftly shaped by hand in two pieces with a well-honed blade, and then bound firmly together with a narrow strip of tin. Sold by itinerant gypsy women from door to door, usually in skeins of a dozen, they were widely bought by housewives in local market towns; for even if they were no great friends to the Romanies, they knew their pegs were cheaper than any they could buy in the shops and would last for years.

Humble products, humble materials, and humble craftsmen and craftswomen, it may well be said. Yet it was often from obscure origins like these that new settlements sprang up, and among them that new customs and new traditions began to

develop. Both Coleshill and Seer Green, for example, lay in one of the early heart-
lands of the Society of Friends, not far from the well-known meeting-house at
Jordans. In the Derbyshire Woodlands, the new industrial hamlets of the eight-
eenth and nineteenth centuries proved fertile seedbeds of the early Methodist
Movement. That is a story that we find repeated in many parts of England, from
Kent and Sussex, through the Midland counties, to Yorkshire and Cumberland.

From the works of writers like George Borrow, George Sturt, Edwin Grey, and
others, moreover, it is clear that the people of the commons were often linked by
a kind of bush-telegraph, connecting wayfaring people of every kind together:
reputable traders, drovers, shepherds, and carriers among them; travelling tin-
kers and potters on their annual beat; chapmen, hawkers, and cheapjacks; hig-
glers, horse-dealers, and gypsies; and, in many places, smugglers, bandits,
brigands, and highwaymen too. Sometimes they met one another in drovers'
inns or hedge-alehouses, sometimes on racecourses or fairgrounds, sometimes
in one of the great London trading inns, frequently in the market places of country towns, and perhaps more often than we know in such fugitive travellers'
haunts as 'Mumpers' Dingle' in Staffordshire, where George Borrow tells us he
met Isopel Berners and fought the Flaming Tinman.

There is a world of interest awaiting discovery in the history of our commons.
The observant reader will have noted already some of the hints and clues to bear
in mind in searching for these areas where they have now been enclosed. Those
with an intimate knowledge of their native heath and its past may wish to pursue
the matter further, and much remains to be done in tracing the local occupations
of the commoners. Otherwise, except where commons still survive at least in
name, the task of recognition is not one to be lightly undertaken. It entails many
complex technical questions of history, topography, and onomastics; it is also
immensely time-consuming, and replete with pitfalls for the unwary. What is
really needed is a thorough scholarly guide or handbook to the subject as a whole.
Nevertheless, it is open to any who will to follow it further in the academic and literary works cited in the bibliography to this chapter. I suspect that some readers,
blessed with the gift of a noticing eye, may come to share my own enthusiasm.

Note: The Royal Commission's figures of commons in the 1950s are conveniently summarized, county by county, in W. G. Hoskins and L. D. Stamp, *The
Common Lands of England and Wales* (1963). The authors were members of the
commission. They emphasize (p. xvi) that their information, 'based largely on
official sources and supplemented by personal inquiries by one of us (L. D. S.), is
given in good faith; but it is not to be regarded as indicating any official opinion,
least of all any legal opinion, about the status of any particular piece of land'. The
figures for the 1950s, cited in the present chapter, are based on those in the same
book, and hence the same qualifications apply. Under the Commons Registra-
tion Act of 1965, any land which was claimed to be a common or a village green
had to be registered by a given date. In the event, the number of claims so greatly
exceeded expectation that the Commons Commissioners set up under the Act
were still at work on the problem in the 1990s. For example, 1,009 claims related
to Dartmoor alone. In this chapter, however, we are not concerned with the current legal position of any common or green, only with the historic situation.

10 Frontier Valleys

Charles Phythian-Adams

Boundary Rivers and Cultural Frontiers

Historic landscapes represent much more than fossilized expressions of past human activity. Following the establishment on the ground of enduring man-made features, as direct reflections of particular cultural needs, the patterns so assembled may themselves help broadly to perpetuate the very functions which they were originally designed to serve. The specialized kind of landscape which is to be discussed here, which has not so far received the attention that it deserves, may furnish a few indicators of some of the ways in which entire countrysides might be read in such cultural terms. To interpret these, however, we need to look at landscape dispositions extensively by including, for example, a generalized view of urban as well as rural settlement patterns.

Two widely differing contemporary comments illustrate immediately the extent to which the 'silence' of the landscape evidence may be broken to reveal how, behind what we see, there may lurk more than we think. Here, first, are the words of a successful, self-made merchant tailor writing retrospectively from London, in 1606 towards the end of his life, to the leading inhabitants of his home-town, Market Harborough, on the southern edge of Leicestershire: 'Remembringe … when I was first fedd in that soile … and remembring alsoe and considering that with my staff I came over that Welland, that I came out of my country and from my father's house with my cuppe empty … I have thought it my duty therefore yet once again to remember that place where I was bredd …' There can be little doubt that, probably like others from the same place, Robert Smyth (who went on to found the Grammar School at Harborough) had for-merly been highly conscious of the facts that his 'country' effectively comprised the townships of the Welland valley, and that beyond them to the south lay an unknown world.

A second observation takes us over the eastern end of the frontier between England and Scotland some 230 years after the final pacification of the Border and the Union of the Crowns, and more than 125 years after parliamentary union. Writing in 1838 from Kelso, its Church of Scotland minster was able to report that although living within 5 miles of England: 'It would indeed seem, that, in proportion as the two countries approach their respective confines, the Scotch and Anglican [*sic*] tongues, instead of gradually losing each its distinctive character …, assume each its harshest and most intractable form; as if for the purpose of keeping their respective *marches* clear and distinct.' The marches in question were specifically 'the east marches' where Scot and Northumbrian faced each other from the 'north' and 'south' banks respectively of the River Tweed.

Albeit representing different spatial levels, the one provincial the other national, the valleys of the Welland and the lower Tweed are to be distinguished from most other valleys in that they share one unusual feature: they are defined to differing degrees in cultural terms by major boundary rivers. As such they are quite unlike those valleys, many of them minor tributary valleys, which function simply as semi-contained arenas for settlement. Most 'frontier' valleys, as it seems best to call them, are (or were) invisibly divided lengthways by a significant boundary running down the central river-bed. These valleys tend to be of considerable extent and to play their role as far as the sea. They therefore reflect a continuing process of cultural definition. The questions that consequently must be asked of such valleys are whether they shared certain particular arrangements on the ground which served to perpetuate a sense of 'frontier' in the minds of both inhabitant and visitor and, if so, how did these features come about?

Since inquiry into this subject is very much at a preliminary stage, in the space available it will thus be necessary to limit discussion to some specific examples of fully developed frontier landscapes of this type (there are of course other types). It is essential to begin then by distinguishing those *major* boundary rivers, at both provincial and national levels, which are open to *extended* chronological inspection from those whose previous boundary functions have been either lost completely or diminished for some time. In these latter cases, landscapes either will have failed to last sufficiently long in this guise to have evolved frontier characteristics, or will have been overlaid by new features perhaps stemming directly from the actual loss of major boundary status and the breakdown of the frontier in question. In the present context, then, we must be sure that we are isolating, and then concentrating on, the unambiguous examples.

Only fifteen or so rivers have represented historic major boundary lines at one time or another, and so might be regarded as potential candidates for examination. They fall into three broad categories. First are those rivers which appear to have once determined boundaries that have long since become extinct like the Ribble in Lancashire or the Wye in Herefordshire. A second category of boundary river comprises those where a previously important dividing line has been reduced to local governmental significance or has always represented that lesser boundary status. At first sight, the former might be thought to include the Thames—as once part of a fluctuating frontier zone between Mercia and Wessex—but because the river was navigable at least until the seventeenth century, from the capital as far as Henley, it may more accurately be seen to have operated in fact as what Dr John Blair has rightly called a 'cultural corridor'. In the present context the grouping of counties coinciding with its drainage basin may thus be said to have represented a 'cultural province' (see insert). A better example than the Thames, therefore, would be the Mersey (Old English for 'boundary river') which, originally representing probably the north-west frontier of Mercia, was

River-drainage patterns
and groupings of
counties: 'cultural
provinces' of England

subsequently relegated to the level of a county boundary. Formerly subdividing earlier political entities, finally, were the Tamar between Devon and Cornwall and the Waveney between Norfolk and Suffolk.

By the time of the Norman Conquest at the latest, and often by the beginning of the tenth century, the valleys belonging to the categories so far mentioned had either forfeited their political frontier status or had been demoted in significance. Those that had not, like the Ribble, were soon to lose their roles thereafter. None of these boundary lines, therefore, would be able to claim continuing political, ecclesiastical, and cultural significances from the conditions of the tenth century through to the later twelfth; this being the crucial period for the establishment of more or less permanent settlement patterns whether rural or urban.

By contrast, the third and last category of river—those valleys that could boast active frontier functions into later times—were either perpetuated in this capacity, being conspicuously renewed as such in the tenth century, or were first elevated separately to frontier status in the tenth or eleventh centuries and so continued thereafter. In the earlier cases there are three 'provincial' frontiers within England dividing clearly definable *groupings* of counties or 'cultural provinces': the Tees, the Welland, and the Suffolk Stour (see Box feature). The newly established boundaries, by contrast, comprise the two international frontiers with Scotland, along the valleys of the Liddel and the Tweed respectively. For these reasons it seems best in what follows to concentrate first on the much longer-established frontier valleys of the provincial interior before seeking to compare and contrast them with the patterns evolved somewhat later on the Scottish border.

'Cultural Provinces'

When seeking an overall view of provincial cultural diversity across pre-modern England, one optional starting point may be furnished by the general coincidence between *groupings* of counties and the major river basins or drainage systems. The area of each grouping tends to have had particular political origins in the Anglo-Saxon period and subsequently to have had its early small territories reorganized into different shires at broadly the same stage. There are many suggestive overlaps between such generalized cultural areas and the underlying physical context, but the 'fit' is too close to have been accidental. The perimeter of each grouping on the map, moreover, is defined culturally (by county boundaries), not geographically. In these contexts, both towns and the greatest rural densities of settlement are apt to cluster in neighbouring or converging valleys and, most particularly, on or towards major navigable rivers. Each 'cultural province'—as such an informal entity might be called for want of a more appropriate description—thus looks outwards eventually and in the first instance at least, for purposes of trade, foreign immigration, and the traffic in ideas and material culture, to very specific maritime neighbours, either in the western parts of the British archipelago or along the continental seaboard. Inland, cultural provinces are most commonly separated either by broad watershed *zones* (not boundary lines) of pastoral countryside characterized by dispersed or lightly nucleated settlement, or—as is discussed in this chapter—by a small number of distinctive valley lines that are independent of the major river systems. It should

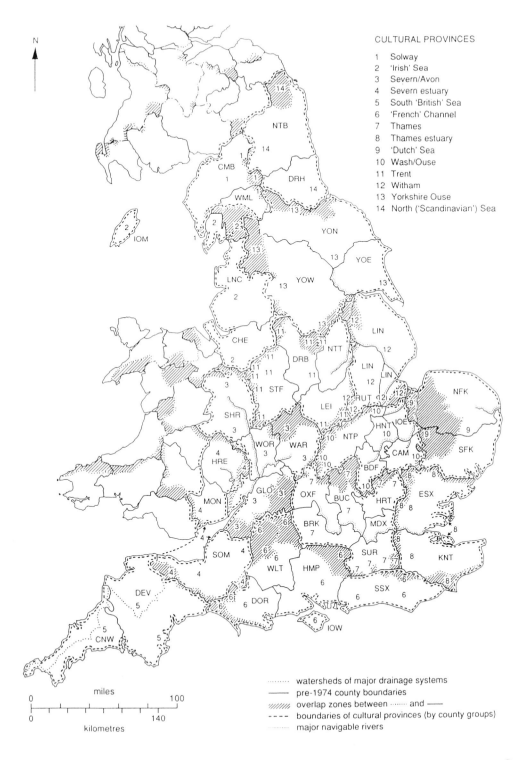

N

CULTURAL PROVINCES

1 Solway
2 'Irish' Sea
3 Severn/Avon
4 Severn estuary
5 South 'British' Sea
6 'French' Channel
7 Thames
8 Thames estuary
9 'Dutch' Sea
10 Wash/Ouse
11 Trent
12 Witham
13 Yorkshire Ouse
14 North ('Scandinavian') Sea

miles
0   100

0   140
kilometres

.... watersheds of major drainage systems
———— pre-1974 county boundaries
////// overlap zones between .... and ————
- - - - boundaries of cultural provinces (by county groups)
.... major navigable rivers

be stressed, finally, that each of these generalized settings thus comprehends a range of *pays*, some of which will overlap the outer edges. Within these 'bounds' a spectrum of local societies, with their own identities and subcultures, is thus to be expected, all of them linked by a regionalized network of towns some of which, however, will of course also look outwards. Such a setting is therefore to be regarded as no more than an overall context within which its component societies will have more in common culturally with each other than with societies in neighbouring provinces beyond any 'edge' or overlap zone.

Provincial Frontier Valleys

Interestingly, and even perhaps partly because of their long-standing boundary role, although the Tees, the Welland, and the Stour lie on the side of England least associated with the survival of British river names, all three of their names may well be Celtic or even pre-Celtic in origin. In each case too the eleventh-century county boundaries running along them were evidently superimposed on much older lines of division. The Tees, for example, is widely accepted as having demarcated the original Northumbrian, and (to judge from the Celtic derivations of their names) very probably earlier, kingdoms of Bernicia and Deira. Between the seventh and ninth centuries, moreover, the Tees seems to have represented the southern edge of the superseded diocese of Hexham. Somewhat later the boundary between the dioceses of York and Durham too was defined by the river, which further broadly delimited the borders of two small, probably archaic, 'shire' units: to its north, Gainfordshire in County Durham, the existence of which is first indicated in the ninth century; to the south, Allertonshire which reached towards and beyond Northallerton.

Similar dispositions are apparent beside the other two rivers. On the Welland (the Leicestershire stretch of which may well have represented the southern frontier line of what Professor Nicholas Brooks has dubbed 'original Mercia'), Rutland could represent another of these ancient small units, and later, in the ninth century, may even have comprised dower land for the queens of an expanded Mercia. Quite possibly it was bordered to the south by a people mentioned in the seventh-century Tribal Hidage, the *Widerigga*, who gave their name to Wittering. A local tradition, twice mentioned by the twelfth-century Geoffrey Gaimar, moreover, associates early Rutland with the erstwhile kingdom of Lindsey based on Lincoln to the north. Such early connections, together with subsequent events, probably help to explain why Stamford could be later carved out of what was obviously the original territory of Rutland and then transferred to the ambit of Lincolnshire north of Welland. A similar modest territory of probable antiquity was that dependent on *Bedericsworth* (later known as Bury St Edmunds (Suffolk)) which likewise was used as royal dower later in the Anglo-Saxon period when its limits had probably been extended to the line of the river Stour. There can be little doubt that it was by this river that both the kingdoms and the dioceses of the East Angles and the East Saxons were divided from each other: the successive dioceses of Dunwich, Elmham, Thetford, and Norwich for the former, and the diocese of London for the latter. It is therefore significant that in earlier times still the parish of Bulmer, immediately west of Sudbury on the Essex bank, contained the heathen site of *Thunreslau*, Thunor's tumulus, which was also a hundredal meeting place. The East Saxons held this god in particular veneration, while such names of ancient gods are often found in close association with territorial frontiers. Thunor has no such associations in East Anglia.

Overlying, yet still broadly perpetuating, all these divisions after the Norman Conquest was the cultural legacy of later Anglo-Scandinavian England. The Tees, for example, then represented the furthest extent of West Saxon shiring arrangements, the north-eastern limit for payments of Peter's Pence to Rome—itself a measure of the expanding territorial claims of the kings of Wessex—and an overlap zone with areas to the north and west that were rich in Celtic customary survivals in renders, services, and the system of justice. It was probably no accident, moreover, that up to 200 years before, the lower Tees valley had been dominated visually from the south by the conical shape of the northernmost monument to Scandinavian paganism, originally called 'Odin's hill' (Roseberry Topping on the northern flank of the Moors). Scandinavianized place-name forms in -bekkr or -by that are common in the territory of the Scandinavian kingdom of York, indeed, spread just north over the river (but not much further). The same is true of that typically Scandinavian funerary monument, the hogback gravestone, the distribution of which, in churchyard contexts east of the Pennines, is concentrated in a region stretching north and north-eastwards of York. North of the significant group in the Tees valley itself, however, there is but one other survivor in all the rest of Northumbria. Significantly the northern flank itself also contained the only wapentake (of Sadberge) north of Yorkshire.

South of Tees by 1100, and over a large tract of eastern England subdivided between three cultural provinces, therefore, lay a heavily Scandinavianized world of local administration, the southern boundary of which north of the Wash was marked by the River Welland. In the tenth century, west of the fens, that river almost certainly comprised the territorial limit of the Danish confederation of the Five Boroughs of Leicester, Nottingham, Derby, Lincoln, and Stamford (Rutland)—and, in those cases where Seven Boroughs were in question, perhaps also included York. In 894, indeed, the West Saxons were in negotiation over the Rutland area with the Danes of York who seem to have 'plundered' and possibly acquired 'large territories in the kingdom of the Mercians, on the western side of the place called Stamford. That is to say between the streams of the Welland and the thickets of the wood called Kesteven.' Between Tees and Welland, all the counties were subsequently divided administratively into wapentakes, while for purposes of tax assessment, the land was subdivided into smaller units, called carucates (each comprising eight yet smaller units, called 'bovates' in Latin, or 'oxgangs' in the vernacular) often calculated according to a duodecimal system of reckoning. South and south-west of Welland, by contrast, was the region of the Anglo-Saxon hundred, the hide (and its subdivision, the virgate), and a decimalized system of tax-reckoning. In turn, the Welland valley as a whole marks the decisive break nationally in the concentration of Scandinavian or hybrid Old English and Scandinavian place names. South-east of Welland, they effectively disappear.

Beyond the river-systems of the Wash at the same period, East Anglia was something of a cultural hybrid as a whole. Here by 1066 the Stour demarcated the north-easternmost edge of hidated country and the West Saxon hundred which together distinguished the wider Thames drainage basin. North of that lay the uniquely East Anglian institution, the leet (a small collectivity of vills answerable for the geld) which might reach back to the days of Guthrum's Danish kingdom

The valley of the Stour

of East Anglia—or even further. It is also noticeable that outside the broad areas so far indicated where Scandinavian place names or hybrids are found, the only other marked distribution in *eastern* England lies scattered, albeit thinly, across this very same territory. Given East Anglia's continuing continental links, moreover, the cultural distinctiveness of the region's late eleventh-and early twelfth-century round, flint, church towers (with minimal overlap into Essex) is also to be remarked. Once again the Stour seems to represent a broad southerly limit to a cultural zone which in this case—thanks to the work of Dr Stephen Heywood—may now be seen to owe its architectural inspiration to the European littorals not only of the Baltic but also of the North Sea.

Frontier Patterns on the Ground

What characteristics did these valleys share? Comparison throws up a surprising number of similarities, in terms of both situation and human occupation, which may begin to suggest how certain consistent patterns could have contributed to their continuing frontier roles whether in political or cultural terms. The underlying factor, of course, is physical. All of these valleys reach eastwards from significant, broadly north/south, watersheds. Westwards beyond the source of each river, therefore, lay territory focused in an opposite or other direction towards a different maritime edge; here lay territory which in early times was also occupied by an ethnically contrasted people, and this fact was later reflected as such in a separate diocesan arrangement. Each valley, therefore, butts onto one cultural province to its west whilst simultaneously demarcating the edges of two others to its north and south.

The Welland valley

These eastward-facing valleys were dominated by single major rivers with no tributaries large enough to rival them in size; they pursued a linear course; and they extended from the broadly north–south watershed barriers of middle or eastern England to the sea: the Tees for about 100 miles; the Welland for more than 80; and the Stour for about half that distance. Although only three valleys are in question, at their fullest extent the areas to be discussed comprehend some 2,000 square miles and the peripheries of eight pre-1974 counties. These valleys thus comprise major punctuation marks on the political and cultural map of eastern England. They do so because in slicing across the eastern flanks of England, the rivers tend to pursue courses that are in marked directional contrast to the wider drainage basins or river systems to their norths and souths.

In terms of its easterly course, it is the gentle valley of the Stour which is the least varied of the three under discussion (Map 1). The river rises over the boundary in Cambridgeshire; cuts south-eastwards across the western tip of Suffolk thus leaving the watershed-zone settlement of Haverhill on its west; and only meets the county boundary of Essex some 3 miles east of the junction between the boundaries of Cambridgeshire and Essex/Suffolk at the 'Stour mere' or 'pool' (Sturmer in Essex) which still covered some 20 acres in 1618. Thereafter, although from around Nayland the countryside so beloved by John Constable begins to open out, especially north of the river whence it is joined by the two tributaries of the Box and the Brett, the terrain remains relatively homogeneous until the estuary itself. The character of this, and the other estuaries, however, lies outside the scope of the inland focus of this discussion.

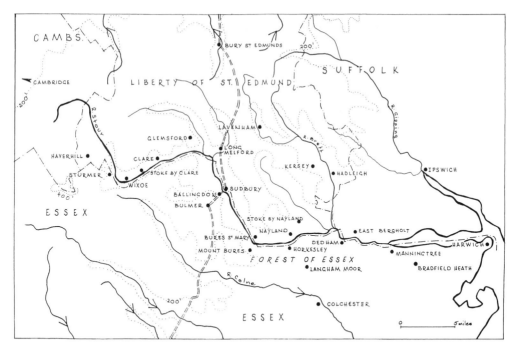

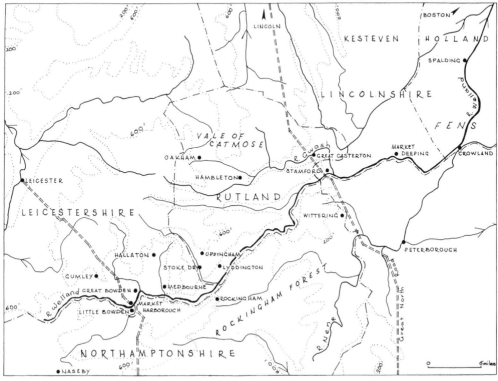

By contrast, the Welland defines both the northern limits of Northampton-shire as far as Crowland, and the southern boundaries in turn of Leicestershire, Rutland, and Lincolnshire again as far as Crowland (Map 2). Effectively from Stamford, but from east of the Deepings at the latest, however, all sense of a broad, physically identifiable valley gives way to fenland across which the water-course simply cleaves its way, sometimes with protective embankments to that point where, centuries ago, the line of the Welland and an earlier digression of the Northamptonshire Nene momentarily converged at Crowland's triple bridge. Thereafter the Welland pursued its very different course towards its original estuary on the Wash where Spalding acted as a river-port. Indeed, despite the fact that reclamation has pushed the coastline further to the north-east, that town was handling vessels of up to 120 tons even at the beginning of this century. It was the old course of the Nene, however, not the Welland, that now demarcated the southern boundary of Lincolnshire Holland from Crowland as far as the sea. As with the estuaries, therefore, the general characteristics of frontier valleys are not properly applicable to fenland areas.

The valley of the Tees is interrupted in a different way (Map 3). The distinc-tively closed nature of Teesdale, which begins at Cross Fell on the old Cumber-land border, only opens out in the vicinity of Barnard Castle some 30 miles from the source of the river. An approximately 10-mile gap between the mouths of the Pennine Dales and the North York Moors then makes possible uninterrupted north–south communications through Piercebridge or Darlington. But east-wards again, the moorland massif intervenes as far as the sea to hedge in the lower reaches of the Tees and its tributaries before the river reaches what once was an estuarine landscape of salt marsh.

The Tees valley

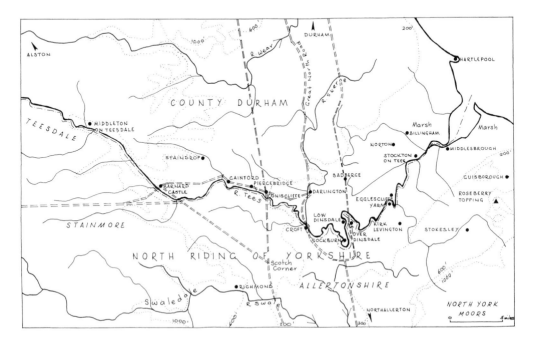

Mention of the Moors brings us to the last preliminary point about the settings for this discussion. If Teesdale is obviously self-contained, the landscape of the northern edge of the moors betrays a characteristic which is shared by the southern flanks of both of the other two valleys. In each case, although the watershed itself is somewhat further south still, the immediate valley landscape is defined on its southern side by a pronounced scarp that separates the valley floor from the watershed zone. In each case, by contrast, the northern flank is a more gentle dip slope and therefore more extensive in the opportunities it provides for settlement and cultivation. In earlier times, the southern scarps of all the valleys, moreover, merged into what had once been rough pastoral country of different kinds. Instead of moorland as in Yorkshire, the top of the south scarp of the Welland valley, for instance, was characterized by woodland. Here lay the medieval royal forest of Rockingham and other scattered areas of tree cover, extending almost as far as Stamford in the east and Market Harborough in the west, and thence through sparsely settled wold countryside south-westwards along the Northamptonshire uplands to Naseby and beyond. In like wise the watershed landscape between the Suffolk Stour and the Essex Colne betrays all the signs of erstwhile heathland: Bradfield Heath, Dedham Heath, Ardleigh Heath, Langham Moor, Boxted Heath, and so on. These marked the northerly skirts of the early medieval royal forest of Essex which, because its extent coincided with the county area, was also bounded by the Stour.

With the major physical barrier along much or all of the south sides of each of these valleys, therefore, the dominance of the generally more ample northern flanks is immediately evident. That reality and the seeds of its establishment in terms of settlement are both apparent early. Roman forts or small towns, for example, controlled road crossings of all three rivers from the north bank: at Piercebridge on the Tees, at Medbourne and Great Casterton on the Welland, and at Long Melford and possibly Wixoe on the Stour. A related function was later performed by important royal townships which acted as administrative centres and meeting points in the Anglo-Saxon period. The upper reaches of the Welland valley were dominated originally from Gumley, some distance to the north of the river and half a dozen miles or so west of the later pre-Conquest royal centre of Great Bowden. In the reigns of Æthelbald and Offa during the years of the Mercian supremacy, assemblies were held at Gumley in 749, 772, and probably 779. Nearer Stamford, and up the tributary of the Gwash, the late Anglo-Saxon queens probably had a residence at Hambleton in Rutland. In the case of the Stour, it is possible that Edmund, the last native king of East Anglia, was consecrated in 856 on a royal estate at Bures St Mary on the north side of the river, and that Guthrum, the Danish king of East Anglia from 879 to 890, was buried in a similarly distinguished site in the latter year at Hadleigh up the Suffolk tributary of the Brett. On the Tees, the former Roman crossing point at Piercebridge was very possibly later controlled from an outlying part of a territory based on Darlington, known as Coniscliffe (a Scandinavianized form of the Old English 'King's cliff') and where in 778, perhaps reflecting a royal administrative status, was killed a high reeve of Northumbria. Significantly, the estate also contained Carlbury, a 'fortified place of the *ceorls*' who may have serviced the royal lands. It is worth adding that other names in *burh* are also found on north banks: Sockburn (originally -bury) in a loop of the Tees, and Sudbury and the lost 'Erbury' (in Stoke-by-Clare) both on the Stour.

This pattern was reinforced by the early siting of significant churches with a frequency that was similarly not paralleled on the south banks. The earliest candidate is probably the British church (*ecles*) that must have given its name to Egglescliffe lower down the Tees from Coniscliffe. To judge from sculptural remains as well as documentary references, before the Viking period a number of Anglian monasteries, churches or cemeteries were also established at a series of central-place sites on the Tees, its tributaries, or the estuary, like Hartlepool, Hart, Billingham,

The Saxo-Norman overlap crossing of the major church of Norton, within whose parish Stockton-on-Tees grew up

Aycliffe, or Stainthorp, let alone Sockburn where a bishop of Lindisfarne was consecrated in 780. In 1083, moreover, when the Cuthbert community at Durham, which included the successors of those who had acted as custodians of the saint's coffin on its travels, was replaced by Benedictine monks, two of the places where the former were relocated were said to have been Darlington (a significant place which had been licensed by King Ethelred himself as a gift to St Cuthbert *c*.1003 and Norton. The latter is of particular interest as a site that earlier had attracted one of the largest pagan cemeteries in the north; and as a place which later boasted both a huge minster-like church, with central tower dating architecturally from the Anglo-Saxon/Norman overlap period of building, and an importance that is to be inferred from its status as the mother church of Stockton. Indeed, it might well seem that before the positions were reversed, and Norton relapsed into being merely the 'north *tun*' of Stockton, Norton could even have been known as 'Stoke', with Stockton thus being named in relation to it rather than vice versa.

Such names in Stoke (Old English *stoc*) often indicate significant religious sites and sometimes may hint consequently at the presence of important residences nearby. We know, for example, that in the tenth and eleventh centuries there was an ancestral mausoleum for the family of one of the East Anglian ealdormen at Stoke-by-Nayland which cannot, therefore, have been situated too far distant— at least originally—from an estate centre. Further up the River Stour at Stoke-by-Clare, again on the Suffolk side, was another secular minster in the mid-eleventh century near to which there also lived a son of a local notable in his own tower. In Rutland, up a tributary of the Welland, finally, lies Stoke Dry, which was tenurially part of the parent estate of Lyddington that belonged to the Bishop of Lincoln in Domesday, but ecclesiastically independent subsequently. Lyddington later attracted an episcopal hunting lodge and thereafter a late medieval episcopal palace. Given the exceptionally elaborate Romanesque carving on the shafts of the chancel arch at Stoke Dry and its unusually spacious original chancel, however, it may well be wondered whether it was Stoke that had comprised the original site of the bishop's residence because, perhaps, of some yet earlier ecclesiastical association which is still echoed in its name.

It may be claimed then that all three north banks and their hinterlands were characterized possibly continuously even from Roman times by a chequered line of variously emergent royal, ecclesiastical, and lordly estates, most of them with centres sited away from, but nevertheless in easy reach of, crossing points over each river (which, in the case of twelfth-century Stockton-on-Tees, was by ferry). The seeds were now therefore sown for more enduring and, in terms of influence, more pervasively important developments still.

The increasing dominance of the three north banks was underpinned by the fact that the rivers in question represented rather more than mere divisions between counties. In all these cases, significant stretches of the north sides also comprised the southern edges of wider territories of exceptionally high privilege, whether royal or franchisal. The precise point at which the various early 'shires' or estates north of Tees coalesced into what became the Palatinate of the prince bishop of Durham is not entirely settled, but it was a process that began perhaps with the secular powers acquired by Bishop Walcher (1071–80) who, for a time, was simultaneously Earl of Northumbria, and was finally accomplished, in territorial terms, but not necessarily power, by Bishop Hugh of Le Puiset with his purchase in 1183 of the Earldom itself and with it the wapentake of Sadberge.

Until that point in the Tees valley the Bishops themselves controlled only the general areas of Darlington and Norton on the north, and as an episcopal liberty, Allertonshire to its immediate south in Yorkshire. Until then much of the northern flank of the Tees—the vaguely defined wapentake of Sadberge—was held by the English or (briefly during the Anarchy) the Scottish crowns, by the Balliol family based on Gainford and Teesdale, and by the de Brus family who held Hart and Harterness towards the coast. In 1066, north of Welland, by contrast, lay the extensive royal soke of Great Bowden, whose dependencies stretched northwards, just over the drainage basin of the Soar. To the east, it shared a boundary with Rutland whose peculiar status was perpetuated not only by continuing royal connections, but also through the granting of its heartland area to Westminster Abbey. In the case of the Stour, there was another great ecclesiastical liberty, the 8½ hundreds of Thingoe, which evolved into the Liberty of St Edmund (to where the martyred king's incorrupt body had been translated between 915 and 925), land which was gifted to the Church by Edward the Confessor. This extensive territory reached over the watershed of the Wash drainage basin to the north-west and as far as Hadleigh and Stoke-by-Nayland on the south-east, only four riverside Suffolk parishes short of the estuary.

It is usually agreed that in contrast to Bury itself, Sudbury was the southern *burh* of the Liberty, its site having formerly been a detached portion of the core area of Thingoe. The location of the *burh* was presumably in the vicinity of the important church of St Gregory which was sited well back from the river-crossing. Settled around this by Domesday was a town of 138 burgesses with a market and a mint. If Sudbury could conceivably have been generated from Bury some 10 miles to the north, a similar process may have taken place in the Rutland area. There, as we have seen, the most likely candidate for the royal *caput* of the district was the hilltop site of Hambleton in the Vale of Catmose. Although overlooking nearby Oakham, which is arguably a quite late urban development, Hambleton contained the mother church of St Peter's, Stamford, some 7 or 8 miles away near the Welland. Stamford in turn was clearly carved out of the territory of Rutland at some point perhaps around 900, so it is distinctly possible that before it became one of the 'Danish' Five Boroughs, Stamford had been the royal *tun*, or administrative centre, for Rutland controlling the Roman ford through the Welland. By 1086, as a King's borough, it could boast over 400 housing plots, six wards, and four churches. Further north, it is also conceivable that after 1083, when Bishop William of St Calais very possibly resettled some of the *congregatio* of St Cuthbert on his manors near the Tees both at Norton and, up the tributary of the Skerne, at Darlington, he may have also laid out two planned settlements as potential 'towns'. As early as 1109, Norton received a grant of a market, that is before any other known in England north of Yorkshire and Lancashire (Newcastle probably excepted). Darlington, however, is recorded as an already fully-fledged 'borough' only in *c.*1183. Nevertheless, given its key position on the communication network, the existence of some sort of pre-Norman ecclesiastical foundation, the presence of a functioning episcopal residence in the twelfth century, and its consistently greater economic importance than Norton then and thereafter, there cannot be much doubt that Darlington was a more significant place than Norton from the beginning. Probably by *c.*1100, therefore, and in two cases perhaps up to two centuries earlier, the future 'capitals' of all three valleys had been planted on the northern flanks of each, all of them on major north–south routes.

In secondary instances of urban encroachment on the north bank, to control trade and vital crossing points, the mother settlement was more readily to hand. By the Stour, with its market and forty-three burgesses, Clare must have been spawned by its neighbouring Stoke (by Clare) well before its Domesday Book entry as a town, and with its new castle was now to become the head of the Honour of Clare which contained no fewer than ninety-five manors in Suffolk and others in neighbouring counties. Nayland—on an island in the Stour—was also carved out of a Stoke (by Nayland), but only received a market grant in 1227/8, although this could well have been no more than a belated recognition of established fact. In the case of the erstwhile royal Leicestershire centre of Great Bowden, we now know that a trading settlement was laid out beside the Welland 2 miles away at what became known as Market Harborough, probably by the Mauduit family in the 1150s, but certainly no later than 1173–4. Within the estate of Norton, a 'borough' at Stockton—appropriately known today as 'on Tees'— was first mentioned in 1283 and is likely to have been of earlier foundation. As its name suggests, Barnard Castle grew up in relation to the fortress established by

The transpontine suburb of a valley capital

Stamford St Martin seen from the north (from the west end of St Mary's church in Lincolnshire), along the Great North Road, and across the Welland into Northamptonshire.

the Balliol family of Gainford to control Teesdale and the crossing of the Tees at this juncture in around 1100, and as the brainchild of Barnard de Balliol in particular. The place is mentioned as an already established borough in *c.*1175.

This process of riverside colonization, usually from the north, further saw settlements invading the opposite south banks of all three rivers both before and after the establishment of county boundary lines. Thus while Great Bowden gave birth to Market Harborough in carucated Leicestershire, just over the river in hidated Northamptonshire in Domesday Book lay Little Bowden, which formerly must have belonged to the wider Bowden complex. Two neighbouring rural parishes of Lubenham and Theddingworth in Leicestershire also had small portions of land on the Northamptonshire side of the Welland. Down river it is well known that in 918, as the Anglo-Saxon Chronicle put it, 'King Edward went with the army to Stamford, and ordered the borough on the south side of the river to be built: and all the people who belonged to the more northern [i.e. 'Danish'] borough submitted to him, and sought to have him as their lord.' It was this transpontine suburb, therefore, that subsequently became Stamford Baron, and the parish of St Martin's. As such it was the only one of Stamford's six wards not to lie in Lincolnshire.

There were even more extraordinary, and probably ancient, arrangements across the Stour. Nayland—north of the river—was taxable in Suffolk but a large portion of it—what are now Great and Little Horkesley—was treated in Domesday as lying in Essex. Upriver on the Suffolk bank lay the probably royal pre-Conquest estate of Bures St Mary (with a later riverside market), but the settlement of Bures Hamlet developed just over the crossing into a transpontine suburb. Further to its south still, and up to the top of the scarp slope of the Essex side of the valley, stretched Mount Bures which clearly had once been a part of the same estate before developing as a separate entity with its own church and motte-and-bailey castle. Above all, there was Sudbury itself where the mother church of St Gregory gave birth to two daughter churches between it and the river-crossing, immediately across which came to lie the Essex suburb of Ballingdon and the hamlet of Brundon.

The same pattern was evident on the Tees. Control of crossings seems to have been established both at Stockton, where one oxgang belonging to it lay on the Yorkshire side of the river perhaps in relation to the ferry; and south of the boundary between Middleton St George and Low Dinsdale where Ponteyse, the former medieval bridge over the Tees, spanned the river to Over Dinsdale—a part of the adjacent Durham parish of Sockburn in Yorkshire. Even where there was no bridgehead, the dominance of the north bank was expressed by the provision of occasional massive bridges (albeit responsibility for upkeep was later divided between the two contiguous counties): by Bishop Skirlaw of Durham *c.*1400 across to Yarm, and, to carry the Great North Road, possibly also to Croft on the further bank of Tees, south of Darlington. Finally, the ancient territorial unit centred on Gainford by the north bank—a name which specifically denotes a 'ford on a direct route'—also overlapped the river, while in Teesdale Barnard Castle eventually developed its own small transpontine suburb in Yorkshire.

What all these processes highlight is the corresponding economic and political weakness of each south bank. Only in the sealed-in gap between the Tees, the sea, and the North York Moors were there signs of independent 'urban' developments: at Stokesley, Guisborough (with its tiny maritime outlet, the so-called 'port of Coatham' for the earlier Middle Ages) and, above all, Yarm. In Teesdale itself there was nothing south of the river. On the Welland there was nothing;

Rockingham never having aspired to being anything more than a market village. On the Stour, apart from Haverhill, on the watershed zone to the west, and the medieval estuarine developments of Manningtree and Harwich to the east, there was only Dedham, a place that simply acquired some sort of urban status with the development of its later medieval clothing industry. What is surely significant here, then, is that both Yarm (its very name denotes fish-weirs) and Dedham seem to have represented the lowest fordable crossing points of their respective rivers and, perhaps, the highest points to which craft could navigate from the sea. In the Welland valley, except for a spell due to silting during the six-teenth and earlier seventeenth centuries, that limit was marked by the bridge connecting the two ends of Stamford, just downstream from which the blocked previous access to a wharf may still be seen on the north bank. In all three valleys, therefore, inland navigable waters were either non-existent or limited.

By contrast, the northern valley flanks developed what might almost be described as local urban networks. For the Tees, there were (eventually) Middle-ton-on-Tees (up Teesdale), Barnard Castle, Staindrop, Darlington, Stockton (which seems to have superseded an earlier 'harbour of trading vessels' at Por-track in the manor of Billingham) and, as the main early maritime outlet, Hartle-pool. For the Welland there were not only Market Harborough, but also for long a more important centre originally, up one of the tributaries, at Hallaton, as well as Uppingham, Oakham (on the Gwash), above all Stamford, together with Mar-ket Deeping and ultimately Spalding. By the Stour, or up its Suffolk tributaries there were Clare, Sudbury, Lavenham, possibly Bures St Mary, Nayland, and Hadleigh, and that is not to mention the important medieval textile manufactur-ing centres at Cavendish, Glemsford, Long Melford, Kersey, and East Bergholt.

Such places reflected the vigorous integrated economies of all these valleys and the employment possibilities that might be generated across county boundaries. Space precludes all but the briefest reference to the underlying general factors com-mon to each. First, and indicative of the climatic advantage of the south-facing northern flanks, was a probably universal fundamental circumstance that may be illustrated from the Stour valley as late as 1906. There, it is recorded, that one old East Bergholt labourer at harvest times had been accustomed to reap for several weeks in Suffolk, before crossing over to Essex to go harvesting again because 'in his day' the harvests of Suffolk were 'almost gathered ere cutting began in Essex'. Sec-ond, ample grazing enabled not only dairy farming or the marketing and fattening of beasts (let alone *ad hoc* fodder for the wayfaring community or even, as before the battle of Naseby, for armies) but also even specialist breeding in the eighteenth century. Third, apart from hides and leather and later malting, other valley resources like horse-hair, flax, or hemp let alone wool (which equally was easily imported), encouraged the development of what Dr Joan Thirsk has helpfully dubbed 'the fibre industries', all of which involved spinning, weaving, and often the further production—according to need—of finished articles like clothing, stock-ings, sail-cloth, or rope. All of these industries were heavily labour intensive, involv-ing a workforce that was often female as well as male, rural as well as urban. The most conspicuous early development here, of course, was the late medieval textile industry of the Stour valley with its quality dyed cloth, skills that were subsequently transferable into the manufacture of the lighter so-called New Draperies, then to silk-weaving (especially at Haverhill, Sudbury, and Glemsford) and, finally in the nineteenth century, to the weaving of horse-hair and imported cocoa-nut fibre.

Both the Tees and the Welland developed woollen industries in the early modern period; both moved later into the manufacture of carpets (Barnard Castle and Market Harborough). In the late eighteenth century, the Darlington area developed a major niche in the linen industry. The upper Welland moved temporarily into silk-spinning and the manufacture of black silk plush for hats and, eventually in 1884, with the establishment of Symington's factory in Market Harborough, into the production of liberty bodices and corsets. Fourth, if all the valleys exploited local clay for their own brickworks from the eighteenth century, only Teesdale also

The economic specialisms of frontier valleys

The quality-dyed cloth industry of the medieval Stour valley brought great prosperity which was expressed in a wave of vernacular, as well as church, building. This shows the superb hall of the guild of Corpus Christi in the market place at Lavenham (c.1520).

developed a major extractive industry following the expansion out of the Alston area of the London Lead Company (founded in 1692), and the establishment of its headquarters at Middleton-in-Teesdale (from where they also exploited the upper end of Weardale). The lead was taken out via Stockton. A wholly different story, however, was represented by the subsequent linking of the more northerly Durham coalfields to the potential created by the exploitation of the iron ore from the Cleveland Hills; by the establishment of the Stockton–Darlington railway; and by the creation of the new Teeside ports and of Middlesbrough.

Two further observations should be made with regard to these valley economies. First, their industries were remarkably self-contained or, if not, usually dominant locally: Teesdale lead mining was spatially distinct from Weardale coal; Stour textiles far eclipsed those of the Colne valley until the days of the New Draperies; the Welland valley remained largely beyond the reach both of the Northamptonshire shoe industry and of Leicester hosiery. Second, as funnels for west–east communication these valleys effectively bypassed the wider regional economies to their norths and souths. Because their rivers were not navigable far inland, goods were drawn from regions to the west, down the valleys towards their riverine and estuarine outlets, and vice versa from the east. The movement of goods and agricultural products may be traced out of Cumberland, via the Stainmore route, from Swaledale, and from Teesdale down to Yarm, in the medieval and early modern periods, though this port was increasingly displaced in the eighteenth century by Stockton, where deeper waters gave access to bigger vessels. The Welland was clearly the major route from medieval Coventry to the great fair at Stamford itself and to the Wash ports of both Boston and Lynn; the Stour was the obvious way from Cambridge or Bury to Manningtree, Harwich, or Colchester, before the Stour Navigation of 1705 brought water transport as far inland as Sudbury.

In social terms, therefore, the north-bank towns were in an unrivalled position to exert influence over the rural districts across the boundaries to their immediate souths. In 1525, for example, on the eve of what nearly turned into a rebellion against the levying of the so-called Amicable Grant, the Suffolk protesters were said to have incited the people of northern Essex when they visited Sudbury 'which is in Suffolk and is their market town'. It was no accident, then, that in subsequent centuries hiring fairs for servants in husbandry were held at all such centres as Darlington, Stockton, Stamford, Market Harborough, and Sudbury. In the case of Market Harborough, nineteenth-century carrier routes, early twentieth-century rural bus-services, and the range of local bicycling all directly linked the countryside up in the Northamptonshire watershed zone to this Leicestershire town.

Nor is it surprising that such urban centres also became responsible for the poor in their vicinities. The workhouses of the New Poor Law of 1834 seem to have been always sited strategically with responsibilities over areas more or less within walking distance: a fair measure, therefore, of a town's catchment area. What is so telling in the present instance, then, is the fact that the Unions based on these towns (which, it should be recalled, subsequently tended to develop into Sanitary Districts and then into Rural Districts for local governmental purposes), now formally breached the county boundaries—and hence the formal boundaries of this discussion—as these had been established for some eight centuries. In such cases, therefore, local social realities and local government were brought together into some sort of 'fit' for the very first time. Unions based on Darlington, Stockton, Market Harborough, and Stamford, or Sudbury and Risbridge, now overlapped extensively into what had always been their natural hinterlands in the valleys concerned.

The moral seems clear. If a frontier valley took its *raison d'être* from the formal administrative boundary running down the river-bed, the informal cultural boundary in each case, was—very broadly speaking—the watershed zone running along the southern edge of the valley in the adjacent county. This zone was largely dominated from the towns on the major river crossings, towns that were sited at the uttermost limits of the cultural province to their north and physically separated from the heartland of their parent territory by a well-defined water-shed in between. In most respects, consequently, such towns and their valley hin-terlands could independently absorb cultural influences from *all* directions in such a way as to create a fissure between the larger cultural blocs to north and south. In their character, therefore, frontier valleys represented in miniature the wider cultural provinces which they served to separate. No wonder then that frontier valleys might act idiosyncratically to mingle dialect elements from either north or south; might operate as invasive cultural edges like that represented by the narrow belt of medieval churches of a peculiarly East Anglian style which

Valley culture

Regionally distinctive Anglo-Scandinavian carved stones from Kirk Leavington church (the mother church of Yarm and therefore on the southern flank of the Tees), showing locally character-istic figures, sometimes with birds.

characterizes the northern fringe of Essex; or might generate their own cultural idioms as did both the Anglo-Scandinavian stone sculptors of the Tees valley or the nineteenth-century headstone carvers of the Welland, or perpetuate their own sense of belonging through such localized organs as the *Darlington and Richmond Herald* or the *Suffolk and Essex Free Press*.

As provincial punctuation marks on the landscape, these 'frontiers' were emphasized by the fact that they also very broadly marked informal *national* subdivisions. On the one hand was the limit set to the direct influence of London, not only via the Great North Road (let alone the London roads through Market Harborough or Sudbury), but also through the coastal contacts with the estuarine ports. In the Middle Ages, the capital largely drew both provisions and migrants (and so some of its dialect eventually) from counties inland from the Wash in particular, but not much further afield than the Welland which, after all, also represented the last long river to the north before the Trent divide. In terms of national culture, it is worth recalling that during the fourteenth century, Stamford—of all the towns under consideration, ever the most metropolitan in its culture—momentarily threatened to become the third university town in the country. In Tudor times, moreover, other than at York and Newcastle, there were no royal residences to be visited north of the Welland line, nor indeed were there any north of the Stour. Both these rivers indeed also came to represent broad northern limits for courtier houses like Kirby Hall, Holdenby, and Burghley in Northamptonshire or Ingatestone or Audley End in Essex, to name but some.

In the north, there was a geographically opposite source of influence: from Scotland. In this case it was the Tees that marked the southern limit on the east of Scottish territorial claims and association. The claims were linked with what was regarded by the Scots as their right to the Earldom of Northumberland (what was left of the old kingdom of Bernicia following the cession of Lothian probably in the tenth century). During part of the reign of David I (1124–53) indeed the Tees line was effectively the international boundary, the battle of the Standard (1138) being fought near Northallerton in part—in Professor Barrow's words—'to save Yorkshire'. Before the Reformation, Durham priory was closely associated with the cult of St Margaret of Scotland and ritually exhibited her cross, one of two black roods of Scotland in the cathedral. Even as late as 1640, it was at the Tees that the Scots halted prior to the Treaty of Ripon. It is then to the problems surrounding the original eastern and western marches with Scotland itself that it will be appropriate now to turn.

National Frontiers: The Anglo-Scottish Border

To the extent that provincial frontier valleys divided cultural blocs from each other, they did so within the same national jurisdiction. By contrast the international frontiers with Scotland reflected the arbitrary subdivision of pre-existing cultural provinces and the subsequent intrusion of a foreign jurisdiction over each area ceded. The interface of the two jurisdictions, therefore, was inevitably characterized both by pockets of no man's land and by codified processes (Marcher Laws) for the settlement of cross-border disputes at customary trysting points. The border line itself, however, needed to represent the defensible limits of incursion, and especially so in a geographical context that eventually spanned the desolate watershed of central Britain between two different cultural regions. Such a line could not continuously respect the courses of rivers, except towards the sea. It needed to exploit the higher watersheds as obstacles to hostile movement. Behind this long frontier, and despite the cross-border landholding patterns of the wealthier elements of the societies involved, at times of international conflict there was additionally a continuing need for in-depth support from the rear, and especially so in the English case where the front line was so distant from national seats of power. To understand such frontier landscapes, therefore, requires not only an appreciation of their making but also a sense of their wider regional contexts. They thus furnish an instructive opportunity for comparison with the older patterns already discussed.

The earlier of the two borders, that on the east, bisected the virtually self-contained drainage basin of the Tweed: the original heartland of the kingdom of Bernicia (which stretched from the Tees to the Forth) and subsequently in part the immediate territory of the Northumbrian earls of the house of Bamburgh. It is possible that the complex process of subdivision was here accomplished in two stages. In the first, a good case has been made for a Scottish expansion of Lothian, south-eastwards from the barrier of the Lammermuir Hills to the Tweed from the 940s at the latest, for formal English recognition of this by 973; for the possible recovery of this area, or some of it, by the English in 1006; and for a possible second cession to the Scots in 1016. In this writer's view, however, the evidence for the area acquired by this date cannot be used to claim that it stretched any further south than the south-eastern edge of the later archdeaconry of Lothian (within the diocese of St Andrews), that is along the Tweed from Gala Water (but excluding the *parochia* of Old Melrose which spanned the river) as far as the sea. South of the river was land that was still regarded as part of the diocese of Durham until the turn of the eleventh century.

It is not inconceivable, therefore, that a second stage of Scottish territorial expansion was effected by the victory in 1018 of an allied force of Scots and Strathclyde Britons (whose own interests may thus have been involved) over specifically the 'people of St Cuthbert' (between 'Tees and Tweed'), eighteen of whose priests were said to have been slain. The battle was at Carham on the south bank of the Tweed, so this could explain why here the boundary between Scotland and England has been apparently redrawn south-eastwards and away from the logical line of the river, initially along the western bound of Carham parish itself, towards the top of the nearby Cheviot. In doing so the former archaic shire of Yetholm (which may well have been Lindisfarne—and thus Durham—property) was arbitrarily bisected, the greater part to the west being absorbed eventually into what would become known as Roxburghshire in—or even before—the early twelfth century. This latter territory (with its southern boundary sharing that of the kingdom along the high ridge of the Cheviot Hills) looked to Roxburgh itself—a fortified place whose political significance surely antedated the days of David I—and included other Lindisfarne dependencies south of Tweed such as Old Melrose and Jedburgh (then Jedworth). In its other guise as Teviotdale, it was precisely this area which was to be transferred ecclesiastically from the diocese of Durham to that of Strathclyde or Glasgow *c*.1100.

Now the fact that Roxburghshire also spilled out of wider Teviotdale over the central watershed of mid-Britain to the south-west, and some distance down upper Liddesdale as far on the left bank as the Kershope Burn, the historic western bound-

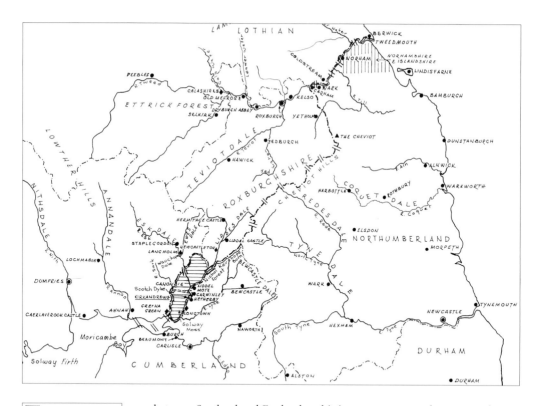

The Anglo-Scottish border (strongholds, settlements, and religious houses not mentioned in the text are not included)

ary between Scotland and England at this juncture, seems to have escaped comment hitherto. Since a twelfth-century charter records the existence in this vicinity of a feature called 'the *fosse* of the Galwegians', however, it may be that this previously marked the easternmost limit of an eleventh-century expansion of Galloway. Certainly as late as 1079 we hear of Galwegian levies used by Malcolm III against the English, getting back from a raid on Hexham to their own border—presumably via North Tynedale—by evening in a single day. It is possible then that the Galwegians had been left free to fill a power vacuum up Liddesdale following the Scottish succession crisis of the 1030s, when the young Malcolm fled from Macbeth, and the Solway kingdom of the Cumbrians was partitioned leaving Cumberland within the area of influence of Anglo-Scandinavian Northumbria. At that point the dividing line with England probably followed the course of the River Lyne, whose confluence with the Esk is then likely to have marked the head of the Solway Firth. It may not have been until after a Galwegian withdrawal and the renewal of direct Scottish control over much of the north Solway littoral, even a decade or so after the Norman occupation of Cumberland in 1092, that lower Liddesdale with the lower Esk were eventually split more or less lengthways between England and Scotland.

As it developed subsequently, the topography of the medieval frontier may only be understood in relation both to emergent major centres of royal power on either side, and to the central upland massif that stretches from the Lowther Hills and the Cheviots southwards to the Tyne/Solway gap: that is, the entire area between Roxburgh and, by the later twelfth century, Dumfries (which, it has been suggested, was deliberately sited defensively on the further side of the Nith from

Galloway), and southwards to the English walled cities which evolved at both Newcastle (from 1080) and Carlisle. Each of these in turn developed downriver outposts to control both access and supply from the sea. Roxburgh looked to the early military and commercial development of Berwick; Newcastle—protectively from the north bank—to the priory and fortress of Tynemouth; Carlisle to castles at Beaumont and Burgh on the Solway (and, in the days of Edward I, to the creation of a supply port for the invasion of Scotland in Moricambe Bay); while Dumfries depended eventually on Caerlaverock. Older patterns of provincial organization were thereby fundamentally altered. In advance of these valley or estuarine lines in the lowland areas, where some distance separated the royal base from the border itself, there was an intermediate valley line, guarded by the strongholds of major tenants-in-chief with exceptional numbers of knights locally dependent on them. Such, on the River Annan, was the de Brus family at Lochmaben castle and burgh and at Annan itself; and, on the Northumbrian River Aln, the de Vescis at the eventually walled town of Alnwick. It was only by or to the rear of these rivers that, with the special exception of Holy Island, religious houses and small urban centres—like Warkworth, Morpeth, or Hexham—and rural nucleated settlements tended to be concentrated. In the lowlands, such inner support zones have therefore to be distinguished from the underpopulated outer zones leading up to the different stretches of the frontier itself.

Of these, that most resembling the valley frontiers already discussed was clearly the lower Tweed to which, in order continuously to reflect the frontier line upstream, have also to be added the valleys of the Teviot and the Jed. Once again it was the northern flank which was dominant along an axis between the royal burgh and provincial capital of Roxburgh and the increasingly prosperous port of Berwick. Controlling the confluence of the Teviot and the Tweed, with a major castle and, eventually, four churches, a mint, and an annual international fair involving the wool clip from the estates of Melrose and Dryburgh abbeys, early medieval Roxburgh also lent its protection to the house of Tironian monks that David removed from Selkirk to Kelso on the north bank in 1128. There too, within forty years, a market centre evolved in the shadow of what was to become one of the wealthiest Scottish abbeys. Up the Jed Water developed the royal burgh of Jedburgh and its refounded religious house; downriver at Coldstream, the twelfth-century house of Cistercian nuns (one of four on the Scottish side), must have provided the initial impetus to the development of what, because of its control of an important river crossing, would also evolve into a small town.

As on the provincial frontiers too (apart from the late emergence of Tweedmouth), the only important new or enlarged settlement on the south bank was the modest English planned borough, planted by Bishop Ranulf Flambard of Durham between 1121 and 1128 in the shadow of his frontier castle at Norham. This stronghold not only guarded a strategic ford; it also served to plug the lowland gap between the Cheviot and the sea, and so to act as some form of safeguard for the two remaining outliers of Durham—Norhamshire and Islandshire—which occupied this district. Between these and the Carham stretch of the Scottish border, finally, a northward tongue of Northumberland was defended by the great castle of Wark overlooking the Tweed.

It was ironic that such 'frontier' patterns, evolved—surely significantly—in relatively peaceful conditions, should nevertheless be unable wholly to withstand the wars which intermittently characterized the period from 1296 to the mid-sixteenth century. Roxburgh was totally abandoned by the Scots in the fifteenth century, having been in English hands for a century previously, and Berwick fell to England for the last time, and so permanently, in 1482. Within another 100 years, the English strongholds on the south bank were both in ruins. Even before the end of the Middle Ages, however, the regional centre of gravity was tilting westwards: burghs developed at Peebles and Selkirk while Jedburgh replaced Roxburgh as county town. With the later industrialization of the dales for textile manufacture, not least at Hawick and Galashiels, and with Kelso's emergence as some sort of social centre, by the nineteenth century, the Tweed basin had rediscovered its role as an undivided 'cultural province', albeit now of Scotland.

A second frontier zone was the central upland region with its deeply incised valleys running up to the major watersheds. The outskirts of this area were ringed by fortresses athwart the only communication lines which performed similar intermediary functions to those on the Annan or the Aln. Such were the castles of Jedburgh, Lochmaben, Bewcastle, Wark on the North Tyne, and Elsdon (later superseded by Harbottle) for Redesdale. For the most part up the dales from these, and inhabiting the valley floors, was a scattered and ever-expanding population, organized (from the end of the thirteenth century at the latest) in ramifying 'Surname' groups, each under a 'heidsman', for the purposes of both seasonal transhumance to their upland shielings and, like the wider Scottish society to the north, the practice of the deadly feud. It now seems possible that these clans were deliberately militarized on the English side by Edward III who encouraged the proliferation of a localized fighting force of light horsemen through grants of advantageous tenures. It was partible inheritance, and hence overpopulation, together with lack of local control that led to the degeneration of this society into cattle-rustling and protection rackets. Whether for war or for cattle raids, however, this 'martial kind of men' evolved an alternative frontier valley landscape, one which exploited the access provided via remote uninhab-ited valley heads abutting the watershed border line along the Cheviot ridge. Across this, accordingly, developed the upland equivalents of fords and bridging points, the tracks known as 'ingates', albeit at local level these connected, rather than separated, the surname societies on either side of the border.

The westernmost strip of frontier was the most complex. As we have seen, there emerged early on a threefold partition of the wider Liddel/lower Esk valley between Scottish Dumfries and Roxburghshire, and the English Cumberland. Before that, however, there is every likelihood that both sides of the lower end of this area had been held as one territory dominated from a strategic position con-trolling the confluence of the two rivers: the Roman *Castra Exploratorum*, near Netherby; or the probable *caer*, or 'fortified place' (Carwinley) of 'the princely' 'Gwenddoleu' who fought at the battle of Arthuret in 573. Even after the creation of the frontier, there was only one stronghold at the foot of the dale, the Anglo-Norman *caput* of Liddel Mote, near Carwinley itself. If the English south flank of Liddel Water now came to contain Nichol Forest, in the thirteenth century the English side of the Esk was still commonable by nearby settlements in Scotland. To the north of the confluence itself, moreover, a twelfth-century cell of Jedburgh Abbey, the Priory of Canonbie, (later described as 'neutral betwixt both the realms'), lay in an internationally intercommonable area reaching for some 10 miles by four from the Solway to the Roxburghshire boundary. Known as the

Debateable Land ('to be occupied' by both sides 'from sunrise to sunset with bit of mouth only'), this in principle could not be permanently settled. Finally, the major through-routes to Dumfries or to Strathclyde bypassed this area to the south-west, the *waths* or fords of the Solway at low tides providing much easier access.

Effectively, then, the local passages out of England lay where the high ground defined both the north-western bound of the Debateable Land and the narrow entry to Eskdale on the one hand and, on the other, the similar narrowing of Liddesdale where the Muir (*mere* or boundary) Burn on the right bank defined the north-eastern bound of the Debateable Land, not far short of the Kershope Burn on the other side. It was consequently only above *these* crossing-points that unambiguously Scottish territory began. Since the English scarp of the Esk and Liddel ran down immediately to the river line, the first stretch of river with both north *and* south banks that might be, and were, settled lay in fact upstream in Scotland: where Liddel Castle and the church of St Martin focused the expanding early medieval population of upper Liddesdale; Hermitage Castle being a later development. Downriver, only to the north, beyond the limits of Canonbie parish, does settlement seem to have concentrated in the Middle Ages: not least at the short-lived feudal burgh of Staplegordon, founded (*c*.1285) in the shadow of Brantalloch Castle, or in the subsequently abandoned steadings of Wauchopedale with its own castle—and access westwards towards Lochmaban—or whatever preceded Langholm nearby. The only possible medieval exception on the south side was a market conceivably located lower down on the left bank of the Esk at Longtown.

The key to the later medieval settlement of all the border dales was the isolated towerhouse or 'pele' in its own enclosure, and subsequently the more compact stronghouse (sometimes since called a 'bastle'). By the sixteenth century, the towerhouses (belonging especially to headsmen of the Surnames) were predominant in Scotland, while the bastles, more expressive of peasant wealth, were common in England. The pacification of the frontier, the construction in 1552 of the Scots Dyke to demarcate the division of the Debateable Land (but only after nearly all buildings on it had been razed to the ground), and the Union of the Crowns in 1603, however, all led to the eventual redevelopment of the landscape as partial reflections of these two differing social emphases. To the south, between the Liddel and the Lyne, for example, the local farmers themselves seem to have stamped a new pattern of settlement on the map. All over this area, largely during the later sixteenth and the seventeenth centuries, we begin to hear of isolated single farms (usually sited on small knolls) boasting a new form of place name, ending not in the older '-ton' but in '-town', and preceded by one of the infamous surnames of the earlier generations: thus Lordstown, Nixonstown, Noblestown, Scotstown, and so on.

On the northern flank of the Liddel/lower Esk line, in total contrast, the wealthy were beginning to plant ambitious nucleated settlements. Up Eskdale itself in 1621–8, for example, Robert Maxwell, Earl of Nithsdale, seems to have founded a small planned town, a burgh of barony, at Langholm where house plots were made available to no fewer than ten members of his own clan. By 1726, this had become a thriving market centre with a developing trade in wool, yarn, and skins. To it, on the other side of the Esk was added New Langholm by the third Duke of Buccleuch in 1778 with 140 houses and subsequently a cotton manufactory. Towards the head of underpopulated Liddesdale in Roxburghshire in 1793, the Duke also planted an extensive planned village with its own market on the north bank at New Castleton. Down by the River Sark, which defined the western edge of the Debateable Land, the village of Gretna (Graitney) Green was wholly rebuilt by its chief resident between 1693 and 1723 as a little market centre and burgh of barony, adjacent to which in 1791 Springfield was added as a settlement for cotton weavers. (Gretna itself was only created during the Great War for the purposes of munitions manufacture.)

The outlines of a familiar pattern of dominance thus began to emerge on the northern and western flanks of the frontier at much later periods than elsewhere, although, apart from Canonbie, the continuing emptiness of the erstwhile Debateable Land as an informal space between the two nations remained marked. In 1771, indeed, Solway Moss burst and swamped the lower end of this area. The emergence in the eighteenth century of the transitory little port of Sarkfoot on that river's east bank to serve the Graham estate, from Netherby to Bewcastledale, and the remodelling of Longtown on the main road north over the Esk, however, also illustrate elements of south-bank development. Designed by Dr Robert Graham between 1753 and 1792, in layout and architectural character, Longtown represents the only example of a classic Scottish planned village/small town to be found in England. As such it thus confirms the cultural influence commonly exerted on south banks from their more populous norths.

Linked as it was also (like Gretna Green and Kirkpatrick to its west) to the nineteenth-century cotton manufactories of Carlisle over 6 miles to its south, however, English Longtown, as in another mode did Berwick, further epitomizes the manner in which such frontier-valley centres functioned most typically as mediators between two much wider national or provincial cultures, whilst yet belonging completely to neither. It was no accident that it was via the south bank of Liddesdale, and then Longtown, that the only Scottish railway company to do so, the North British, penetrated the Border en route from Edinburgh to Carlisle and eventually Silloth on the Solway, nor that the thinly populated Cumberland countryside, between Longtown and Carlisle, came to be characterized by numerous nineteenth- and early twentieth-century single-storey, lime-washed cottages of a vernacular type which emanated directly from Scotland. By this period, indeed, the development of a thriving Solway economy, integrated not only by sea but also by the viaduct of the Solway Junction Railway of 1864–9, had led, as in the case of the Tweed basin, to the renewal of cultural provincial realities that straddled the formalities of the international boundary.

Provincial and National Frontier Valleys Compared

The contrast between English provincial and national Scottish–English frontier valleys is thus most marked. Contrary to what might have been expected from their relative significances, the former were unambiguously defined in physical terms, uninterrupted and extensive, and, above all, enduring in their informal function; the latter were exploited for *ad hoc* purposes, represented only sections of a discontinuous frontier line, and were, when measured in historical time, only temporary. The one continued to separate cultural blocs in areas largely unaffected by urbanization; the formal role of the other was displaced as soon as the underlying identity of the cultural province which it had bisected was reasserted as an undivided whole.

Part II

Cameos of Landscape

11 Stonor:
A Chilterns Landscape

Leslie W. Hepple and Alison M. Doggett

Introduction

The small village of Stonor sits in a dry valley in the south Chilterns, Oxfordshire. The motorist, driving north out of Henley-on-Thames on the Oxford road, first travels up the famous Fair Mile before climbing out of the valley towards Bix and Nettlebed. A minor road continues up the Assendon valley to Stonor, 5 miles north of Henley. The valley road has fields on either side, and the white of the chalk and flints shows through the ploughed arable. On the steeper and higher slopes there are woods. Stonor village (known as Upper Assendon until this century) is little more than a hamlet, with its brick-and-flint houses and ancient barns. Just beyond the village are the lodge gates of Stonor Park, which lies in a combe off the main valley, and cannot be seen properly from the road. The Stonor family have lived here since the twelfth century, and they have played an important role in the local landscape history.

Our territory is the parish of Stonor and Pishill, and it encompasses all we can see as we stand in this upper Assendon valley: the arable and pasture fields of the valley sides together with the wooded horseshoe of surrounding hills. This small landscape encapsulates much of what is distinctive about the Chilterns region as a whole. Leland wrote in the 1530s of 'plenty of wood and corne about Henley. The soyle chalky and hillinge', and he captures much of the Chilterns character. The Chiltern Hills are chalk, with their scarp-edge facing north-west and forming one of the strongest lines of division across the English landscape. The dip slope is deeply dissected by valleys running up from the Thames and its northern tributaries, and these have eaten back almost to the scarp crest in places. On the higher ground the chalk geology is overlain by clay-with-flints and Reading Beds, which yield much heavier soils mainly unsuitable for arable farming. The result is a landscape with much local detail, a patchwork of chalk valley, steep slopes, localized clays, and a corresponding chequerboard of arable fields, pasture, woods, and commons.

• • • • •

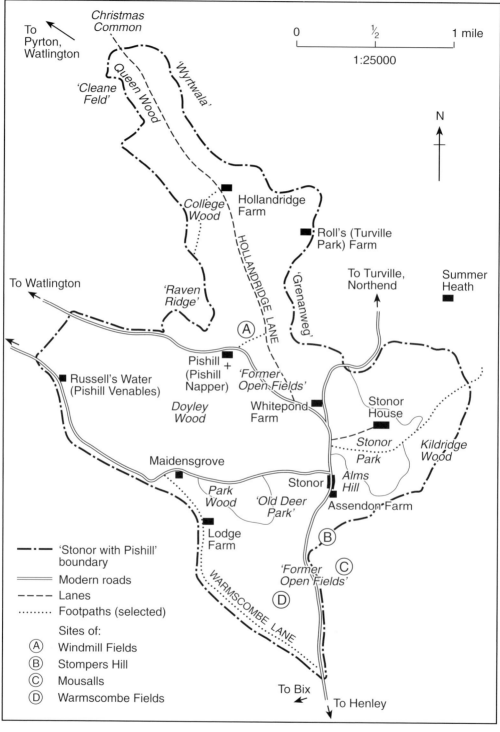

To Pyrton, Watlington

Christmas Common

'Wyrtwala'

'Cleane Feld'

Queen Wood

0   ½   1 mile

1:25000

N

College Wood

Hollandridge Farm

Roll's (Turville Park) Farm

HOLLANDRIDGE LANE

'Grenanweg'

To Turville, Northend

Summer Heath

To Watlington

'Raven Ridge'

Ⓐ

Pishill (Pishill Napper) +

'Former Open Fields'

Russell's Water (Pishill Venables)

Doyley Wood

Whitepond Farm

Stonor House

Stonor Park

Kildridge Wood

Maidensgrove

Park Wood

'Old Deer Park'

Stonor

Alms Hill

Assendon Farm

Lodge Farm

Ⓑ

'Former Open Fields'

Ⓒ

Ⓓ

WARMSCOMBE LANE

To Bix

To Henley

—··— 'Stonor with Pishill' boundary
═══ Modern roads
--- Lanes
········ Footpaths (selected)
Sites of:
Ⓐ Windmill Fields
Ⓑ Stompers Hill
Ⓒ Mousalls
Ⓓ Warmscombe Fields

The Assendon valley is one such dip-slope dry valley. Today's visitor, travelling by car, can easily think the valley ends just beyond Stonor Park, for the road forks and the two prongs climb steeply up to the Chiltern plateau, westwards towards Watlington and north-east to Turville Heath and Northend. In fact, the dry valleys extend much deeper into the hills, penetrating to within a few hundred yards of the Chiltern scarp-crest itself at Christmas Common. Numerous routes are accessible only by footpath or bridleway, and you need to walk to appreciate the nooks-and-crannies of this countryside.

The valleys were formed during the periglacial conditions of the late ice age and when the water table was higher, and during historical times the valley farms have had to rely on wells. Just north of Stonor at Lower Pishill a 'wellfield' is marked on a map of 1725, and there is still a well on the modern Ordnance Survey map. On the higher ground farmers had to rely on ponds set in the heavy clays of the summits, and in these parts it is waterlogging rather than water-shortage that can be a problem. In the lower valley closer to Henley, the famous Assendon Spring sometimes flowed after heavy rains, as Robert Plot noted for 1674 'with that Violence, that several Mills might have been driven with the Current; and had not the Town of Henley made some diversion for them, their Fair-Mile must have been drowned for a considerable time'. Such events recurred and gutters constructed along the Fair Mile to cope with the problem can still be seen today.

The parish of Stonor with Pishill is modern, dating from 1922, and brings together three separate land parcels—Upper Assendon (Stonor), Pishill, and Warmscombe—that have a distinctive and largely joint history. For although the Assendon valley runs down to the Thames at Henley, the history of this area is linked northwards across the Chilterns summit to the scarp-foot settlements of Pyrton and Watlington. It is the Chiltern or 'hill-land' portion of a strip land-unit. These long, skinny parishes are typical of chalk scarps, combining the resources of fertile Vale soils and meadows with the downland grazing and woods of the hills. In many instances the hill sections formed detached portions, physically separate from the rest of the strip, and in the Stonor case the hill parts reach beyond and include the upper parts of the more fertile dip-slope valleys on the other side of the summit.

It is likely that all of this area was part of an early Anglo-Saxon land unit called *Readanora*, the red *ora*, and approximating the later Pyrton Hundred. *Ora* was the precise term used for a hill or slope shaped like an upturned canoe; here the *ora* is the scarp slope, and *Readanora* probably refers to the red subsoil that is common on the summit. This land unit was later split lengthwise into the two strip-parishes of Pyrton and Watlington, possibly at the time the *-ton* settlements were established. The division of the hill portions was unusual in that, for both parishes, they became detached and the upper and lower parts of both parishes join at a point on the Chiltern crest at Christmas Common and cross over: Pyrton is north of Watlington, but 'Upper Pyrton'—our Upper Assendon or Stonor—is south of Watlington's hill-land of Christmas Common. This odd arrangement probably arose from the line of an ancient routeway called 'Knightsbridge Lane' or 'Ruggeway' (Ridgeway) which runs right through lower Pyrton, up the escarpment and along the narrow ridge of Hollandridge until it descends into the Assendon valley just north of Stonor, and thence to the Thames. Such a route would probably have led to historic landholding links between the areas. A small triangle of the west side of the Assendon valley, known

Map of Stonor and the immediate locality

● ● ● ● ●

as Warmscombe, became a detached portion of Watlington, and Pishill parish on the north-west side was probably formed after 1066 from detached portions of both Pyrton and Watlington, later identifiable as the manors of Pishill Venables (later the hamlet of Russell's Water) and Pishill Napper (later Lower Pishill).

The Stonor valley, looking south towards Henley from the hillside north of Whitepond Farm

The hedgerow marking 'the inclined boundary bank' of the Anglo-Saxon charter lies immediately below the camera location, with Whitepond Farm beyond. To the left are the woods of the present Stonor Park; in the middle distance is Stonor village and the road to Henley. On the immediate right is Hollandridge Lane, descending into the valley, and then Pishill. Further away are the old woodlands of the former deer park, including Park Wood.

The Anglo-Saxon Charter Boundary

Our knowledge of the early land units is inevitably somewhat speculative, but a charter for Pyrton, dating from the eleventh century but claiming to refer to 774, gives a detailed boundary for the split land unit and its detached Chiltern portion. The document can be seen in the British Library and the 'boundary of the woodland holding' begins on Christmas Common, recording the outline in a clockwise circuit.

Much of this charter boundary can still be traced today, certainly all the way along its northern and eastern sides, which continue to form the parish boundary, and much of it is also the Oxfordshire–Buckinghamshire boundary. From its apex at Christmas Common the boundary follows a narrow valley as it descends from the Chiltern top, proceeding 'ever along the woodland verge' (*wyrtwalan*). The indications are that *Wyrtwala* was a special form of woodland edge, possibly with a timbered hedgebank and a perimeter track, and an old bank can still be seen along parts of the same track. The valley becomes the 'timber slade' and then drops to the *grenanweg* (Green Way) of the valley by what is now Turville Park Farm, but used to be Rolls Farm and is still within Stonor. The boundary follows

the narrow valley bottom, but not quite to the junction near Whitepond Farm. A short distance before this it suddenly climbs across the valley shoulder eastwards 'along the inclined boundary bank'. This is still the boundary, visible from the hillside following the curving hedgerow. The line then winds up 'wild cat valley' around the back of Stonor, encompassing the woodland (*Stanora lege*), before descending again to the valley south of Stonor, to a 'small ash tree', leaving it again 'by a maple tree on the west side of Assenden'. The western ascent may well have been by way of Warmscombe Lane, now a tree-tunnel climbing up to Maidens-grove Common. It is uncertain whether the boundary circuited Pishill, or just the later post-Conquest bounds of Pyrton, but it can certainly be followed along the magnificent boundary hedge from Pishill Bottom to 'raven ridge' and then back up to the 'cleane feld' or 'open country' of Christmas Common.

The charter tells us more about the character of the area: it is called the 'woodland holding' of Pyrton, but the valley parts were undoubtedly cleared and open, as is shown by the use of single-tree markers in the main Assenden valley (the ash and maple trees) and also by the 'foul slough, above which stands a solitary apple tree' in the valley above (Lower) Pishill.

Our area includes the detached portion of the medieval parish of Pyrton—that part known as Upper Assenden or Stonor—together with Pishill parish, probably formed after the Conquest out of parts of detached Pyrton and Watlington, and the small detached triangle of Warmscombe, part of the western side of the Assenden valley. Both Warmscombe and (parts of) Pishill may have been within the charter bounds. Certainly they are intimately connected with the medieval and later life of Stonor, and the rejigging of parish boundaries in 1922 brought them into one, and may have restored the Anglo-Saxon land unit.

The Medieval Landscape

The Anglo-Saxon charter provides the boundaries within which to construct our landscape history, but it is the medieval documents which start to give it real life. Domesday Book and early taxation records are not as illuminating as we would like, partly because the detached areas are treated with their larger parishes. It is the local manorial and estate records that give us glimpses of the rural economy. Here the history of the Stonor family is central. The Stonors of Stonor first emerge in the records in the late twelfth century as holders of a free tenement under Pyrton manor. This vigorous family subsequently built up major land-holdings, controlling most of our locality and creating what came to be accepted as their own manor across several parishes, as well as acquiring numerous other estates across southern England. A major force in this expansion was Sir John Stonor (1280–1354) who became Chief Justice of England and whose tomb-chest effigy can still be seen in the Abbey at Dorchester-on-Thames, 12 miles from Stonor. At the end of the fifteenth century all the extant family papers and records were seized by the courts and have remained in the public records. These provide a major insight into gentry life, comparable with the Paston Letters for Norfolk, but they also provide us with detail on the local landscape.

The late Anglo-Saxon and medieval landscape had many of the lineaments we can see today. The woodland was probably somewhat more extensive, especially on the plateau tops, but in the main Assenden valley, and where the upper valleys come together around Lower Pishill, the land was cultivated (already in evidence in the Anglo-Saxon charter) and common fields developed. The name Pishill derives from 'hill where peas are grown'. Medieval and later documents show that the common field systems developed across the earlier parish boundaries: in the upper valley the fields lie across both Upper Assendon and Pishill, and in the lower valley the fields cross Assenden, Warmscombe, and parts of Fawley (so also crossing the county boundary). Many of the medieval fields can be located today, using a very detailed map of the Stonor estates in 1725. This map which can be seen on the study-wall at Stonor Park names all the tenant-farms, fields and woods, providing a historical bridge between the present Ordnance Survey map and the medieval documents.

Stonor manorial accounts for 1387–8 and 1476 mention several fields, and in 1387 the demesne sowed 84 acres with barley, wheat, oats, pulse, and maslin (*mixtum*), a mixture of wheat and rye. Stompars-feld in 1387 (Stamperhyll in 1476 and Stompers Hill, subdivided into three enclosures, in 1725) was the sloping hillside field behind (east of) Stonor village. Molsows of 1476, also divided into three closes by 1725 (Great and Little Mousalls and Mousalls Hill), was the east field just further down the valley. Facing this on the west of the road was Warmyscombe field, also later subdivided. North of Stonor the land on either side of lower Hollandridge Lane made up Pishill open-fields, and the present Whitepond Farm is largely constituted out of the east portion of those fields.

Payments in the manorial accounts for 1336–7 show some local occupations and activities: 'to two servants assisting the shepherd of Stonor and Bix during the lambing-time'; 'food for the haywards in autumn'. Sales of sheep-skins (both with wool and as hides) are recorded, as well as swineherds, repairs to dovecotes and all the activities of a busy, mixed economy. A *Stonor Letter* of *c*.1475 notes that a dyker had been brought to Stonor 'for to make yowr dykes in … feld between the hy way and the ew tre'. There was uncertainty about whether a single or double ditch was wanted, but the top was to be set with whitethorn and the ditch had to be a yard deep, all priced by the perch-length. Possibly we see here the beginnings of enclosure in an open field, though it may be associated with the deer-park, for a later letter of 1480 notes 'your parke gothe welle onewarde in dyking and pale, your husbondrie like wise'. The 1475 letter concludes 'And ye shud have a Monday next comyng a xl plowys in Pyssyll felde', but a 1482 letter argues 'youre husbondrie is note welle gydide', having too many workers to only one plough.

As well as providing farming land, both open and enclosed, the Stonor locality also served its role as hill-land for the scarp-foot residents of Pyrton. It provided summer and autumn grazing, and such transhumance is probably commemorated in the name of 'Summer Heath' behind Stonor. In the mid-twelfth century a grant to Hurley Priory (which held land in Pyrton) allowed feed for twenty pigs in Breton-heth (now unidentifiable) in Assendon. A charter from Ralph Stonor to two Pyrton residents in 1394 includes a note of 'a right of way through the grantor's field called Millefeld for driving their sheep and cattle to Pishill'. The extensive commons and scrub in upper Pishill, now known as Maidensgrove and Russell's Water Commons, were an important resource, and William Esynton would take his animals up Knightsbridge Lane and along Hollandridge Lane, then crossing Pishill Bottom onto the commons. The 'Windmill Fields' are named and located here on the 1725 map, and a right-of-way still crosses them today at this point.

Pyrton also possessed isolated farms, with their own enclosed fields cut out of the woodland. Little documentary evidence exists to show such colonization in this locality, but plenty of evidence is found of early medieval hill-land settlement in neighbouring Watlington and Shirburn parishes. Hollandridge Farm is first mentioned in 1282. It is tempting to derive the name as hill-land-ridge, which fits the situation exactly, but the first known form is Herlingerugge (Hollingdridge in 1725). In the valley east of Hollandridge, what became known as Rolls Farm (now Turville Park Farm) was also always separate. On the surrounding heaths and woods, cottagers had also settled on the commons side and farmed their own small fields.

The Stonors' house, Stonor Park, lies back from the road, and one has to enter the grounds to gain a view. Today the house presents an Elizabethan E-shaped design with eighteenth-century alterations to the gables and windows, but the chapel and core of the present house date from about 1280, with extensive rebuilding by Sir John Stonor in c.1349. Later building work in 1416–17 is recorded in the *Stonor Letters & Papers*: the local clays of the Reading Beds at Nettlebed and Crocker End were good for making bricks, and the accounts note payments of £40 to Michael Warwick for making '200,000 de Brykes' and £15 for transporting them from Crocker End to Stonor.

The house is today set in an elegant deer park, with groups of fallow deer grazing under the trees. Medieval Stonor also had its deer park, and amongst the family correspondence in the *Stonor Letters* is one written by Elizabeth Stonor in London just before Christmas 1476 to her husband William at Stonor: 'as ye wryte that ye will sende me of a wylde bore and other venison ayenst Sonday, truly I thanke yow'. Leland noted of Stonor: 'Ther is a fayre parke, and a waren of connes [rabbits], and fayre woods. The mansion place standithe clyminge on an hille, and hathe 2. courtes buyldyd withe tymbar, brike and flynte.' However, it emerges that the medieval deer park was not next the house, but to the west of Stonor village, rising up to Maidensgrove. John Steane has traced its typical convex outline in the field boundaries, and its legacy is marked by Park Wood, Park Lane, and Lodge Farm up at Maidensgrove, together with old field names called 'laundes' near Lodge Farm.

The historic character of many of the present-day woodlands is revealed in their names, which reach back to the medieval period. Queen Wood commemorates, not Elizabeth II or even Elizabeth I, but Elizabeth Woodville, Edward IV's queen in the 1460s. College Wood makes the local connection to the Dean and Chapter (or College) of Windsor, important landholders here from the late fifteenth century to the nineteenth. Doyley Wood recalls the d'Oilly family, present in the area at the time of Domesday Book; Park Wood locates the former Stonor deer park to the west of the hamlet, whilst Kilridge Wood behind Stonor occurs frequently in fifteenth century deeds. A 'Wodewardi de Kyllyngryghe' is recorded in a Court Roll for 1414.

The woods and their products were an important and valuable resource, and rights of ownership and use were jealously protected. Medieval records show the varied role of the woods' products. Most were coppice-with-standards, less dominated by beech than they were later, but making coppice growth the major yield. Manorial owners usually 'sold' a woodland for a specified number of years, with limitations on sizes of trees that could be cut, to protect both young growth and high, standard timber. A deed of 1525, still preserved in the archives at Windsor, indicates the range of species: it sells part of Kilridge Wood, with 'beeches ashes withese maples apssis and whitebemys' (beech, ash, withies or willow, maple, aspen, and whitebeam), but only 'whereof the trees be above 25 inches above the ground at brest hight of a man for 3 years for £40'. The *Stonor Letters & Papers* for the 1480s include several wood-sale deeds and accounts. In 1480 William Stonor sold Herre (Harry) Chone 'all the wood, oak and ash, except unto the assise of xxvi inches, bounding on Kilrygge and Litelbowettess and on the highway to Somer-heth [Summer Heath] … with free ingress and egress to and from the wood to carry coals etc'. A wood account of 1482 gives the most detailed picture, with many references to 'Saunders', probably the woodward. It includes 'wode the wch my master sold for tile' and 'XXX qrs cole' that were sold. Firewood was also being sent from Stonor to the household in London, probably by barge from Henley.

Although most of the enclosed woodland was a manorial preserve exploited commercially, tenants often had specified rights to collect fallen and small wood. Thus in 1496 Robert Rolffe was allowed 'wood sufficient for haybote, cartebote and plowbote' in the woods belonging to Windsor. Wider communal rights were exercised in the larger wooded and scrub area that extended from Bix across upper Pishill and into upper Watlington. This was the wood of 'Minigrove' or Maidensgrove, deriving from the Anglo-Saxon for 'common grove', and shared by Bix, Pishill, Pyrton, Watlington, and Swincombe. Here there were common rights of pasture and 'estovers' (wood-collection).

The Landscape after 1500

As in much of the Chilterns, the common fields of the upper Assendon valley disappeared quite early. By the time of the 1725 map the whole landscape was composed of separate, enclosed fields, but the change greatly predates this time. A lease of 1584 refers to former common fields at Pishill as 'closes', and it is likely that consolidation and enclosure were associated with the Stonors' control of the lands and their interest in sheep-farming. The *Stonor Letters* reveal the extensive wool and sheep interests of the family in the fifteenth century, linked to their wider estates in the Cotswolds and elsewhere as well as local estates, and Thomas Stonor was certainly enclosing at Rotherfield (west of Henley) around 1500. Any major switch to sheep in the Stonor locality was, however, substantially reversed in the sixteenth to seventeenth centuries as the general growth of population and especially of London increased demand for cereals. This can be seen in the probate inventories for Hollandridge Farm: Nicholas Smith's inventory for 1628 listed 120 sheep and lambs with only £16 worth of 'corn, hay and vetches', whereas Edward Drew's inventory in 1692 records corn worth £88 in the ground, and only 10 sheep and 8 lambs. The corn was a mixture of barley—possibly for malting at Henley—oats, peas, and wheat.

The later seventeenth and early eighteenth century saw significant agricultural improvements. The Stonor family had remained stubbornly Catholic after Henry VIII's Reformation—as they do to this day—and suffered both persecution and fines. Their estates shrank from their medieval extent, but they retained their local dominance and focused on farming advances here. The 1693 lease for Hollandridge included 'parcels of wood ground that was lately grubbed and converted into tillage', including Ballasdree [Bullistree] coppice, and these pieces (on

The great barns at Whitepond Farm, north of Stonor village, as drawn by W. Fairclough in a wartime portrayal of the English landscape

Similar barns can be found on other Stonor estate farms and date from the period of Thomas Stonor (1677–1724) and his son, also Thomas (1710–72).

the west side of Hollandridge Lane) have remained as fields. From this period, or the early eighteenth century, date the solid brick, timber, and flint farmhouses and the substantial barns with their tiled roofs and enormous timber doors, which can still be seen at Assendon, Hollandridge, and Whitepond Farms. With these additions, much of the rural landscape visible today was already present by 1750, but this did not mean that the locality stood aside from later improve- ments. Thus Sewell Read, in his survey of Oxfordshire agriculture in 1854, com- mended Assendon Farm at Stonor: 'It would be difficult to find a hill farm more advantageously cultivated,' with its rotations of corn (wheat or oats), turnips and vetches, and use of the new imported fertilizers. The tithe award maps show the dominance of such arable, fertilized by sheep folded on the stubble. Perhaps the major price of this era of High Farming was the removal of many hedges, amal- gamating smaller fields that were shown on the 1725 map.

During this period the woods became even more valuable, because of the expansion of the Thames river-trade and the London market for firewood. Details can be found in wood-accounts in the Stonor archives. Thus in 1748–9 oak and beech timber and poles were sold, plus oak bark for tanning ('1 load 14 yards in Stonor Park: £3.-5s.-0d.'), and a whole range of sizes of wood for fuel. Some was sent up to London, both for family use and as commercial sales from Henley. The same accounts note

long faggotts 100 in Stonor Park	0-15-0
669 in Horsells Bottom	5-19-8
Bavins 1230 in Stonor Park	9-04-6
830 in Stompers Hill	6-04-6
5560 in Kildrig	41-14-0

All of these places can still be located on the map (Horsells Bottom is now the lower portion of Longhill Hanging Wood).

Samuel Rockall turning a chair-leg on his wheel-lathe in the 1930s

Rockall was one of the very last of the Chiltern 'wood bodgers', and his cottage and workshop were at Summer Heath, just to the east of Stonor Park. Specimens of Rockall's craftsmanship and tools were collected by H. J. Massingham and are now in the Reading Museum of English Rural Life.

A long-standing irritation to the Stonor family was the fact that Kilridge Wood behind Stonor was owned by the Dean and Chapter of Windsor, and the Stonors only managed to lease it for substantial periods. It was allowed to grow high and dark, and when the poet Alexander Pope visited Stonor in 1717 (Thomas Stonor was a subscriber to Pope's Homer) he noted 'the gloomy verdure of Stonor', although his view was probably jaundiced by Thomas being absent and his having to travel on all the way to reach Oxford at 11 at night!

By 1800 the demand for wood was changing—wood for household fuel was facing rivalry from coal, whilst wood for furniture-making, and especially chair-making in the Chilterns region, was increasing. Larger poles and timber were now in greater demand. The 1871 Census records a number of woodmen, chair-makers, and turners located around the common in Pishill parish. Eventually furniture-making moved further to factory-production and imported timber, but one of the last of the Chiltern bodgers worked in his cottage on Summer Heath, behind Stonor Park. Samuel Rockall was much extolled by H. J. Massingham in the 1940s as the archetypal craftsman. Rockall's father-in-law was a gamekeeper on the Stonor estate 'living all his life in the same cottage in the woods', and Samuel's wife claimed that her husband had 'never been anywhere in

46 years except to visit her mother and cut his timber' (and the mother had only had three visits!).

The changing wood market also saw the introduction of fir-tree plantations in the late nineteenth century. On the borders of our locality both Jubilee Plantation and Kitchener's Firs give the game away. On the west of the Stonor valley, 'The Firfields' can be seen, occupying part of what was Warmscombe, whilst the firs of 'Row Dow', just to the north, are 'Rowdown' or 'Rough Down' on the 1725 map, within the bounds of the old deer park. Much of the older woodland has been cut down and replanted with firs, like Balham's and Kildridge Woods, managed by the Forestry Commission behind cosmetic screens of beech. In contrast Park and Doyley Woods retain their 'high beech' character.

On the upland common at the hamlet of Pishill Venables a brick-making industry, like that in Nettlebed, was started, also using the local clay, and was certainly active from c.1665 to the mid-nineteenth century. Russell's kiln is mentioned in 1695, and the hamlet is now known as Russell's Water, named after the brickworks pool there. Both pool and former kilnhouse can be seen today. This works, like that at Nettlebed, was a consumer of wood for the kilns and kept up local demand for coppice after the general decline of firewood sales; it may well account for the survival of old coppice-woods at Maidensgrove and in the Warburg Nature Reserve across the boundary into Bix.

Twentieth-Century Landscapes

The long-running agricultural depression after 1870, combined with the changing wood industry, had its impact on the Stonor landscape. As arable incomes and wood sales fell, jobs on the land and in the woods shrank. These parts of the south Chilterns suffered less than some others because they were able to change to dairy farming to provide milk both by train to the London market and to the milk-processing plants in Reading. The unprofitable woodlands were subsequently left to grow overly tall, making the woods vulnerable to disease and wind. Writing immediately after the Second World War, Massingham noted: 'Now the old Catholic house of Stonor is strangled with the ivy of finance; the fields are heavy with sow-thistle, fat hen and charlock.'

Coincidental with these changes was the rise of motoring, which brought neighbourhoods such as Stonor within the orbit of Londoners, who came for day visits and bought country cottages. Some roads were surfaced, and others such as the old Hollandridge road left aside, so that now the visitor by car would not know this old road existed, and it is left for the walker, rider, and cyclist. For a period the steep Alms Hill (behind the Alms Houses) became a challenging competition hill-climb for enthusiastic early motorists. The popular novelist and playwright Cecil Roberts bought a cottage at Lower Assendon, and observed 'a large van-like car, with curious brightly varnished wooden sides, a drop end, and a hooded driving seat.... It was Lord Camoys's pantechnicon from Stonor Park, and what it was for or what it contained I was never quite sure. Sometimes it was crammed full of young people, lively shoots from the Stonor family tree, sometimes a dark-eyed girl drove it, or a youth in a gay check suit half-leaned out of it ...'

However, the Stonor area escaped most of the land sales for building that scarred more accessible parts of the Chilterns in the 1920s and 1930s, and instead it was extolled in guidebooks as the most remote and unspoilt part of the Chilterns. Cecil Roberts's best-selling *Pilgrim Cottage* sequence is partly set in the valley, and his non-fictional trilogy *Gone Rustic*, *Gone Rambling*, and *Gone Afield*, published in the mid-1930s, made the area known to a wide reading public, as did H. J. Massingham's *Chiltern Country* in 1940. Visitors explored the area, and 'incomers' began buying up and refurbishing the decaying cottages around the commons, a process that has continued since 1945, making the locality one where property prices are high.

After 1945 agricultural policy and new technology revitalized arable production, and Massingham's down-at-heel landscape vanished, replaced by mechanized, efficient farming practices. The price here has been comparatively low, involving further removal of field boundaries and the construction of an ugly farm silo right at the valley junction. Fortunately the landscape has not been devastated like some others, and the 1725 map of old farms, woods, and enclosed fields is still largely reflected in the countryside of today. At Stonor Park itself, the family has successfully restored the house and estate (without any resort to theme-parks), and a quite remarkable line of continuity is being maintained.

The heritage of this landscape was recognized in the 1994 Chilterns Management Plan, where the Stonor area is noted as a prime example of 'ancient countryside' to be conserved. It is a landscape that can be enjoyed through its dense network of public footpaths and bridleways, typical of much of the Chilterns. To walk these paths is to tread a historic landscape, and imprints of those who have gone before still linger. From Summer Heath a path leads down through conifer woods into Stonor Park, following the footsteps of Samuel Rockall on one of his visits to his mother-in-law. Out of the conifers, a rise and a twist in the path suddenly presents the best panorama of Stonor House, a view which must have been admired by generations of local woodmen and the Stonor family. For despite their very different social positions, they have all shared deep roots and long attachments to this locality.

Walking from Christmas Common along the ancient route to Stonor and Henley, just before Hollandridge Farm itself, a stile and a path lead off to College Wood. Here in August 1534 John Ap Powell, a servant of Sir Walter Stonor, lay in wait for Walter's brother Edmund, who farmed Hollandridge. He assaulted Edmund (possibly over a sheep-farming dispute), but came off worse himself and died. Sir Walter's letter to Henry VIII allows the scene to be visualized today: 'There is a narrow lane by the house, which is on the highway to Stonor. He [Ap Powell] set there a sparrow-hawk that he kept of mine upon a stile, and tied her fast, and at my brother's coming called him: "Alight, coward knave, for I will have a piece of thy flesh." '

On a hot summer's day, we can still climb from the Assendon valley up the tree-tunnel of Warmscombe Lane, with its vistas down onto the arable fields that were Warmscombe's, and across to sheep grazing the upper fields. Up this same lane Michael Patrick (or Partidge), a villein in the 1279 Hundred Roll, must often have climbed. Michael cultivated his own 15 acres, for which he paid 10 shillings rent, but he also owed three days' work, with two men, on the manorial holdings at harvest time and he had to help shear and lift the sheep. In this luscious piece of Chiltern countryside the continuities in the rural landscape are tangible and through the fortunes of its unique heritage we can reach out and touch the past.

12 Hook Norton, Oxfordshire:
An Open Village

Kate Tiller

Hook Norton in the 1920s, looking west from the church tower

The linear settlement pattern can be seen, with a long street flanked by the houses, crofts, orchards and barns of the sixteenth-, seventeenth-, and eighteenth-century farmers of the open fields. At the far end of the village is its nineteenth-century brewery, which brought new diversity to the village economy.

The traveller does not stumble casually upon Hook Norton. Rather it has to be sought out, along by-roads leading from the ancient ridgeway, which runs north-east to south-west across the parish a mile and a half north of the village, or by branching off the modern A-road between Banbury and Chipping Norton, whose course lies just south of the parish boundary and some 2 miles from the village. Approaching from the south-east and the Chipping Norton road the way is undulating, across the small valley of the river Swere, and twisting, around sometimes right-angled bends, to South Hill overlooking Hook Norton village. Settlement is focused here, apart from some outlying farms, most of which post-date the parliamentary enclosure of 1774. The village is long and straggling, following the slopes of a steep-sided stream valley, and bracketed at its western end by the tower of the nineteenth-century Hook Norton Brewery. The villagescape is dominated by a rich stock of domestic, vernacular buildings, several of which

Hook Norton in the 1920s, looking south-east from the church tower

Immediately below lies the 'market place', with smithy, shops, and inns. The unregulated buildings of the twin settlements of Hook Norton and Southrop lie on either side of the valley with the railway viaduct (opened in 1887) and the open countryside of the north Oxfordshire redlands beyond.

signal in their red and cream striped walls Hook Norton's position on the boundary of the north Oxfordshire Redlands (to Arthur Young 'the glory of the county') and Cotswold oolitic limestones.

At the centre of the village, on the north bank of the valley, stands the parish church. From its tower settlement and landscape become clearer. To the west the long street is flanked by the stone and thatched or tiled homes of the sixteenth-, seventeenth-, and eighteenth-century farmers of open-field agriculture. This pattern of houses, crofts, orchards, and sometimes barns can still be unravelled from later infill and twentieth-century 'refurbishment'. The scene is empty of a 'big house' or park, the nearest thing being the gabled building now called the Manor House, but built by the local Austin family in 1636. At the far, Scotland End of the village is the tower of the brewery ('an extraordinary essay in brick, ironstone, slate, weatherboarding, half-timber and cast iron', according to John

Piper), built in 1897–1900. It is a successor to the brewery begun on a local farm by John Harris in 1849 which helped inject new diversity into the village economy during the difficult years of agricultural depression.

Turning to look south and east from the church the street below opens out into a vestigial market place. Hook Norton was and is a large, diverse, and vigorous village society, but one that never developed into a market town, despite the fact that in 1437 the Earl of Suffolk, the lord of the manor, had a Charter to hold a market and two fairs. Then, as subsequently, seigneurial influence failed to make a lasting mark on Hook Norton. Rather the market place was the location of some of the many businesses and events that figured in village life. In 1871 there were no fewer than 133 crafts and tradesmen, over 10 per cent of the population, in the village. Around the market place in the late nineteenth century were a smithy, an ironmonger's shop, a tailor and outfitters, and the Sun and Bell inns. Stock sales and the annual Club Day fair on the Tuesday before Whitsun were held here. Buildings tumble in an unregulated way down Bridge Hill to the stream and it is here that settlement spills over from the main village onto the south slope of the valley, creating a subsidiary hamlet of Southrop, first mentioned in 1316. Archaeological evidence suggests earlier activity in the area, which may subsequently have been amalgamated into the main area of village settlement. At the bottom of the hill was one of four village tites, places where water was collected in buckets on a yoke by those who did not have their own wells. Mains water came to Hook Norton only in 1955, and sewerage in 1965.

Beyond the roofs of Southrop the landscape opens out, mixed farming country in a parish (large for Oxfordshire) of 5,340 acres, rising from 450 to 650 feet above sea level on the ridges to north and south of the village. In the far distance the tall piers of a viaduct carry the Banbury to Cheltenham Railway. This connection came late to Hook Norton. The line was opened in 1887 and closed in 1962. Plans had been made as early as 1845, but technical problems and the cost of overcoming the terrain meant that nothing came of them until 1883–7 when two sections of viaduct and a tunnel were built. The work is said to have taken 400 men four years, and six workers were killed. In August 1883 police had to be brought from Banbury to Chipping Norton to restore order when disgruntled navvies laid siege to the manager after their ganger had disappeared with the wages. A navvy was sentenced to hard labour for stealing chickens from a local farm, whilst another had his leg broken, fighting at the Bell at Christmas 1886. Once the railway opened, in 1887, the longer-term impact began to emerge—a monthly cattle market near the new station was started, the first seaside excursion (to Portsmouth) was oversubscribed, and above all the low-grade ironstone of the area became a commercial proposition with direct access to lines to South Wales and Staffordshire. This new ability to exploit a natural resource added another source of local employment for the village.

To walk through the village east of the church is to see the signs of different phases of Hook Norton's fortunes. There are more of the rich yeoman houses of the seventeenth and eighteenth centuries, several with distinctive staircase turrets built from the local red ironstone. There are small cottages and extended roof spaces into which the expanding population of the late eighteenth and early nineteenth centuries was fitted. The 1,351 people recorded in Hook Norton at the 1821 Census represented a 31 per cent rise in twenty years, an increase which was said 'to consist only of paupers'. Then beyond East End we reach the Railway

Inn, the station site, and shortly, incongruous in surrounding fields, the remains of the Brymbo Ironstone workings and their accompanying terraced housing. Finally, the minor road heads back towards the 'beaten track', the main road to Banbury, passing on its way Manor Farm and Butter Hill, a rich grassland, the field-name of which can be traced back to 1154 when it is mentioned in Oseney Abbey cartularies. Here, as throughout Hook Norton's landscape, land, buildings, field and place names, artefacts, and documentary sources when brought together yield a rich and lengthy story.

Hook Norton is in many respects a classic 'open' village. Its parish area is large and its population high by rural standards, undergoing a particularly rapid rise (from 1,032 in 1801 to 1,525 in 1841) during the period when contemporaries began to employ the distinction between 'open' and 'closed' communities. It was a village with numerous farmers, high poor rates, a wide range of rural industries and crafts, many shops and pubs, a large housing stock in diverse ownership, an absence of large estates or resident gentry, a strong tradition of religious nonconformity dating back to the seventeenth century, and a reputation for independence. This distinctive character can be discerned long before the term 'open village' was coined in the nineteenth century. Perhaps it underlies a number of unflattering rhymes dating from the sixteenth century on (e.g. 'Hogs Norton, where pigs play on the organ'), which identify the village with rusticity and boorishness. The name of the place appears as 'Hogesnorton', in the Close Rolls, as early as 1381.

Industry in the countryside

The opening of the Ban-bury–Cheltenham Railway in 1887 made it viable to work the local, low-grade ironstone, which was dug out of the fields to the east of the village (see Map 12.4)

The Brymbo works

Here the calcining kilns stand amidst fields, with railway wagons on the embankment to the right, waiting to take the processed ironstone to South Wales and Staffordshire. Ironstone working was short-lived (lasting some forty years), but important to the village at a time of agricultural depression.

The first, written reference to Hook Norton comes in the Anglo-Saxon Chronicle and describes a battle there in 913. The form of the name is 'Hocneratune', rendered by the place-name historian Margaret Gelling as 'the tūn of the people at Hocca's hill slope'. Thus at the very outset we are faced with the challenge of relating people and events to the surviving physical terrain (Map 1). As we have seen, the present village certainly lies on a slope, along the stream valley at the centre of the parish. However, recent work by John Blair suggests that the events of 913 may have been focused some 2 miles north-east. The Chronicle (John of Worcester's text) describes how, 'After Easter the pagan army from Northampton and Leicester plundered Oxfordshire, and killed many men in the royal vill (*regia villa*) Hook Norton and in many other places . . .' This was one of the abortive Viking counter-attacks which punctuated the reconquest of the Danelaw. It was the practice of Saxon writers to locate such military events in relation to royal vills. Blair notes that the ridgeway running south-west from Banbury towards Gloucestershire and through the north of Hook Norton parish would have been the major route along which Vikings from Danelaw areas to the east could cross the Cotswold uplands into southern Mercia. Moreover, near to the point where the ridgeway enters Hook Norton parish no fewer than five later parishes meet in an area close to a striking concentration of earlier sites—a holy well, a pagan Anglo-Saxon burial, the major Iron Age hill fort at Tadmarton Camp, and two smaller polygonal enclosures. Using the evidence of later field names it has also

•••••

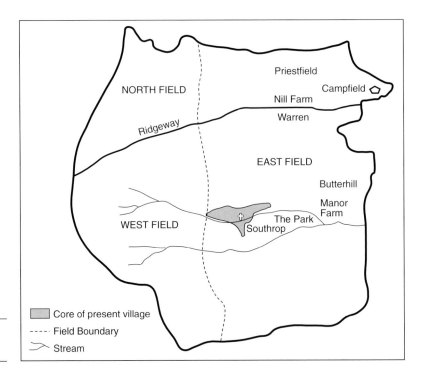

Hook Norton: the medieval landscape—principal features

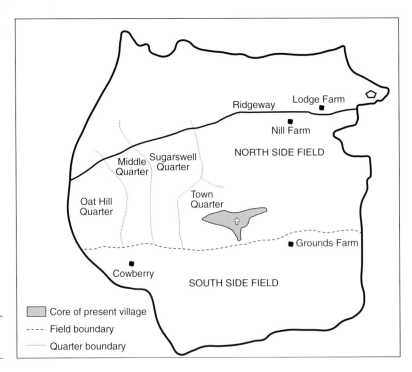

Hook Norton: post-medieval fields and farms

been possible to locate Priestfield as an area astride the ridgeway and one which, according to twelfth-century documents recording the gift of the Hook Norton church by the post-Conquest lords of Hook Norton to endow Oseney Abbey, had previously 'pertained to the church of that vill'. Taking all these clues of strategic, political, and ritual importance Blair has suggested that the Saxon royal vill of Hook Norton may have been sited on the ridge west of Tadmarton Camp, in the north-east of the present parish and not in the modern village.

There is no doubt that shifts in settlement are a feature of English landscape history and that this part of the Hook Norton landscape saw considerable activity in pre-documentary periods. The application of modern fieldwork techniques to one field demonstrates this. The author, together with students of a local history extra-mural class held in Hook Norton, field-walked Campfield, adjoining the ridgeway and near the meeting of boundaries of five parishes. Once again field names provided an initial clue. An early task in the analysis of the Hook Norton landscape was to compile a field-name index, starting with the fields shown on a post-war 6-inch Ordnance Survey map, then collecting currently known names, and then working back through successive historic periods, helped by maps only as far as the eighteenth century but by documentary sources as far back as monastic cartularies of the twelfth century. Using this painstaking compilation of data the patterns of the medieval and post-medieval field systems began to emerge. Other sources, including always (as we shall see at Campfield) looking at the landscape on the ground, were then brought into play. Through such a process the reality is apparent of W. G. Hoskins's notion of landscape as a palimpsest, as a fabric used and reused many times, but the layers of which, by dint of careful investigation can gradually be separated and peeled back. Here too the particular value to landscape history of regressive analysis, as propounded by Marc Bloch and others, becomes clear.

Hook Norton's field names suggested many things; land use (Cowpasture), size (Five Acres), soil conditions (Pudding Furlong), open field features (Butts, Long Cut Furlong), ownership (Parson's Hill), crops (Old Sandfoin ground), and archaeological features (Campfield). Using the field-name index and map, Campfield was located. References to Hook Norton Camp were found in antiquarian sources. Robert Plot's *Natural History of Oxfordshire* (1677) mentions a 'quinquangle', near to Tadmarton Camp, which he associates with the campaign of 913. By the nineteenth century Alfred Beesley, quoted in a commercial directory of 1852, was still linking the Camp with 'the terrible slaughter of the English … driven by the Danes from the camp of Tadmarton', but the banks of the camp were reported to be reduced by the plough 'almost to the level of the soil'. Nevertheless, twentieth-century aerial photographs in the Oxfordshire County Sites and Monuments Record revealed the irregular pentagonal outline of the Camp as a soil mark and it was decided to field-walk the site after autumn ploughing. Differences in soil colour still marked parts of the single bank of the enclosure, and within its area thirty-four worked flints were found and subsequently identified as prehistoric. Apart from a solitary earlier Neolithic example, eighteen of these were Bronze Age, including a fine middle Bronze Age skinning knife, and equally distributed between early, middle, and late periods, probably indicating use of the site by a small group, possibly seasonally, but over a long period.

By 1086, when Hook Norton is recorded in Domesday Book, it had ceased to be a royal vill. Had Hook Norton lost its previous status when a block of royal

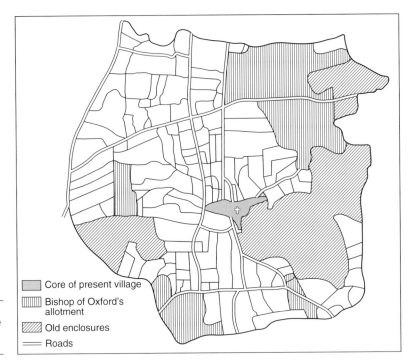

Hook Norton: parliamentary enclosure 1774—old and new enclosures

Core of present village

Bishop of Oxford's allotment

Old enclosures

Roads

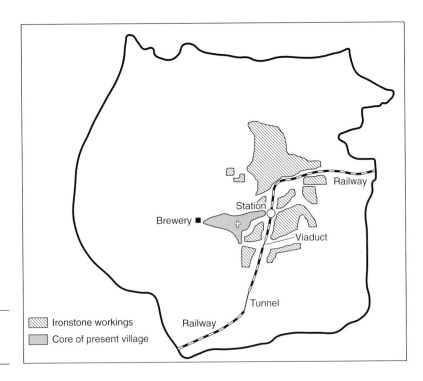

Hook Norton: the nineteenth century— brewery, railway, and ironstone workings

Railway

Station

Brewery

Viaduct

Tunnel

Railway

Ironstone workings

Core of present village

land in north Oxfordshire had been split into separate manors in the tenth century? Within Hook Norton had settlement shifted in focus to the present village site at the end of the Saxon period from a previous centre on the ridge to the north? Only some pieces of the jigsaw are available to us. Other archaeological finds in the area of the parish suggest settlement and land use to the south, east, and west of the present village in the Romano-British and Iron Age periods. In the area of the village itself evidence of earlier settlement has also been uncovered. The oldest surviving fabric of the parish church, previously thought to date from the early to mid-twelfth century, has proved to be late Saxon with the discovery, during work in 1987, of long-and-short quoins on the eastern angles of the nave, part of what was a substantial Anglo-Saxon building. A short distance away, in Southrop just the other side of the stream, a burial accompanied by a hoard of silver coins has taken the story back into the ninth century. The find was first made in 1848, but a recent reconsideration of the manner of the burial and the coins found with it, has identified this as a warrior burial, and coin hoard, characteristic of a soldier in the Viking army and datable to 875 or shortly afterwards. If there was still a place of strategic importance up on the hill to the north in 913 it seems increasingly likely that a significant settlement was also already in existence in the valley. This was to be the central place in the medieval and later landscape of Hook Norton.

At thirty hides Hook Norton was the largest of Robert d'Oilly's manors in 1086, and the centre of his barony. Its landscape, as to be expected of a subsistence agriculture, was dominated by arable cultivation. There was land for thirty ploughs, but with five ploughs on the five-hide demesne and a further thirty ploughs on tenant land, the arable must have been fully exploited. Hook Norton's rich resources of pasture also figure, with 140 acres of meadow, and pasture five furlongs by two furlongs. There was also a spinney, extending to two furlongs by half a furlong.

The arable land was managed through a common-field system fully developed in its complexity. For example in 1260 a half-yardland holding consisted of thirty-four separate pieces together with one acre of meadow per year in the common meadow. At this stage Hook Norton had two great fields, East and West, but by the mid-fourteenth century, a third, North, had appeared (Map 1). Much of what we know of the medieval landscape and its workings comes from the rich surviving records of Oseney Abbey, the Augustinian house in Oxford founded by Robert d'Oilly in 1129.

Robert gave Oseney not only Hook Norton church, its advowson, and rectorial rights (the living subsequently became a perpetual curacy), but also substantial areas of glebeland amounting to three hides. These holdings later became the basis of Nill and Manor Farms. The effect was that Hook Norton had two manors, that of the d'Oillys and their successors (the longest lasting of whom were the de Plessets, and the Chaucer-de la Pole dynasty, Earls and Dukes of Suffolk) and a second abbey manor. Each lordship had some of the landscape accoutrements of its status, near Nill Farm a dovecote and a rabbit warren, and to the east of the village, Hook Norton Park, first mentioned in an *inquisition post mortem* of 1301 as part of the holding of Hugh de Plesset.

The relationship between Oseney Abbey and the local parish is apparent today in the marked contrast between the chancel and nave of St Peter's church. The strikingly plain and largely unaltered twelfth-century chancel was the responsibility of the abbey, the rectorial appropriators, whilst the nave was the concern of the parish. It, and the tall west tower, are altogether grander in scale and, in their Decorated and Perpendicular style, betoken the prosperity of the village in the fourteenth and fifteenth centuries. Inside the church the only seigneurial monument is a thirteenth-century grave slab of Isabel de Plesset, tucked away in the corner of the north transept. Thereafter the seats of power of the medieval lords of Hook Norton manor lay elsewhere. There were continuing contacts with absentee lords, as shown by the charter of 1437 in the Earl of Suffolk's attempt to establish, or more likely control existing market activity. However, this already reduced link ended in the early sixteenth century when first (in 1513) the fortunes of the de la Poles foundered and then (in 1539) Oseney Abbey was dissolved. The old patterns of landholding broke up and, although some vestiges of manorial jurisdiction remained through the Bishop of Oxford as successor to Oseney, the real momentum of the place rested with the local families who are memorialized in the yards of the parish church and the Baptist chapel, the Goffes and Lampetts, the Austins and Wilmots and their like.

The village took on its now characteristic appearance in the early modern period. During the late sixteenth, and especially the seventeenth and early eighteenth centuries it underwent a great rebuilding of its vernacular houses and farm buildings. Hook Norton was rebuilt in stone, predominantly the fissile, coursed rubble walling of richly coloured orangey-red ironstone. Except for the few grander houses with stone slate roofs, the majority of buildings were thatched. The results of this rebuilding stretch from East End to Scotland End and over the stream into Southrop. They encompass a small number of large houses; in the Hearth Tax returns of 1665 the Crokers, lessees of the Bishop's Manor, held the Parsonage House (described in 1650 as 'a Fair Stone built House . . . containing many fair and useful Rooms with Necessary Barns Stables outhousing and Yards') with thirteen hearths. The Austins, at the 'Manor House' already mentioned, had eight. Otherwise the overwhelming majority of Hook Norton taxpayers had homes with between one and four hearths. These represented the typical houses with a ground plan of three rooms in line, usually parlour, hall, and kitchen, and chambers over, or smaller two-room cottages. Something of the interior lives of these homes is revealed by probate inventories. For example, the Calcotts were an established village family. Alexander Calcott (1616–82) was a village baker. His probate inventory describes moveable goods in a two-storey, six-roomed Hook Norton house. The total worth of his goods and chattels was £72. 15s. 0d. Of this over £39 was due to him in debts, and some £32 accounted for by furniture and household belongings. Because he ran his business in his home, domestic cooking was done in the hall, whilst the kitchen became the bakehouse, with a furnace, dough troughs, and moulding boards. In the stable-cum-fuelhouse was 500 of furze, fuel for baking. In the cellar was brewing equipment and 'two dozen of hemp'. The showpiece of the house was obviously the chamber over the parlour, its contents worth more than any other room and showing the level of comfort attainable in a village home by the second half of the seventeenth century. There was a feather bed with scarlet curtains and counterpane, a red rug, a table and eight red leather chairs, four pictures, and fire irons.

Building was going on outside the village too. In 1646 Lodge Farm, a smart new yeoman farm house, was built 2 miles north-east of Hook Norton on the old ridgeway (Map 2). This was probably created in connection with enclosure for sheep-farming, and is just one element in major changes which Hook Norton's farming and thus its landscape underwent in the early modern period. Other enclosure was taking place. A rare surviving agreement of 1672 to enclose Cowberry field and meadows shows the process at work. The parties were William Horwood, grazier of Hook Norton, James Beal, mercer of Hook Norton, Joseph Davis, mercer of Chipping Norton, and Richard Archer, cooper of Battersea, Surrey, an alliance of local farming expertise, local tradesmen's capital, and London business interests set to exploit Hook Norton's grazing to the full. Clearly the village's agriculture was no longer a matter of local subsistence or static common-field organization.

Robert Allen, in a study of agricultural development in the South Midlands (including Oxfordshire) between 1450 and 1850, has recently re-emphasized the capacity of local agriculture to improve and modernize without recourse to wholesale enclosure by parliamentary act. He looks rather to 'a yeoman revolution' of the sixteenth and seventeenth centuries. Taking Hook Norton in the same period one can see the strengths of such an argument, for enclosure by agreement, consolidation of holdings, improved crop rotation and specialization for the market all seem to be in evidence. At some point before 1700 the common fields were reorganized into North Side and South Side Fields. As we know from a sale document of 1709 North Side Field was further subdivided into quarters (bottom map p. 282), making greater flexibility in cropping and management possible. It was this considerably evolved landscape that eventually saw parliamentary enclosure in 1774.

Early modern Hook Norton has all the marks of a successful and generally prosperous community. The exact size of the population is not recorded, but an estimate, based on the Compton Census returns of 1676, would put the total then at around 720. This census was undertaken for the Archbishop of Canterbury to determine numbers of communicants and nonconformists in each parish. The results show that Hook Norton had 338 conformists, 6 papists, and 90 nonconformists, more than any other parish in the diocese and proportionately extremely high. (The next greatest concentration of dissenters was at neighbouring Bloxham, with 80 nonconformists to 800 conformists.) The dissenting presence was to be one of the hallmarks of Hook Norton's open village character. North Oxfordshire had always been a stronghold of Puritanism. During the Civil War in 1644 a Particular Baptist Church was founded at Hook Norton, under the pastorate of James Wilmot and Charles Archer. In 1655 Wilmot signed the Articles of Faith and Order when Hook Norton became part of an Association of seven Baptist churches in the Midlands. With the enactment of the Clarendon Code after the Restoration he paid the price of his faith and was imprisoned and his goods confiscated. These were offered for sale at Chipping Norton market but no one would buy them. Eventually the goods were returned to Hook Norton, where a friend of Wilmot bought them and returned them to their original owner on his release from gaol. Despite all of this, and excommunication, Wilmot continued in his ministry and the Return of Conventicles in 1669 shows a monthly meeting of sixty Anabaptists in Wilmot's house in the village. When he died in 1681 he left a cottage and inventoried goods worth £40. 3s. 6d. His ministry passed to his son Daniel who was to serve until 1741. During that time the curate of Hook Norton was suspended by the Bishop (1682) because of the level of dissent in the parish and, after the Toleration Act of 1689, Hook Norton's Baptists became publicly established. In 1718 they moved from worshipping in a private house to build their own meeting-house (on the site of the present 1787 building on the main street of the village) and were able to do so because of the support of substantial benefactors from Hook Norton and surrounding villages, notably William Harwood, who died in 1720. They built not only a meeting-house, but also a minister's house and three almshouses, and land was provided for their own burial ground, a substantial presence indeed.

In some ways the parliamentary enclosure of 1774 was for Hook Norton part of a continuum of agrarian and landscape change. Nevertheless it raised tensions, between local landholders and the principal land- and titheowner, the Bishop of Oxford, and on the part of the smallest players on the scene, those with only rights of use—grazing or the gathering of fuel—on the old customary system. Little of these rumblings penetrated the formal parliamentary procedures. Rather they were expressed in an earlier petition querying the Bishop's right to all the tithes being claimed. A local agent investigated the petition and reported to the Bishop (January 1773) that the principal signatories 'have retracted as they did not know what they had signed', whilst the smaller ones 'are those who now have an opportunity of committing trespasses on their Neighbours' Property with their sheep, which in so great a Field cannot be altogether prevented. The poor people who have no property have usually cut a few Furze upon the Greensward Part of the Field, and these poor creatures are the only people that seem to merit your Lordship's Consideration.' He went on to propose an allotment of land, the rents from which would provide a dole 'to the Honest and Industrious poor at Christmas … although (they) have no pretence to claim a right'.

The Act was duly passed and in September 1774 the Award was made (Map 3). Of 116 allottees the Bishop received some 835 acres, Nathaniel Appletree 217, and six others between 100 and 157 acres. Four years later dissatisfaction continued and a leading local farmer wrote to the Bishop that 'some discontented souls' would probably renew their petition despite his efforts, for 'obstinacy has no ears'. By this time quickset thorn hedges surrounded new fields, various exchanges of land had taken place between the principal allottees, roadways had been laid to prescribed widths, watercourses diverted, new farmsteads outside the village (like Appletree's at Belleisle) built, and an allotment of 40 acres for the poor placed in the hands of Trustees. How far was the social structure of Hook Norton affected by parliamentary enclosure? The year 1774–5 saw record poor relief expenditures and annual sums spent continued to rise thereafter, probably part of a broader trend of rising population and low wages. A listing of the Bishop's copyhold tenants in 1774 and 1808 revealed 46 per cent of the same individual or family names and 54 per cent new names. A parallel list of tithe payers, mostly small freeholders, shows 80 per cent of new names, indicating high levels of post-enclosure sales. Land tax assessments, sadly available only after 1785, indicate that by then and into the 1830s, Hook Norton had a steady pattern of 25–30 per cent owner-occupiers and 70–5 per cent tenants. By the mid-nineteenth century the Bishop's estate was being dispersed by sale.

Victorian Hook Norton reached a peak of growth (1,525 people) in 1841. It remained dependent on farming, with twenty-one farmers in 1871, but to this was added the brewery from 1849 and the ironstone workings from 1889 (Map 4). The other distinctive feature of the village economy and social mix was the large number of crafts and tradespeople. A native of the place, leaving Hook Norton school just after the First World War, got his first job gardening for the brewery owner, another brother washed bottles at the brewery, whilst a third worked in a local bakery shop, which had an extensive delivery round in the parish and small adjoining villages, Swerford, Wiggington, Whichford, and Ascott, which looked to Hook Norton for services. These economic options buffered the village from the worst effects of agricultural depression after 1873. Hook Norton retained its robust openness: for example, despite the arrival of a resident and energetic Anglican incumbent in 1841, more people attended Baptist, Quaker, Wesleyan, or Primitive Methodist worship on Census Sunday 1851 than at the parish church. In 1875 the Rector reported to the Bishop that at least half the population were habitually absent from church and that in the last three years things had got worse, 'especially since the formation of the Agricultural Labourers' Union', a member of whose Executive Committee lived in the village. Organizations like the Friendly Society flourished.

Despite good fortune compared with many villages, population growth was not sustainable. Numbers fell to 1,232 in 1881, only to recover partially, to 1,346 by 1901. The inter-war years were tough, the population falling again to 1,153 by 1931. Fred Beale, whose interview is quoted above, emigrated to Canada at the age of 19 in 1927 to work on a Prairie farm. He returned, for family reasons, in 1936, and took a job in Alcan's new aluminium factory, out of the village, in Banbury.

In 1943, as part of the preparation for post-war reconstruction, a survey, *Country Planning*, was made of rural English life, together with a film, *Twenty-Four Square Miles*. The area chosen was north Oxfordshire around Hook Norton. The resulting picture showed the relative deprivation of the country dweller, in terms of standard of living, housing conditions, educational and medical services, and social opportunities. Here was a landscape of declining population, dilapidated agricultural buildings, and farming methods, fossilized by economic depression in a traditional mixed husbandry, and carried out in units (average farm size 109 acres) and field patterns derived from a history long past. Rural crafts had declined to the point that only three smiths worked in the area. There were twenty jobs at the brewery. Community activities were marked by apathy and lack of leadership. The centre of village life was the pub; Hook Norton had seven and a beer shop. The authors argued that mains water, gas, electricity, metalled roads, a bus past the door, and active citizenship should be brought to these neglected areas. Now, fifty years on, hardly 2 per cent of Hook Norton's population works in agriculture and people live, but do not work, in the village. In the 1980s 136 new households were added. The school, the post office shop, the brewery, the pubs, the band, the Baptist church, and the local historical society all flourish. There are few dilapidated buildings to be seen in a landscape which is still recognizably the product of its long history.

13 Eccleshall, Staffordshire:
A Bishop's Estate of Dairymen, Dairy Wives, and the Poor

Margaret Spufford

Diligent travellers, of a historical turn of mind, should logically approach the little town of Eccleshall and its huge parish from the south-east, and travel through it on the old coach road to the north-west, the route that John Eardley, the carrier, took for his weekly cart from London in the 1690s. This arterial road was a by-way of one of the branches of the great 'North-West' road from London to Chester.

The snag about this travellers' route on the main road is that its line is so well chosen: it runs very roughly along the division between the heavy soils to its north-east, and the lighter and less fertile soils to its south-west. The landscape on the clay is gently rolling, and not unlike the Cheshire Plain (map on p. 301). The light soils of the south-west always were, and still are, much more heavily wooded, with one or two dramatic sandstone gorges. Really discerning travellers will notice that the larger farmhouses, with the superimposed eighteenth-century three-storey brick façades, typical of Staffordshire, mainly lie to the north-east of the old main road, and that few older houses survive to the south-west.

However, the most diligent explorers of all will benefit much more from crossing the 20,000 acres of the parish on the minor road from the north-east to the town itself, and then exploring the maze of lanes on the poor soil of the sands and pebble-beds in the south-west of the parish. That way, they will bisect the parish, and get a full cross-section of the different types of landscape there. The multiple township names force themselves on the travellers' attention, immediately on entering the parish at Cotes, and then crossing into Millmeece before even reaching the township of Slindon on this minor road. Even in so gentle a landscape, some features are very apparent, like the old holloway leading away from Slindon to the tiny township of Aspley, which is off the road. The nature and use of these heavy soils is evident too, and gives some clue to their possible past uses. The heavy clays of Slindon and Aspley are always either under the plough, or, more commonly, being grazed by cattle. It is already clear that this is predominantly dairying country.

Despite the lack of drama in this rolling landscape, a gentle climb is needed from Slindon to Eccleshall taking us to the top of a hill, which our enquiring

The hollow-way from Slindon to Aspley

Even in gentle landscapes specific features, such as this old hollow-way, provide dramatic clues as to how the land was once used. In this case, the heavy clay-based land was grazed by cattle.

Eccleshall Church

The church nestles amongst the trees which surround the site of the castle. It became the frequent residence of the bishops of Lichfield from 1200 onwards, and was their principal seat from the sixteenth to the nineteenth centuries, despite its destruction in the Civil War.

travellers will learn to call 'Catshill'. From there a view opens out to the little town in the shallow valley of the Sow below. This view, and the town itself, are dominated by the church tower. The church contains suitably magnificent monuments to three of the sixteenth-century bishops who made Eccleshall their main residence, one of whom was President of the Council for Wales and the Marches. It can be no accident that the bishops were more frequently in residence after 1200, and that the church of the Holy Trinity at Eccleshall is described by Pevsner as 'one of the most perfect thirteenth-century churches in Staffordshire'. The very first view of it emphasizes its importance, although our travellers have as yet no knowledge of just how great that importance was. Nor do they yet know, as they cross the valley floor and pass the lodge to the main gate of the Bishop's Castle that Leland passed this way in the late 1530s or early 1540s, and noted the 'vj faire pooles' lying before it. A mill was already here at Domesday.

Bishop's Wood

This, along with the common of 'Great Wood', lies on the small hills of the Upper Sow valley and was integral to the village economy because of the wood, ironstone, and soil (for glass-making) it provided.

The main road of the little town is sufficiently arresting to tell the traveller, without further documentary evidence, of its importance as a staging point, from its numerous inns with their eighteenth-century façades, and as a market, from the arches of the old Butter Market. But if our travellers are determined, or wish to get to the source of the tiny Sow, they have to travel up the main road to the north-west and then plunge off it. They immediately find themselves in a much more complex landscape. The contours are still gentle, but rise to small hills each side of the Sow. The outcrops are sandstone, and the whole landscape is still heavily wooded and criss-crossed by a bewildering network of small lanes. Along the hill at one side of the upper Sow valley runs a suspicious-looking dyke, detectable by the sharp-eyed.

To interpret this confusing palimpsest, travellers must turn to the documents. Neither landscape alone, nor documents alone, will tell a full enough story. Put together, they make a rich interpretation possible, and both become much more readable.

Eccleshall in the Eleventh Century

Eccleshall is fortunate in its Domesday entry in 1086, which takes the historian further back into the past than is normally possible. The element *ecles* in the place name itself records not the Saxon, but the Romano-British word for a church, and the second element *halh* means 'land by a river'. So the place name means 'land of the church by the river', but since it incorporates a pre-Saxon word for the church, it implies that a Romano-British Christian church with a settlement around it already existed when the Anglo-Saxon invaders from Mercia reached the area in the late sixth or seventh century. This church was served by a priest in the eleventh century, for his existence is recorded in 1086. A few fragments of a stone cross, which was probably a preaching cross, survived and can be dated to the tenth century. The Romano-British element *ecles*, testifying to an early tradition of Christianity here, is also supported by another tradition recorded in Domesday, that St Chad, the great Celtic evangelist of Mercia, had held the whole estate of Eccleshall. We cannot be sure, in the absence of a charter, that this is true. We can be sure, however, that by the time of Domesday, the bishops of Lichfield had held Eccleshall so long that it was assumed that they had always held it. If there was a Romano-British *ecles* here when the Mercians arrived, it may be that the estate seemed a peculiarly suitable gift for St Chad himself.

Other signs of a Romano-British presence still survived in 1086. One of the immediately neighbouring hamlets to the settlement of Eccleshall itself was 'Waleton', the 'ton' of the 'Welsh', as the Saxons contemptuously called their predecessors. It was very near the little river Sow, which also kept its Celtic name.

As well as giving clues to the distant Celtic and early Mercian past, the Domesday entry bears witness to the eleventh-century present. Not only was Staffordshire a poor county, but Eccleshall was a poor place within it. The main village itself had only fourteen villeins, or unfree farmers, two cottagers, and two millers, making nineteen households in all. There are unlikely to have been more than 100 people all told. Many of the hamlets were 'waste' and the others were very small. Many of their names, like Aspley, told of settlement in the woods, from the 'ley' element in Aspley, Bromley, and Offley, to the description of the remaining woodland: 'the woodland of this manor of Ecceshelle has a length of four leagues and a breadth of two leagues'. Travellers are introduced to the part of the parish, the heart of which was later known as the 'Bishop's Wood', at its western edge. The Bishop's Wood, and its neighbour, the common of 'Great Wood' which lay next to it on the small hills of the upper Sow valley, were to be extremely important in the household economy not only of the lord of the manor, but of his very poorest tenants. They were also to provide a very important resource for early industry, for they contained both wood and suitable sandy soil for glass-working, as well as deposits of ironstone, and employment for both brick-makers and charcoal-burners.

Eccleshall in the Twelfth and Thirteenth Centuries

Not only is the Domesday entry for 1086 unusually helpful, but we also have a detailed survey carried out of the Bishop's estate 200 years later, in 1298. When this was updated, in 1320, only a few extra acres of assarts, or clearings, were recorded, so the 1298 survey seems to have been made, conveniently, at the height of the population explosion. Apart from the number of householders recorded, this survey, full of unknown place names and references, is initially baffling. However, Victor Skipp's early work, which was published in pamphlets on the Forest of Arden, inspired us to take the surviving tithe maps of nineteenth-century Eccleshall, and see if we could make sense of the survey of 1298. An extramural class mapped all these nineteenth-century field names for the 20,000 acres of the parish, and indexed them. To our astonishment, we discovered that these nineteenth-century field names had preserved, for over six centuries, the memory of their thirteenth-century predecessors. We were then able to analyse and map the Bishop's Survey of 1298 in a quite different, and more exciting, manner.

Eccleshall had, indeed, had a population explosion in the two centuries since Domesday. The village itself in 1298 not only contained four or five times as many people as in 1086, but had turned into a market town with an annual fair. It had also turned into a borough, at the instigation of its lords of the manor who had fortified their castle as well. The hamlets had grown likewise. The impact of this growth on the area of landscape under cultivation was dramatic. Yet another, new, township with a woodland name, Horsley, appeared. And everywhere the tenants were engaged in pushing their way into the wood, and clearing new 'assarts' in it. It is possible to map the directions from which the tenants were clearing. The men and women of Meece were clearing land southwards up to the top of the Catshill, where they were likely to meet the tenants of Eccleshall itself, who were clearing northwards. Informed by the documents, we can stand on the top of Catshill, and be aware of our thirteenth-century predecessors clearing assarts both to north and south of us. We also understand the importance of the glimpse we get of the church of Eccleshall below amongst the trees, which we now know has a continuous history back to a Romano-British church. Adam, son of Isolde of 'Meece', may be taken as a 'typical' unfree tenant in 1298, despite having the largest villein holding in the survey. He had nearly 15½ acres of assart, as well as his virgate of 21 acres of arable and an acre of meadow. But the whole community in the parish, from the biggest free tenants who were the gentry, to the smallest cottagers, were engaged in transforming the landscape and pushing back the agricultural 'frontier'. The Bishop himself had contrived to get a whole new field with a better aspect brought into cultivation.

**Assarting and the
Bishop's Park in 1298**

Much attention was being focused in 1298 on clearing Great Wood down in the south-west of the parish. Men from Broughton, Croxton, Sugnal, and Bishops Offley were all engaged in this. Already six cottages and a house in Great Wood itself had been built, so the rise of population was leading to settlement on very poor land. The Bishop's Wood was, of course, exempt from encroachment: but the survey of 1298 casts light on the mysterious 'dyke' still to be seen. The wood, in fact, was already enclosed in 1298, for a villein had to keep up the palings round one area of it, and the north end left the name 'Pallys Moss' behind in the tithe award. The wood itself supplied sport, a substantial amount of meat for the Bishop's table, and sales of brushwood. Hunting there was obviously important, for the free tenants had to provide as many as 84 beaters for three-day hunts for the Bishop, three times a year.

The combination of evidence from the nineteenth-century tithe maps and the survey of 1298 make it clear that Eccleshall, and all its hamlets, each had a core of common field, often called 'Town Fields' round them, which was subject to common control, according to the sixteenth-century manor court regulations. At Aspley, in 1576, the tenants were forbidden to tie beasts in the 'winter corne field or lentle corne field till all the corne be carried forth'. But the 'new land' of the thirteenth-century assarts was not necessarily added to this core of common fields at all, for the fifteenth-century court rolls show that a good deal of land was held in small enclosures outside the open fields. Hence, the open fields were inconsiderable enough not to be mentioned in later enclosure acts, which only affected the commons. Eccleshall was, therefore, the sort of place where historians used to think that open fields did not exist.

The Fourteenth-Century Slump

The harvest disasters and famine of the fourteenth century, followed by epidemics, inflicted on an already-malnourished population, did their worst in Eccleshall. The pressure on land of the large thirteenth-century population went into reverse, and rents fell dramatically. By the 1470s, they were down by at least a third from their late thirteenth-century levels, and some marginal land reverted to waste. For the Bishop's officials, watching a sinking rent roll, the situation was gloomy. But for men of enterprise, whether they were gentry or small yeomen and their wives, the outlook was quite different. The opportunity to build up their holdings loomed, and was seized by such men as Richard Broughton, a gentleman, on the one hand, and the Tildesley and Key families who were yeomen of Aspley, on the other. In Broughton, at the edge of the Bishop's Wood, eight inhabitants had been taxed in 1327. But by 1524, the Great Subsidy recorded only the single name of Richard Broughton, taxed in the hamlet on the substantial sum, for Eccleshall, of £13. 6s. 8d. After that date, Broughton Hall stood in solitary dignity to represent the houses of the original settlement. Indeed, it is the only substantial gentry house to survive in the parish now. Broughton Hall was described by Pevsner as 'the most spectacular piece of black and white in the country' of Staffordshire. Across the road from the Hall stands the charming family chapel of the Broughtons, built in the 1630s, complete with its original box-pews, and the proud display of armorial bearings illustrating the family's expedient and sensible marriages.

At Aspley, over the same period, the number of householders shrank from the eight villeins and bordars of Domesday Book, to only four taxpayers in 1524–5. Three of these were labourers, dependent on the household of William Tilsley, or Tildesley, who was a very unusual yeoman for Eccleshall, though not for eastern or lowland England, taxed on as much as £4. Aspley remained a hamlet dominated by yeomen. The next major tax, the Hearth Tax, showed that hamlets tended to be dominated either by superior gentry with their huge houses containing as many as ten hearths, like Bromley Hall, home of the noble Gerard family, from which other tenants had nearly, or quite disappeared, or by yeomanry, with their more modest, but for Eccleshall substantial, houses with as many as three or four hearths. It appeared that the two social systems, exemplified by Broughton and Bromley, ruled over by grandees, on the one hand, and Aspley, with its yeomen, on the other, could not easily co-exist. Really prosperous gentry had a tendency to clear away their neighbours. They also, incidentally, liked to have their seats in good hunting country, so there was a tendency for these major gentlemen to dominate in the hamlets fringing on the Bishop's Wood, and Great Wood.

The most striking feature of the Hearth Tax in Eccleshall, however, was the level of poverty it revealed. The tax showed, in its proportion of residents with only one hearth and those exempt from paying any tax at all, that Eccleshall was poorer than any other rural area we yet know in the 1660s and 1670s: 83 per cent of its taxpayers paid on one hearth only. A disturbing number of 200 probate inventories, showing people's property at death between 1660 and 1700, covered many of the very poor in Eccleshall, and spanned the Hearth Tax in date; they listed no fire-irons or indications of a grate at all. There seem to have been a great many lodgers living in single rooms.

Agriculture in Eccleshall

Twenty years after the Hearth Tax of 1673, between 1693 and 1697, the Bishop set in hand his own extraordinary, and minute, survey of his manor. Analysis of this gives a remarkably full picture of Eccleshall in the late seventeenth century, and particularly of its poor. The survey demonstrates very fully the different agricultural and occupational structure of the hamlets to the north and east of the main highway through the parish, and the old common land to the south and west of it. Its contents are, however, much less full on the hamlets to the north and east and farming practice there, and these, therefore, have to be filled in from other sources. So an accurate picture of the agricultural hinterland of this huge parish, which must have fed the people of the market town and sold its surplus goods there, could only be made by analysing all the agricultural inventories surviving from 1665 to 1738. This was done in three periods, 1665–89, 1690–1714, and 1715–39, looking for any perceptible swing in the use of land from arable to pasture, or major changes in farming practice. The area is defined by the agricultural historians for the period 1640–1750 as 'subsistence corn with cattle rearing, dairying and grazing'. By surveying a similarly long period, we covered the whole possible lifetimes of anyone who either appeared in the Bishop's Survey, or handed on land to those who did. We divided the moveables of these people into their arable produce, that is crops, horses and gearing, 'husbandry ware' with 'raw' linen, hemp, flax, and wool, their livestock of all kinds except horses, fodder, cheese and butter, and dairy implements. We were also interested in the proportion of the yeomen or 'gentleman's' goods invested in household furnishings and linens and 'napery-ware'.

Despite the survival of stray fragments of common 'Town' fields round the hamlets of Eccleshall parish into the seventeenth century, leases made in that century commonly restricted the lessee to ploughing only one-third of the arable. We know, therefore, that two-thirds of the arable area on many farms in these hamlets was normally under temporary or permanent grass. Obviously arable farming was mainly for subsistence. The inventories of 113 'yeomen' and 'gentlemen' made between 1665 and 1739 show that only one single man, William Wright of the Fieldhouse, had crops that were more highly valued than his livestock. He was the second most prosperous yeoman whose inventory survived from between 1665 and 1689, even after the vast debts of £1,320 owed to him were deducted: 70 per cent of his £347 remaining were in crops, horses, and tools of husbandry. He was an oddity in this area. Yet anywhere but in a pastoral landscape, his inventory would have stood out because of his stock. His considerable milking herd of ten cows, a bull, and five calves would have distinguished him. Yet in Eccleshall he appeared as an 'arable farmer'.

Looking at all farmers in Eccleshall between 1665 and 1740, there appeared to be no change over time and no progressive movement towards more grassland farming in this period, according to the inventories. Nevertheless, the value of the livestock, fodder, cheese, butter, and dairy implements was uniformly higher as a ratio of a man's goods at death than the value of his arable crops and tools. It is likely that this emphasis on pasture as against arable farming existed in Staffordshire at the very least by the later Middle Ages, when, for instance, neighbouring Warwickshire emerged as a specialist pastoral farming region. This emphasis may have increased in the later seventeenth century. By 1737, there were fifteen English and Welsh cheese-producing counties, amongst which a great north-western belt, including Lancashire, Derbyshire, Cheshire, Staffordshire, and Shropshire figured largely. Joan Thirsk suggests that the importance of dairy production in the national market was enhanced during the period after the Restoration, and this development was reflected in a geographical extension of the dairying regions and greater specialization in the traditional ones.

The problem of how the Eccleshall cheese was marketed is probably solved by one laconic entry in the Bishop's Survey. John Stevenson, tenant of the Unicorn Inn, later the King's Arms, was also a 'cheesemonger', as well as a 'tipler', disapproved of by his landlord. This means that the local cheeses were almost certainly taken to the Unicorn, and then disposed of by means of John Eardley's carrier's cart, either up the road to Chester, thence to be carried coastwise to London, or down to London direct to the Castle and Falcon in Aldersgate Street. It might seem improbable that cheese was carted to Chester; yet in 1722, John Prickett, master of the *King's Fisher* had evaded duty on 215 tons of cheese, Isaac Dove, of the *Dove* had evaded paying on 330 tons, and Francis Cooke on 236 tons of cheese, all from Chester. All in all, examination of the inventories gives us a very clear impression of the pastoral, dairying hinterland to the little market town of Eccleshall and its hamlets to the north-west and north-east. The assarts of the thirteenth century were being used as pasture.

Any farmer not retired was likely, and by 1715–39, practically certain, to have cheese and butter in store, and dairying equipment. On first reading, the inventories are misleading, for the value of cheese was low. However, some considerable light is thrown on the reality by three inventories. William Smith of Eccleshall, who died in the 1670s, had forty-nine cheeses in his store chambers, valued at £3. 15s. 0d. John North of Wootton had a 'hundred' (that is, a hundredweight) of cheese, priced at only £1 6s. 0d. in April 1685. If the appraisers of North's inventory were being consistent, William Smith's cheeses must have equalled 300 lb of cheese, so each cheese must have weighed about 6 lb. This puts the activities of John North's father, Robert, who had died ten years before, leaving 'cheese old and new' valued in August at £8. 0s. 0d. (perhaps 6 cwt?), in their proper, somewhat large-scale perspective. It also points up the unusual degree of specialization practised by Thomas Hadderton of Acton Hill, whose inventory of 1707 listed in his store chamber '23 hundred of Cheese, part old and the other part new of the first make £18 and of the later make 110 cheeses at £5'. The whole were valued at £23, no less than 15.7 per cent of all Thomas Hadderton's farming goods. He had twelve cows and three weaning calves. He also had £300 owed to him in debts and a lease worth £400. Although his household furnishings were by no means luxurious, and no window curtains, long-case clocks, or coffee-pots decorated his house or impressed his neighbours, the furnishings were still worth a fifth of his goods, excluding the debts and the lease. This was an unusually high proportion of household comforts for Eccleshall. Even without coffee-pot or curtains, Thomas Hadderton had made himself more comfortable than most Eccleshall yeomen. Furthermore, his daughters had dowries of £200 each, providing always that they married with the consent of the executors. He had also committed the ultimate act of conspicuous consumption of the late seventeenth century. His son Ralph was up at University. Specialization in dairying had served him very well.

The hamlet of Aspley was dominated by two yeomen families, the Tildesleys and the Keys. Both William Tildesley and John Key died in 1695. William Tildesley was, no doubt, a descendant of the dominant yeoman of 1524, also named Tildesley. His moveable goods were worth £149, almost double the median average, which was £74 for yeomen in the parish. He leased land worth £72 a year, the Bishop tells us, and the key to his farming is given by his 11 cows, 9 calves, and a bull. After that, it comes as no surprise that his cheeses in store were valued at £4. 5s. 0d. Thomas Key was an even more prosperous man, with goods worth £255, well above both the median and the average for the east of England. He, too, specialized in dairying: his 54 cheeses in the chamber over the cellar were worth £4. 0s. 0d., and he had 8 cows and their calves at Aspley, as well as another 7 on a holding elsewhere. His will bequeathed 'my new purchased land in Aspley' to one son, and 'my aunchient land' to another. His inventory, with its chairs, cushions, and iron back to the fire in a parlour from which the bed had been banished, showed an unusual degree of comfort for Eccleshall. His wife, Mary, must have been a proud woman.

Either William Tildesley or Thomas Key must have lived in the timber and plaster house which was pulled down just before the First World War but of which a photograph survives. If this house was in any way typical of those of prosperous yeomen in Staffordshire, it is important, for it demonstrates just how deceptive are the wholesale eighteenth-century three-storey red brick façades, and how they conceal from us the 'real' face of the county in previous centuries. Aspley House Farm had a stone carved with the words 'Cheese Room' which marked the upper window of the store from which the cheese was lowered by pulley to the waiting wagon.

Stability both of farming practice and family has been pronounced in Aspley since the seventeenth century. Mr Jim Warrington who was farming there with his father from the 1950s to his own retirement, is descended from a Tildesley, so Tildesleys and Keys formed part of that stable core of rural society to which Professor Hey draws our attention. Mr Warrington himself reckons the size of a farm not in terms of the acreage, but of the milk yield, so that a 'small farm' to him would be one which produces 10–20 gallons a day, a 'median farm' 30–50 gallons a day, and a 'large farm' 60–100 gallons a day. His wife told me that they had 300 acres in the 1940s, which gave 100 gallons a day. Mr Warrington also emphasized (in 1996) the role of the farmer's wife as cheesemaker: she would manage all cheese production on a small farm, might have help on a median farm, and would certainly employ a cheesemaker on a large one. Thirty years before, in the 1960s, Mr Warrington's mother told me the way her own mother-in-law had taught her to make cheese when she came to Aspley as a young bride. The farmer's wife was an essential partner in the business, as indeed anyone who has read *Adam Bede* (1859) will have noticed. Not only was the dairy the hub of the farmhouse in that novel, even without the charms of the dairymaid, but when Mr Poyser on his way to church 'turned a keen eye on the crops and stock as they went along ... Mrs Poyser was ready to supply a running commentary on them all. The woman who manages a dairy has a large share in making the rent, so she may well be allowed to have her opinion on stock and their "Keep"'. George Eliot's Loamshire may really have been Warwickshire, but its economy was not very far removed from that of Staffordshire.

The Poor of Eccleshall

Bishop Lloyd's survey of the 1690s was weakest and least helpful on the agricultural hinterland of the market town, and on just such yeomen as William Tildesley and Thomas Key. The only inhabitants he neglected even more were his own social peers or dining companions, the nobility and gentry. Despite the pre-eminence in Staffordshire of the Gerards of Bromley, and indeed their suitability as a subject of moral gossip of just the type the Bishop most enjoyed, his informants' pens passed over that vast household. But on the subject of the poor, the Bishop was driven by a combination of moral anxiety and anxiety about the poor rates to pay close attention. As a result, his survey is one of the fullest seventeenth-century documents known which deals with the very poor and their wives, the women who went out to work for wages. The contemporary Mrs Poysers, Mrs Mary Key and her gossips, did not interest him, but Margaret 'Suds' Owbury, the charwoman, and her like, did. Margaret's labour might keep her off the poor rates.

Interestingly, the Bishop's preoccupation with the poor, their industry and employment, may well have been shared by the substantial farmers about whom he said so little. A detailed church rate of 1687 survives. It shows Thomas Hadderton of Acton Hill paying 10s. 7½d., and Thomas Key of Aspley paying 15s. 8¼d. for poor relief. William Tildesley of Aspley paid as much as £1. 7s. 7¼d. The size of the problem raised by the poor in Eccleshall can be gauged by comparing this list of 188 ratepayers with a list of the heads of 568 households of all Eccleshall parish, made by the Bishop six years later, in 1693. One-third of the householders were supporting the remaining two-thirds. Thomas Key and William Tildesley might well feel resentful. Moreover, in assessing the rate of 1687, the Overseers of the Poor had included all possible candidates from whom money might be squeezed. John Croft, an agricultural labourer, lived in the town itself in a burgage (on a yearly rent). He was the poorest man for whom we have an inventory. All his goods were worth only £2. 1s. 0d., and his cottage, burgage though it was, had no table, no chair, and no bench in it. His whole wardrobe was a pair of breeches, a waistcoat, and two coats. He left his pair of shoes in his will to a woman cousin, so they were precious; new shoes usually cost 2s. 0d. or less in the seventeenth century. Yet John Croft had had to pay 6d. towards the poor rate. The

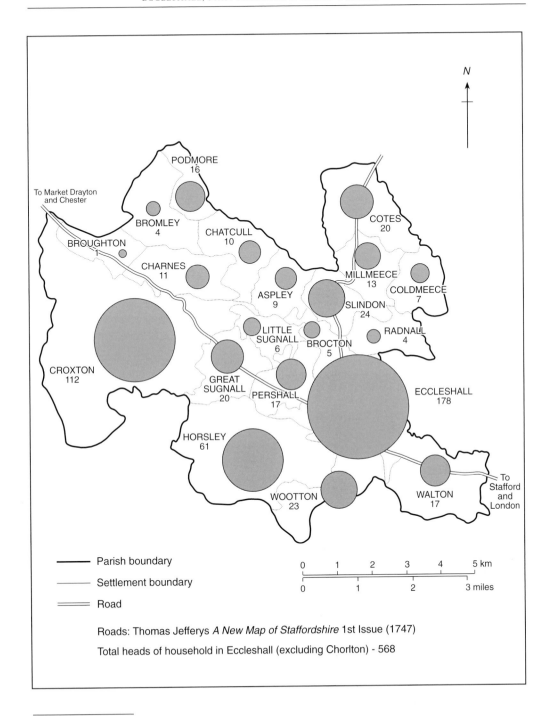

N

To Market Drayton
and Chester

PODMORE
16

BROMLEY
4

CHATCULL
10

BROUGHTON
1

CHARNES
11

ASPLEY
9

COTES
20

MILLMEECE
13

COLDMEECE
7

SLINDON
24

LITTLE
SUGNALL
6

BROCTON
5

RADNALL
4

CROXTON
112

GREAT
SUGNALL
20

PERSHALL
17

ECCLESHALL
178

HORSLEY
61

WOOTTON
23

WALTON
17

To
Stafford
and
London

Parish boundary

Settlement boundary

Road

| 0 | | 1 | | 2 | | 3 | | 4 | | 5 km |

| 0 | | | 1 | | | 2 | | | 3 miles |

Roads: Thomas Jefferys *A New Map of Staffordshire* 1st Issue (1747)

Total heads of household in Eccleshall (excluding Chorlton) - 568

**Size of Eccleshall town
and hamlets in 1693**

man who paid the lowest rate of only 3d. was the parish clerk. He had a by-employment, killing hogs at Christmas, and his wife was a midwife, so they made ends meet. But these poor people, paying rates for those even poorer than themselves, might well also have been resentful. So might the more successful craftsmen and retailers who lived in Eccleshall town, which acted as a magnet for retailers and craftsmen. A fifth of the Survey, and its occupations, is missing, which may undermine our calculations. However, a quarter of the other heads of household were craftsmen, including the shopkeepers, for Eccleshall was both the market and established shopping centre for its area. The mercer had assorted stock which we might well think suitable for both a grocer and a bookshop, as well as for a haberdasher. An apothecary, a wig-maker, and the all-important innkeepers in this road-town, were also living in Eccleshall itself. So were most of the other craftsmen, including the big group of leatherworkers, except those whose craft was linked to the raw materials of the Bishop's Wood. It is no accident that the public house on the minor road leading past the wood in the upper Sow valley is called the 'Mason's Arms'! A small group of stonemasons, brick-makers, sawyers, charcoal-burners, and, above all, glass workers, were to be found in Croxton. The intrepid historian can still fight through the jungle of undergrowth in the Bishop's Wood, and find the remains of the only sixteenth-century glass furnace still in existence. It was in use from 1580, when Bishop Overton brought glass workers with him from Hampshire, to about 1615, when legislation prohibiting the use of wood in the manufacture of glass came into effect.

Eccleshall itself in 1697 had 178 heads of household, out of the 568 listed by the Bishop for the whole parish. The market town was by far the largest settlement: but the hamlets on the poor sandy, pebbly soils to the south-west of the great London to Chester road formed an ideal habitat for trees, commons, and squatters. Greatwood Heath, where the Bishop's rabbit warren lay, had been described in a somewhat denigrating way by the parliamentary surveyors as 'fitt Soyle for Berryes and Connyes'. In the later seventeenth century it was also fit soil for the desperate. The area of the common wood accommodated at least nine separate settlements, Fairoak, Blackwater, and Wetwood among them. The township of Croxton whose land comprehended the wood had in all 112 scattered houses or shacks. The maze of lanes that bewilder travellers in this area is explained by the history of the settlement of this common.

Alehouses, cheap drinking houses, always flourished near commons like this, as far away from the eye of authority as they could manage. Both the 'idle' and the 'industrious' poor needed their thirst quenching, and some temporary light relief. After the town of Eccleshall itself, Croxton township had more alehouses where food, drink, and music could be supplied than anywhere else in Eccleshall parish. Alehouses were also, or could be, centres of immorality. There were at least six alehouses in Croxton, Fairoak, and Blackwater. Robert Dodd, who was, in Bishop Lloyd's words, a 'mad, swearing fellow', kept one of them in a cottage at the side of Croxton Green. He worked as a labourer as well: this was a typical pattern of by-employments. It is indicated that he had been witness to a murder, and was also said to keep a bawdy-house. His wife had prostituted herself for a gentleman; his eldest daughter had had a child by one John Powell, for the upkeep of which the Overseers of the Poor had had to pay out 16s.; John Powell himself was described as 'an impudent whoring rogue'. He had fled as far as London, where he was 'nobbled', again in the Bishop's words, by a local officer, who searched for him and ran him down at The Angel, Islington. What is remarkable is that we have yet further information on him: he did marry, although not Elizabeth Dodd, who had borne his child, and he did manage to obtain a cottage in Fairoak, despite his past. It even had two rooms, for the inventory survives, made after his death in 1727. It is the next to poorest of all the inventories we have: his goods were worth £2. 14s. 2d. in all. He did have a bed to sleep on: he also had one table, one chair and a bench, and six wooden plates. This poverty-stricken cottage had one candlestick to light it. His livestock was a single pig: the only clue to his livelihood was two axes. He was probably a woodsman, like Oliver Barnaby, the charcoal-burner who lived in Simon Hall's alehouse at Blackwater.

There were respectable alehouse keepers, though: the Bishop had not a word to say against Robert Hodge of Croxton Bank, who ran an alehouse, was a musician, and so must have played whatever ballads were the current rage. The 'Poor Man's Complaint', registered in 1695, at just this time, seems likely to have been popular, and potentially incendiary, in this environment. One of the many verses, which complained of central taxation, rent-rises, lack of charity, and the whole state of the kingdom, ran

> The times they are hard,
> Yet those that have Treasure,
> and Wealth out of measure,
> They little regard
> poor Labouring Men,
> Who are out of Employ,
> Whose Children cry,
> which troubles them sore.

The ballad had a refrain which went to a very catchy tune,

> I weep when I think of
> I weep when I think of
> the Cry of the Poor.

Robert Hodge is likely to have had a sympathetic audience, and been himself a sympathizer. He eked out his living by gardening at Bromley. He needed to: the entries in the parish register revealed that he must have known all about crying children. He himself had nine, all of whom lived. His inventory, too, reveals that keeping an alehouse was not a rapid road to prosperity: his goods were worth £4. 16s. 10d. in all at his death, although in his inventory we are relieved to find he had a grate and a pair of tongs in the 'house' of his three or four roomed cottage, and we are also startled to find he possessed one 'luxury', a 'looking glass', which was pedlar's ware, worth 8d. He brewed his own ale of course: he had two pieces of brewing equipment, and three barrels.

Gregory King in Eccleshall

The existence of Bromley Hall was not only important to the parish of Eccleshall itself, but to the gentry and noble circles of the county, as well as supplying employment to the poor. We have already seen that where superior gentry or nobility lived, substantial yeomen tended to disappear. In this particular case, the magnificent house of the Gerards provided employment, not only for the poor, but for a very promising young man indeed. The noble house, which was sometimes likened to Hardwick Hall (Derbyshire), has disappeared completely, with the exception of the entrance gate, and the stable. The stable was where Gregory King probably left his horse. He was one of their employees as steward, auditor, and secretary to Lady Jane Gerard for two and a half years from 1669 to 1672; he was a very rare bird indeed. Gregory King was still a young man at 21 when he was appointed to the post. He had already acted as assistant, making topographical notes and drawings for William Dugdale on his Heralds' Visitation of the Northern Province. The young man boarded with Lady Jane's father elsewhere, and rode to work daily. He will, therefore, have stabled his horse in the only original building which survives at Bromley Hall. Inside the splendid gateway is the stable, which has an equally splendid Renaissance manger. Whichever way Gregory rode to work must have led him past one of the poorest squatters' settlements. His training in topographical drawing and observation, as well as his insatiable

Renaissance gateway at Bromley Hall

All that remain today of Eccleshall's grandest property, Bromley Hall, home to Gregory King in the late seventeenth century, are the entrance gateway and the stable.

curiosity, will have meant that he took careful note of what he saw. But we are not pushed to prove it. As a man in his early 30s he came back. His 'Staffordshire Notebook' records his comments on Broughton, Wetwood, Charnes, and Fairoak in August 1680. He also updated his notes on the gentry of Charnes and Chatcull in that year. In 1680 he even recorded that 'Eccleshall water raises in the Woods a quarter of a mile beyond Fairoak'. So he had been up the Sow valley. He went on visiting the parish, for he updated his notes on the gentle families of Great Sugnall and Broughton yet again in 1690. His notes on the spring in the woods beyond Fairoak are of particular importance.

It is possible, thanks to the Bishop, to do a social anatomy of Eccleshall's poor a decade or so after King drew up his table of 'profitable' and 'unprofitable' families in England in 1688. We know, now, that King had ridden past the hovels of these people many times over. While he was an acute observer, obsessed by measurement, certainly, his experience of the poverty of real human beings that he had seen in his youth informed his estimates when he quantified them on paper. The problem of poverty in Eccleshall's hamlets was extreme. In his 'Scheme of the Income and Expense of the several Families of England ... for the year 1688' King had a total of 764,000 labouring people and outservants, cottagers and paupers (people like those who lived in Wetwood and Fairoak, in brief) decreasing the wealth of the kingdom. 'Sparse as King's official sources were', wrote Professor Holmes, 'they were enough to allow him a more penetrating and dramatic insight into the frightening extent of the problem of poverty than any previous economist or political arithmetician had achieved.' 'We need many more social anatomies of provincial communities in pre-Industrial England,' he added. In this parish where Gregory King had personally worked, we have just such an anatomy. I suggest here that Gregory King gained his insight not only from his 'sparse' official sources, but primarily from his own observations. Colin Brookes summarizes what he describes as Professor Holmes's 'devastating critical scrutiny' of Gregory King, and sums up his work thus: 'Holmes, following Charles Davenant (1771), believes Gregory King's revelation of the extent of poverty and its demographic and economic implications to have been his most valuable achievement.' 'By placing King's political arithmetic within the context of the course of King's life, Holmes taught us what we can and cannot learn from it.' We can now go further. King plainly had first-hand knowledge of the extent of poverty in one of its most extreme forms, in a rural 'horn and thorn' parish, where he had spent two-and-a-half working years. He was a man who, like all good fieldworkers, got 'mud on his boots'. That mud was acquired riding the lanes to work at Gerard's Bromley, and up through Fairoak to find the sources of the stream. His experience is at least as relevant as his documentary sources. It is not irrelevant that he even estimated, among his 364,000 'Day Labourers and Outservants' who decreased the wealth of the kingdom, that some 300,000 were in rural employment. Wetwood, Fairoak, and Greatwood seem to have had a considerable effect upon an impressionable and observant youth.

Even Gregory King's official work at Bromley Hall itself might well have brought him into direct contact with individuals, members of whose families became paupers. This is revealed because in the 1690s the Bishop wrote a set of notes on a woman called Jane Shelley, care for whom partly fell on the overseers of the poor because she had cerebral palsy. She would have been born during Gregory King's period at the Hall. Her father was dead but his three brothers were still living. Two of them worked at Bromley Hall; perhaps Gregory King met them face to face. The eldest was indeed the 'prop of the house', but he had a somewhat dubious wife who was an ale-wife. She was suspected of incest with the next brother, another Bromley Hall workman. This man was unmarried and crippled. Jane's own eldest brother was described as an 'arch-thief'. Jane's cousin, Thomas, was living in 1698 with his wife, who had been a servant, and three children, the youngest of whom was a baby 6 weeks old, in the kitchen of the third Shelley brother. When this brother married, his new wife took exception to the arrangement and bundled the lot of them out. The parish register recalls that the young couple had two more children, and on the birth of the last, Jane Shelley's cousin was recorded as a 'pauper' in Croxton, like so many others. These indeed were Gregory's King's people, observed by him both as flesh and blood, living in their shacks in Fairoak, Croxton, and Greatwood. He might even have paid some of them, in his capacity as steward and auditor to Lady Jane Gerard. They were certainly not merely desiccated numerals, taken from central government documents, who, according to him 'decreased the wealth of the kingdom'.

Gregory King's economic concerns, and those of Bishop Lloyd when he compiled his Survey, may have been similar. It seems almost certain that anxiety over the poor lay behind the enclosure of 'Gratwood Heath' in Eccleshall in 1719. This was the earliest Enclosure Act for Staffordshire. Greatwood was then estimated to contain 1,000 acres, of which the Bishop was allotted a sixth, as compensation for the loss of his rents and his rabbit warren. It is true that cottagers who had a legal right of settlement were allowed to stay, but where did the rest of that choked and crowded community go? Only the maze of lanes survive to tell the traveller a little of the past of this area. But it seems possible that behind the neatly manicured present-day landscape lies an episode possibly as tragic as the Highland Clearances in miniature. Eccleshall offers a tantalizing glimpse of the history that lies embedded in our landscape.

14 Staintondale, North Yorkshire:
A Moorland Estate of the Knights Hospitaller

Barry Harrison

The Physical Landscape and Early Settlement

Staintondale is a parish of 3,145 acres sandwiched between the North York Moors and the North Sea coast. Until recent times it was merely a 'township'—a distinct community within its own boundaries—lying within the great ecclesiastical parish of Scalby. It is bounded on the north and east by towering cliffs, on the south by the deep valley of Hayburn and Thorny Beck, and on the west by open moorland. The land falls regularly from over 800 feet in the north to about 400 feet in the south, providing excellent views of almost the entire township from both sides. From the north the vista is rather bleak: large rectangular fields, thin hedgerows, conifer plantations, shelter-belts, and square Victorian farmhouses all suggest an upland landscape of recent enclosure. From the south however, hidden depths are revealed: wooded valleys flanked by small and irregular fields, mature hedgerows, and irregular farmsteads blending into an apparently ancient landscape. Seen from this side the bleaker northern half of the township now appears as an attractive and inviting backdrop, framed by hummocky cliff-tops to the east and moorland heights to the west. Seen from either side the settlement pattern is one of scattered farms, many of them occupying prominent positions near the cliff-tops and on the ridges between the streams. The area now regarded as the 'village' is a loose straggle of buildings beside the main road through the parish, none of which existed before the last century.

In agricultural terms Staintondale has always been more attractive than its geographical situation would suggest. The Middle Jurassic sandstones of the North York Moors extend right to the coast, where they appear as a thick band just below the cliff-tops. They are generally overlain by thin, stony, and acid soils, but during the last glaciation the great North Sea glacier, pressing in from Scandinavia, forced its way up the then shallow valleys of the becks between Scarborough and Fylingdales, depositing thick layers of fertile boulder clay. Beyond the valleys, however, the terrain remained unglaciated and suitable only for rough grazing until modern times. The great line of cliffs, some 500 feet high in

A view over Staintondale from the western edge of the pre-1829 moor, looking south-eastwards

In the middle ground, to the right, lie the wooded upper reaches of Stainton-dale Beck, with Bell Hill—the site of the Hospitallers' hall—to the left. Beyond lie the Riggs, dividing the valleys from the high cliffs above the North Sea.

the north diminishing to some 200 feet in the south, form the most spectacular feature of the local landscape. Here the hard Jurassic sandstones overlie a thick band of soft Upper Liassic shales (the alum shale), the erosion of which has caused massive collapses, and created a broad undercliff resting on harder jet and ironstone bands below. This platform has constituted an important resource for the inhabitants of Staintondale until quite recent times, affording good rough grazing, supplies of building stone, jet, and ironstone from the collapsed strata, and suitable locations for a major industry: the manufacture of alum, essential to the cloth-dyeing process.

Like many other parts of the North York Moors, the Staintondale area was quite heavily settled in Bronze Age times and then again, after a long gap, in the late Iron Age and Romano-British periods. Bronze Age barrows and cairns, often in large clusters, proliferate around the northern and western boundaries of Staintondale and on Cloughton Moor to the south, although no settlements have been located. From the Iron Age, however, there are several cross-ridge dykes—thought to have defined the outer boundaries of 'estates' or small territories—with some evidence of settlement. The north-western boundary of the parish is formed by the Green Dyke which linked the cliff-top with the shallow valley of Pye Rigg Slack. Further south the earthwork known as War Dyke linked the valley of Staintondale Beck with the coast and, incidentally, formed the northern boundary of cultivation in the medieval period. Another dyke spans a ridge on Cloughton Moor and may be related to a small settlement on better land a little to the south. There is, of course, unlikely to be any continuity between such early settlements and those of the Anglo-Saxon period which, to judge by the forms of

The Parish of Scalby

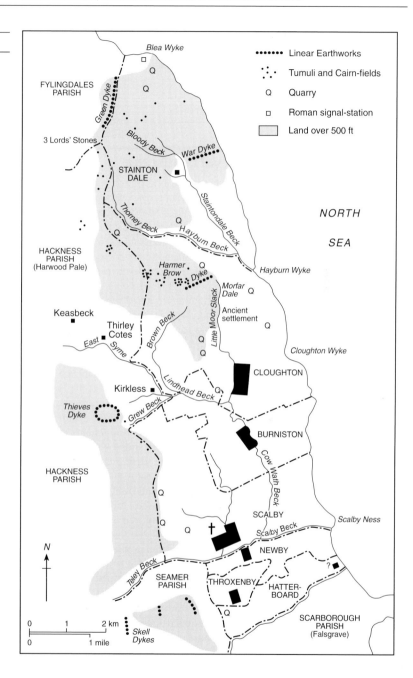

Blea Wyke

FYLINGDALES
PARISH

Green Dyke

3 Lords' Stones

Bloody Beck

War Dyke

STAINTON
DALE

Thorney Beck

Staintondale Beck

NORTH

SEA

HACKNESS
PARISH
(Harwood Pale)

Harmer
Brow

Dyke

Hayburn Beck

Hayburn Wyke

Morfar
Dale

Keasbeck

Ancient
settlement

Thirley
Cotes

Brown Beck

Little Moor Slack

Cloughton Wyke

East Syme

Kirkless

Lindhead Beck

CLOUGHTON

Thieves
Dyke

Grew Beck

BURNISTON

HACKNESS
PARISH

Cow Wath Beck

SCALBY

Scalby Ness

Scalby Beck

NEWBY

?

N

Teley Beck

SEAMER
PARISH

THROXENBY

HATTER-
BOARD

0   1   2 km

0   1 mile

Skell
Dykes

SCARBOROUGH
PARISH
(Falsgrave)

Linear Earthworks

Tumuli and Cairn-fields

Q   Quarry

□   Roman signal-station

Land over 500 ft

place names in the area—mostly compounded with '-tun' and '-by'—seem to belong to the mid-Saxon and Scandinavian periods. Just to the west of Stainton-dale, however, the parish of Fylingdales bears a name compounded with the relatively early '-ingas' element, and the discovery of a pagan Saxon cemetery there may indicate that the boulder clay embayments of the area were not unattractive to early Germanic settlers.

The Medieval Landscape

Staintondale, which appears as *Steintun* in the Domesday Survey, was one of over twenty sokelands attached to the great federal estate of Falsgrave, near Scarborough, held by Earl Tostig of Northumbria before the Conquest and by the King in 1086. These sokelands were heavily settled and populated on the eve of the Conquest: 108 sokemen (near-freeholders) with 46 ploughs tenanted this part of the estate in 1066. Nearly all had vanished twenty years later, following the ravaging of the North by William the Conqueror, but they were to re-emerge soon afterwards. The integrity of the Falsgrave estate had already been disrupted by the formation of pre-Conquest holdings that had been granted to thegns or lesser lords, but the northern part of the complex comprising the Domesday townships of Scalby, Burniston, Cloughton, and Staintondale, together with a group of small vills south of Scalby (Newby, Throxenby, and Hatterboard), perhaps of rather later foundation, retained a degree of unity for several centuries. Together they formed the ecclesiastical parish and the 'Soke' of Scalby, and most of the East Ward of the royal Forest of Pickering which was also known as the Forest of Scalby. Although under Forest jurisdiction by the early twelfth century, the king's tenants here continued to be classed as relatively free 'sokemen' who owed money rents and regular taxes (tallage) to the Crown, but who were not encumbered with menial services. There were, in fact, no royal demesne lands in the area other than a series of great 'hays' or forest enclosures which were used primarily as game reserves and cattle pastures. Of these only the extensive Scalby Hay remained by the late thirteenth century, although the names of Hayburn and Swinesty Hagg (*Swinestschage*) just to the south of Staintondale may suggest earlier sites.

Intercommoning rights enjoyed by the inhabitants of most of the townships within the Soke of Scalby suggest that the farmers of the whole area could graze their stock and take wood and peat for fuel on a fairly free basis. The area known as Fullwood, which embraced all the moorlands to the west and north of Cloughton, together with the peat mosses along the western boundary of Staintondale, were intercommoned by the villagers of Cloughton, Burniston, Scalby, and Newby until the eighteenth century. It is probable that Staintondale itself was originally included in these arrangements, but the policy of the Knights Hospitaller, who became manorial lords in the mid-twelfth century, seems to have been to detach the township as far as possible from the Soke. Even so the rector of the parish church of Scalby and the incumbent of its subsidiary chapel of Cloughton continued to enjoy free common throughout Staintondale. In fact, it is possible that Staintondale was once used for transhumant grazing by the inhabitants of the Soke as a whole. A charter of 1262 refers to the reclamation of land in the southern part of Staintondale Moor 'around the old *Herdewik*', a name suggesting the site of an earlier shieling.

The Parish of Scalby:
Medieval Landscape

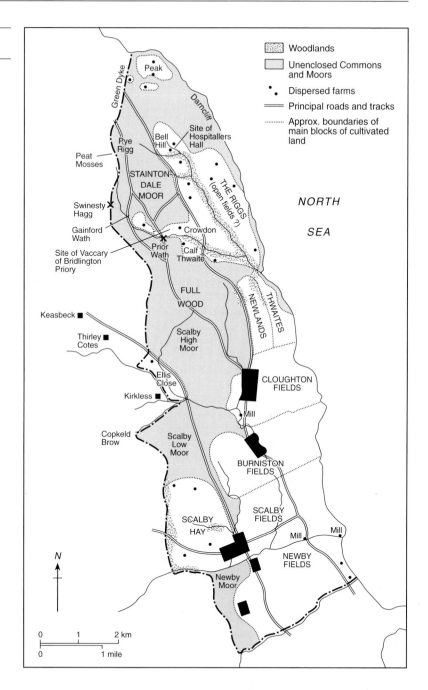

Green Dyke

Peak

Darncliff

Peat
Mosses

Pye
Rigg

Bell
Hill

Site of
Hospitallers
Hall

STAINTON-
DALE
MOOR

THE RIGGS
(open fields ?)

Swinesty
Hagg

Gainford
Wath

Crowdon

Site of Vaccary
of Bridlington
Priory

Prior
Wath

Calf
Thwaite

FULL
WOOD

NEWLANDS

THWAITES

NORTH

SEA

Keasbeck

Thirley
Cotes

Scalby
High
Moor

CLOUGHTON
FIELDS

Ellis
Close

Kirkless

Mill

Copkeld
Brow

Scalby
Low
Moor

BURNISTON
FIELDS

SCALBY
HAY

SCALBY
FIELDS

Mill

Mill

N

Newby
Moor

NEWBY
FIELDS

0   1   2 km

0   1 mile

Legend:
- Woodlands
- Unenclosed Commons and Moors
- Dispersed farms
- Principal roads and tracks
- Approx. boundaries of main blocks of cultivated land

311 ● ● ● ● ●

The township may not have become fully detached from its southern neighbours until quite a late date. The western boundary, from Old Blea Wyke (*Blawych*) via the Green Dyke (*Grenedic*) to Swinesty Hagg (*Swinestschage*) was already established by 1109–14, but the later southern boundary, along the line of Hayburn and Thorny Beck, may not have been so clear. A vaccary or cattle-farm at Hayburn, together with grazing rights throughout the moors and woods of Staintondale, granted to Bridlington Priory in the early twelfth century, actually included land on both sides of the beck, all of which was regarded as part of Cloughton. Gradually, however, the Hospitallers were able to detach Staintondale from the rest of the Soke. It remained part of the Forest of Pickering, and fines were occasionally levied for assarting and other forest offences during the thirteenth and early fourteenth centuries, but the paucity of references suggests that the Crown and later the Duchy of Lancaster, as chief landlords, had few interests left there. This did not, however, prevent the Crown from successfully claiming an allotment when the moors were enclosed in 1829!

The earliest surviving charter relating to any land in Staintondale is a confirmation, made between 1138 and 1154, of a vaccary given to Bridlington Priory in the time of Walter de Gant, who had founded the Augustinian House about 1125. The core of the vaccary lay 'in the moor of Stainton next to Hayburn', and the irregular hedged enclosures still seen around Crowdon Farm, and the significantly-named Prior Wath (ford) nearby, almost certainly represent the site of this early holding. In 1262 the vaccary was extended eastwards as far as the 'great gill' called *Staindicgrip*, which can only have been the deep and narrow valley of Staintondale Beck. The Priory, as we have seen, enjoyed grazing rights throughout the common pastures of Staintondale, extending even to the under-cliff (*Derneclive*) in the far north of the township.

The Knights Hospitaller appeared on the scene around the middle of the twelfth century. What appears to have been their first acquisition was granted them by one Henry, son of Ralph or Ulf, probably one of the sokemen of the area. It seems to have involved quite a small area of land—certainly not more than 50 acres—around Bell Hill (*Balladehow*) in the upper reaches of the valley. The grant included a *Hallstede* where the Hospitallers' hall or 'camera' was later built, and its boundaries extended to another house some distance away, suggesting that enclosed and scattered farmsteads were already a feature of Staintondale. There may have been a small nucleated settlement as well: a rental of 1542 mentions a close called *Aldres Tonge* which may well be a corruption of *Aldetoftes* (Old Tofts), a place name that features when one Geoffrey de Stainton, son of Nigel de *Aldetoftes* granted land in Cloughton to Bridlington Priory in the thirteenth century.

Why the Hospitallers should have been interested in such a remote location as Staintondale is something of a mystery. At the time their nearest preceptory (headquarters manor) was over 30 miles away, on the edge of the Vale of York. Perhaps proximity to the major routeway linking the important towns of Beverley, Scarborough, Whitby, and Guisborough, which is shown on the fourteenth-century Gough map and which certainly ran through Staintondale, was the key factor. Their impact was considerable. In place of the hitherto largely pastoral economy of the area, their main interest seems to have been in reclaiming the commons and moors for arable cultivation, a policy which alarmed the Prior of Bridlington, the Forest authorities, and no doubt the local farmers as well. By the end of the twelfth century King Richard I had extended the Hospitallers' holding to embrace the northern two-thirds of the township, extending from the sea on the east at least as far as the main road on the west. Within a few years, during the reign of King John, the Hospitallers were accused of making enclosures for the cultivation of wheat and oats right out on the coast 'next to Darncliff'.

Later in the thirteenth century the assarting (reclamation) activities of the Hospitallers shifted further south, fortified by a charter of King Henry III (1253) granting the Brethren leave to enclose land and cultivate it without interference from the royal bailiffs. The main scenes of activity were the slopes on either side of Staintondale Beck, extending eastwards to the sea, westwards to the Scarborough/Whitby highway, and southwards as far as the lower reaches of Hayburn Beck. Here the Brethren came into contact and conflict with the Prior of Bridlington, concerned for the erosion of the grazing lands attached to his vaccary. The problem was resolved in 1262 when the Prior granted the Hospitallers leave to assart one carucate of land—probably 100 acres or more—in Staintondale Moor, in return for an enlargement of the Prior's vaccary and an undertaking not to reclaim any further land without the Prior's permission. The newly reclaimed land in this area does not appear to have been added to the Hospitallers' demesne which, according to a survey of 1338, consisted only of 100 acres of arable, 20 acres of meadow, and 10 acres of underwood. Instead the land was organized into tenant farms and included a small block of open fields in the area known as 'The Riggs' in the south-eastern corner of the parish, where tracts of good rigg-and-furrow can still be seen. The sinuous outer boundaries of the new assart farms, such as White Hall, Planetree House, Riggs Hall, and Prospect House, can still be detected on the modern map and in the field. Further assarting was taking place in the far north of the parish where a small group of farms around the Peak probably emerged during this period, although they do not appear in the records until the early sixteenth century.

Intense assarting activity in Staintondale reflects what was going on elsewhere in the Soke of Scalby during the central Middle Ages, as population increased and more land was needed to feed newcomers. Just to the south, the great forest tract known as Fulwood in Cloughton was reported, in 1334, as having been largely denuded of timber through the actions of the inhabitants of Cloughton, Burniston, Scalby, and Newby. In the previous century the farmers of Cloughton had been busy extending their open fields and creating new enclosed holdings in the areas still known as Cloughton Thwaites and Cloughton Newlands, great blocks of land which extended almost to Hayburn Beck and effectively cut off Fulwood from the sea. These clearances appear to have been a collective enterprise: a thirteenth-century grant of one bovate of land in the fields of Cloughton adds that 'if any land shall have been or shall hereafter be broken up in the moor and waste the grantee and his heirs shall take and cultivate from this land as much as belongs to the said bovate'. In other words, everyone with a bovate of land was entitled to a standard share of the new assarts. The assarts of Thwaites can still be seen from the old Whitby/Scarborough railway line which bisects them. The enclosures are bounded not by walls and hedges but by broad and near-impenetrable swathes of bracken and furze, remarkable vestiges of the moorlands which once covered the area.

By the end of the thirteenth century the Hospitallers appear to have been in control of the whole of Staintondale except for the Bridlington Priory holding in

Medieval ridge-and-furrow on the Riggs

Here a small common field was located among a group of assart farms, one of which (Plane Tree Farm) can be seen on the left. In the foreground is the wooded valley of Staintondale Beck.

the south-west. Their position was temporarily challenged in 1341 when King Edward III seized the property on the spurious pretext that it had originally been granted to the Knights Templar on condition that a chapel be maintained and hospitality provided for poor travellers—services which, it was claimed, were not being provided. However, a jury appointed to enquire into the claim found for the Hospitallers and no more is heard of the matter. This incident is of some interest since it indicates that the Hospitallers ran their Staintondale estate purely for its farming income. The holding was described in 1338 as a 'camera' (i.e a hall with its attached farm). There is no evidence that any of the Brethren ever resided there and the usual monastic-type buildings, including a chapel and guest-house, would not therefore be needed. The estate was probably run by a salaried bailiff who occupied the 'camera' or hall, organized the work of a small demesne labour force, and collected rents from the tenant farmers.

Following the Black Death (1348–9), lower grain prices and higher wages diminished the enthusiasm of landowners generally for direct farming, and encouraged the leasing-out of demesne lands. At Staintondale the Hospitallers at some stage let out their home farm to tenants, so that by 1542 the whole manor was in the hands of seventeen tenant farmers, including *Le Olde Hall* and *Le Parke*, which must once have been at the heart of the demesne. Furthermore, the balance of agriculture had shifted back from arable to pasture farming: of the thirty or so closes shared among the farmers, only seven were described as arable in 1542.

Rural Society and the Landscape from the Sixteenth to the Eighteenth Centuries

The Hospitallers summarily withdrew from Staintondale when the Order was dissolved in 1540 and its properties were seized by the Crown. An account drawn up in 1542, and the tax assessments of 1543, provide us with our earliest picture of the people farming in Staintondale. Certain families had by this time established themselves firmly in the dale: the seventeen customary tenements listed in 1542 were in the hands of just seven families. The Harrisons and Hogesons held ten holdings between them; they paid nearly two-thirds of the total rental and a similar proportion of the tax. Most of the individual holdings consisted of messuages (houses) with unmeasured closes of land attached, which were valued more or less equally: six paid between thirty and forty shillings per annum, and another four paid between thirteen and twenty shillings. The remaining holdings were cottages, each with at least one close of land, paying rents of between six and eleven shillings. All were sufficiently well off to contribute to the 1543 tax, nine of them paying above the minimum rate. The picture is thus one of a fairly equal society, without extremes of wealth and poverty. The farmers evidently had both the means and the determination to acquire control of their dale and to free it from outside interference. Thus a consortium of local farmers was able to purchase the estate from the Crown in 1553, paying just over £773 for it—nearly fifty times the rental of 1542. This consortium then set about selling off the seventeen messuages with their lands in fourteen lots, mostly to the sitting tenants. Finally, in 1562, a group of four of the purchasers bought the manorial rights for £40. These rights were important since they included the holding of manor courts and gave their owners some control over common grazing, peat and mineral extraction, hunting, and the disposal of wrecks—a significant asset on this rugged coastline. The other purchasers, who were now effectively freeholders, expected the manorial rights to be passed on to them as a body, but the four 'lords' and their descendants thought otherwise and the dispute was not finally resolved until 1662 when the freeholders bought out the one remaining claimant. The only other source of external interference, the payment of tithes demanded by the tithe-owner of Cloughton in 1628, was successfully challenged in the ecclesiastical courts ten years later. Thus was achieved a remarkable freedom from squire and parson, which was no doubt an important factor in the early development of Quakerism in Staintondale. By 1669 meetings were being held in the house of William Warforke, on the site of the present Meeting House Farm, and a meeting-house was built nearby in the 1690s.

The 'freeholders' republic' which had emerged in Staintondale seems to have encouraged the development of a vigorous land market. The farmers were now free to round off and expand their holdings by purchase and mortgage, and it was not long before some of them began to acquire freehold land in adjacent parishes. Thus in 1602 William Hay of Old Hall was able to bequeath another farm, at Thorny Beck, to his daughters, and a few years later his son left substantial property in Cloughton. In 1690 James Harrison was able to set up all five of his sons respectively with his farm in Staintondale, land in Cloughton, a house and land in Fylingdales, Calf Thwaite Farm and other lands in Cloughton, and a house in Scarborough. Conversely, the farmers were also able to sell and mortgage land, thus allowing a number of outsiders to gain a foothold in the dale. The North Riding of Yorkshire was still largely a country of great estates in the sixteenth and seventeenth centuries, including those of the Cholmleys of Whitby and the Hobys of Hackness to the immediate north and west of Staintondale, so that freehold land was at a premium and much sought after. Although Staintondale was hardly prime agricultural land, the development of sheep-farming (see below) may have made it an attractive investment at a time of booming cloth-production. Even in the 1550s, five of the Staintondale purchasers do not seem to have been resident in the parish, and in the course of the seventeenth century many of the earlier family names disappeared.

The Hearth Tax assessments of 1673 suggest that considerable prosperity pervaded the whole dale, and a fairly even distribution of wealth is indicated by the fact that no less than ten of the twenty-two contributors occupied houses with two or more hearths—an unusually high proportion for north-east Yorkshire—although none had more than four hearths. The personnel, however, continued to change: only eight of the family names of 1673 are found in a list of freeholders made only sixteen years previously. To judge by wills and inventories, another trend was also setting in during the late seventeenth and early eighteenth centuries: the leasing out of property by both resident and non-resident freeholders to tenants was introducing another social layer into local farming society. Many of the inventories made during the period from 1690 to 1730 were of low overall value, usually under £25, and record rather small farms with, typically, a horse, a couple of cows, between twenty and forty sheep, and a little corn and hay. By the late eighteenth century tenant-farmers were probably in a majority, though a substantial minority of eight to ten resident freehold farmers always remained in the parish.

The appearance of the dale during this period is hard to gauge. The number of farms—between seventeen and twenty—remained remarkably constant throughout. Many of the farms are named in sixteenth- and seventeenth-century wills and they appear to be in the same locations as their nineteenth-century successors. There is very little evidence of expansion into the adjacent commons before the enclosure of 1829. Farmers acquired extra land by purchases in adjacent parishes, particularly of grazing rights in Cloughton Newlands, which had become a stinted pasture by 1600. The main change in the landscape was effected by the internal subdivision of the thirty-odd closes recorded in 1542, probably to afford more sheltered pasture for animals and more systematic manuring of the arable land by sheepfolding. The lands attached to Peak House, for example, embracing only a couple of closes in 1542, had been divided into at least one dozen fields by 1659. The former open field on the Riggs (*Ould Rigges*) was well on the way to enclosure by 1602 when it was evidently held in four still undivided blocks. The pattern of internal enclosure can still be observed in some parts of the township, particularly in the south-eastern quarter, where a number of farms show a pattern of small but regular enclosures within the sinuous medieval outer boundaries.

One might expect the prosperity of the Staintondale farmers to be reflected in the quality of their housing, particularly in view of the Hearth Tax evidence. However, pre-eighteenth-century rural housing in north-east Yorkshire was generally of a very poor standard, and any superiority in Staintondale was only relative. Probate inventories, listing personal property at death, are only available from 1688 onwards and only a minority of them provide information about rooms, but the overall impression is one of rather basic accommodation. Even the wealthiest farmers seem to have lived in rather humble surroundings. John Armyn of the Peak (1716) had 200 sheep, twenty-eight cattle, four horses, and large quantities of corn and hay, together worth £115. His house boasted a kitchen, dairy, and a couple of parlours but there were no upstairs rooms and the total contents were valued at only £4 10s. 0d. Most of the houses recorded before 1730 consisted only of a living-room (forehouse) and a ground-floor parlour used for sleeping.

View looking northwards from the south side of Hayburn Beck, above Bridge Farm (in the trees to the left)

The barn in the middle ground is a former long-house, bearing the date 1655. Note the relatively luxuriant hedgerows in this part of the township, and the woodlands to the north and east, around the valley of Staintondale Beck.

Unfortunately very little seems to survive on the ground. A survey by the Royal Commission covered six farmhouses in the parish of which only one appears to predate the nineteenth century. The exception is a stone cowhouse to the east of Bridge Farm which was once a dwelling. It is a single-storey and cruck-built structure of four bays, two at the 'house' end, probably for forehouse and parlour, and two at the 'low end' with ventilation-slots suggesting use as a byre. The door-way leading into the cross-passage dividing the two ends of the house has the date 1655 inscribed on its lintel. The building is a typical North York Moors 'longhouse' which, in spite of its humble character, is an early example of the 'Great Rebuilding' in this area. We know from documentary sources that earlier cruck buildings in the district had outer walls of mud or, at the best, dry stone walling, rather than the good mortared stonework that we find in this structure.

The Industrial Landscape

One might have expected farming to have dominated the life of Staintondale throughout its history, but, in fact, the rugged cliffs in the parish shared in the development of one of the new large-scale industries of Elizabethan and Jacobean England which were designed to ensure national self-sufficiency in materials formerly imported from Catholic Europe. Alum, used as a mordant in the dyeing of cloth, was vital for the greatest of all national industries, and it was found at the Peak in Staintondale. Works were first constructed here by Sir Brian Cooke, around 1615. They remained in the Cooke family until final closure in 1862, although usually operated by lessees from outside the area who occasion-ally took up residence for a time at the nearby Peak House.

View to the east from Ravenscar village over the Intakes above the edge of the Peak alum quarries, which lie to the right

Beyond the enclosures lies a still unenclosed moorland tract. Some of the enclosures are medieval, but the small 17th- and 18th-century farmsteads were occupied by alum workers.

A tenement known as *Le Peeke House*, one of the largest in Staintondale, was first mentioned in 1542 and the existence of at least one cottage nearby suggests that a small community had developed in this location, separated from the rest of the parish by a great tract of moorland. The settlement must have expanded considerably with the onset of alum working—most of the larger alum works on the coast, of which the Peak was certainly one, employed over 100 workers when in full production. Although nothing resembling an industrial village developed before the nineteenth century, a group of very small and irregular enclosures carved out of the moor near to the alum quarries, around smallholdings such as Crag Hall and Black Head, are probably survivals from seventeenth-century 'intakes' within which workers' cottages once stood.

Peak House, with its extensive farm, came to be regarded as a desirable property in its own right, attracting a succession of moneyed men from the 1650s onwards, due partly no doubt to its dramatic situation overlooking the alum works and the great sweep of Robin Hood's Bay, and partly to the ready market for agricultural produce that lay on its doorstep when the alum workers came to live nearby. By the nineteenth century the Peak House estate embraced three farms, ten cottages, and 200 acres of land, to which a further 125 acres were added at the enclosure of 1829. In 1774 the house itself was rebuilt as a Georgian villa by the Childs, a London banking family, who had no known connection with the alum business.

The alum workers themselves are difficult to identify. They often doubled as agricultural labourers, and were described as such in the records. This may be due to the intermittency of production: the Peak, like other works, was frequently laid down for long periods when overseas competition lowered prices. In addition, much of the work itself was irregular: only a dozen or so skilled workers were required and the rest of the labour force was mostly engaged in barrowing shale from the quarries to the clamps in which it was burnt. When sufficient supplies had been stock-piled, further labour might not be needed for a long time, and the labourers would be thrown back on such agricultural work as the developing estate could provide. There always seems to have been a close relationship between the alum workers and the owners of the Peak estate. In the only alum workers' will so far discovered (1715), William Smith left absolutely nothing, but he requested Peter Lindley of Peak House to provide 'a good drink' for the alum workers in return for the great trouble they would have in carrying his body the 6 miles to Scalby church for burial.

In the early nineteenth century, the poor and scattered workers' cottages on the Peak estate appear to have been largely replaced by better housing located in a new hamlet known as Peak Hill, a few hundred yards to the south of Peak House. In 1841, a time when the alum works were probably lying idle, the hamlet was inhabited almost entirely by agricultural labourers, yet most of them were described as alum workers ten years later. Other cottages were built in and around the alum works themselves, probably as part of a major redevelopment scheme for the works undertaken in about 1830. Eighteen cottages were sold along with the alum works in 1862 which, together with the Peak Hill dwellings, were probably adequate for the relatively small labour force then employed. The 1851 Census lists only 38 alum workers located in 24 dwellings. A pair of ruined cottages at the works are currently being cleared and restored by the National Trust.

As for the remains of the works themselves, they are the subject of an ongoing programme of excavation and restoration by the Trust. Near the quarries are the calcining sites where the shale was burnt and the steeping pits, with their reservoirs, from which the alum salts were extracted in liquid form and fed, via long stone-lined channels, to the 'alum house' on the cliff-top. Here can be seen the remains of the boiling house, where the liquors were heated, and the roaching houses in which they were left to crystallize, together with warehouses, cottages, and many ancillary buildings. On the cliff-edge are the spectacular remains of machinery to winch coal and other supplies up from the beach and to send manufactured alum down, while at the bottom of the cliff a succession of harbour-works, cut into the rocks, can be studied at low tide. Although the present remains are mostly of nineteenth-century date, it is important to recognize that all this plant would have been necessary from the very beginning. It involved a huge investment of capital, initially made by the Crown—the Cookes were only 'farmers' or lessees before privatization of the works in the late seventeenth century.

The Nineteenth Century and Beyond

The extensive medieval commons of Staintondale survived largely intact until the nineteenth century, long after the commons of neighbouring parishes had been enclosed. Perhaps the difficulty of securing agreement among a large number of freeholders hampered any action. When parliamentary enclosure was finally achieved, in 1829, it embraced no less than 1,531 acres of common pasture and moorland. Within this area seventy separate allotments were made to eighteen parties, most of whom received between three and eight blocks of land in different areas of the parish. Many of the beneficiaries of the enclosure appear to have been outsiders: only eight of their number appear as residents in the 1841 Census. Furthermore, the outsiders obtained most of the larger allotments; according to the 1851 Census resident farmers occupied a total of only 1,510 acres of land within the parish, of which the new enclosures accounted for not more than 20 per cent. But in spite of this significant shift in the balance of property in the parish, the local freeholders remained an important element. At least eight of the nineteen farmers listed in the 1851 Census were descendants of the resident freeholders of 1829.

Staintondale in the 18th and 19th centuries

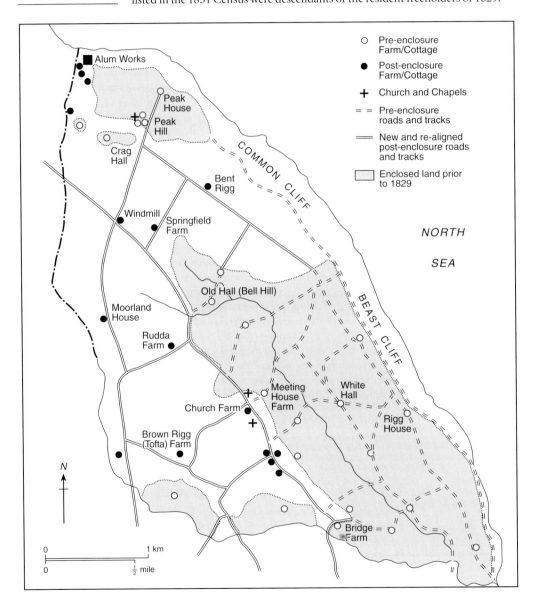

The actual process of enclosure was spread over a long period. The first edition Ordnance Survey 6-inch map, published in 1854, still shows large unenclosed areas in the north and west of the parish. Only three new farms had been built since 1829, at Bent Rigg and Springfield to the north and at Brown Rigg in the south-west. Along the line of the main road the present 'village' was beginning to appear: an inn, school, smithy, and one or two cottages, later supplemented by a church, a farmhouse, and a few more cottages.

The topographical aspects of enclosure are difficult to assess since no map survives with the Award and there is not even the usual Tithe Award to help out. The enclosure landscape can, however, be largely reconstructed from the text of the Award. The new allotments lay in two main areas: in the south-west, above the ancient enclosures along Hayburn and Thorny Beck, and in the north-east around Bent Rigg. It is interesting to note that the commons had formerly extended one field to the east of the main road, on its current alignment, so that all the old farms west of the Beck, from Meeting House Farm to the cruck long-house above Bridge Farm, had been located on common-edge sites. The new allotments on Bent Rigg were laid out on either side of the new Bent Rigg Lane, which continued the line of the old War Dike Lane all the way to the Peak. They extended from the cliffs on the north-east to the ancient Hospitaller enclosures around Bell Hill on the south-west, covering an area of about 320 acres. Eight parties to the Award received allotments in this area, but the bulk went to Mrs Willis of Peak House (125 acres) and to Sir William Bryan Cooke, owner of the alum works (75 acres), a significant accretion of property at the north end of Staintondale which was rapidly becoming the main focus of settlement in the parish.

The enclosure coincided with a general resurgence of arable cultivation, a return—as far as Staintondale was concerned—to something like the conditions of the earlier Middle Ages, after several centuries of pastoralism. Although of relatively poor quality, even the fields newly enclosed from the moors were put under the plough. Thus when the Candlers leased out a large block of property in the south-west of the parish, the farmer was instructed to pursue a five-course rotation of fallow; turnips or potatoes; wheat; oats or legumes; and grass, sustained by liberal applications of lime. Limestone outcrops occur in the northern cliffs and a limekiln near the alum works may have supplied local needs.

At the death of W. H. Hammond, who owned Raven Hall (the former Peak House) from 1841 to 1885, the *Scarborough Gazette* proclaimed that 'he created farms out of the wild moorlands and changed heather into turnips, corn and pasture'. In 1858 he also built the windmill which still stands on the high ground to the south-west of the Peak.

W. H. Hammond was perhaps the nearest approach to a squire that Staintondale had experienced since the days of the Hospitallers, providing a chapel, schoolroom, shop, and inn. He also enthusiastically promoted the Scarborough and Whitby Railway, with stations at opposite ends of the parish, which was begun in 1872 but not completed until 1885. It was, however, a previous owner of the Peak estate, the Revd Dr Richard Child Willis, who in 1831 rebuilt and greatly extended the 1774 house. The Palladian villa gave way to the rather bleak neo-classical building which still forms the core of the present Hotel.

On Hammond's death in 1885 his four daughters and their husbands, along with other speculators, promoted the Ravenscar Estate Company to develop a new seaside watering place focused on Raven Hall, which was now to be turned into a luxury hotel. Initial optimism knew no bounds and an extensive grid of streets for terraces of houses and of serpentine cliff-walks for villas and pleasure grounds were staked out. In practice not more than thirty of some 1,700 plots found purchasers and less than a dozen of forty or more planned streets were laid out. Most of them were left as muddy tracks disappearing into the fields. A few solitary buildings around the impressively named Station Square, intended as the centre of the new town, are about all that remain of this grandiose scheme. The Hotel, however, taken over by a group of Bradford businessmen in 1898, has enjoyed a near-unbroken existence until the present day. It is now the Raven Hall Country House Hotel affording an impressive range of facilities for its guests.

Conclusion

From at least the twelfth century Staintondale has been a rather remote place, cut off from its immediate neighbours not only geographically, by great commons and moors, but socially by its independence from the great neighbouring property complexes of the Abbot of Whitby and his successors, the Cholmleys and the Hobys. Even to the south, the large nucleated and open-field villages along the coast imply a different lifestyle from that prevailing among the scattered farmsteads of Staintondale, little bothered by squire and parson. The emergence of a 'freeholders' republic' reinforced these differences and although there was considerable mobility out of and into the parish, with much land changing hands, a significant proportion of resident freehold farmers always remained. Developments at the Peak, in the far north of the parish, always a separate community, associated with the growth of the alum industry and the emergence of a gentry estate, had very little impact on Staintondale as a whole.

The railway was closed in 1965, but the popularity of 'Ravenscar', a name invented by the Estate Company, has increased considerably in recent years. Apart from visitors to the Hotel, the magnificent view and walks have made it a popular centre for the weekend motorist and for serious walkers following the Cleveland Way along the coast. The National Trust, which now owns much of the undercliff and the site of the old industrial workings, has a coastal centre near the hotel where a small but growing stream of visitors are directed towards the geological features, natural history, and industrial remains. The rest of Stainton is, thankfully perhaps, still largely unknown. The main road between Whitby and Scarborough, crowded with traffic in the summer months, has long been diverted well to the west of the parish leaving the old road—which now comes to a dead end at Ravenscar—mostly to local traffic. Even for walkers, the attractions of the paths along the coast and along the old railway line, mean that the many ancient tracks linking the scattered farms are nowadays little frequented. It is here that one can still recover the early farming history of the dale and sense something of the geographical and social remoteness which has characterized Staintondale through the ages.

15 Fen Drayton, Cambridgeshire:
An Estate of the Land Settlement Association

Pamela Dearlove

Introduction

Fen Drayton in Cambridgeshire, lies about 8 miles north-west of Cambridge between the main Huntingdon to Cambridge Road (A14) and the River Great Ouse. It sits in a mainly flat, featureless landscape of large open arable fields, where, with little else to catch the eye, motorists on the A14, in the vicinity of Fenstanton, are likely to notice the group of glasshouses partly hidden by a windbreak of poplar trees, hinting at a different kind of landscape. Drivers who turn into Mill Road, will see more greenhouses, accompanying pairs of uniform, unremarkable, semi-detached houses, set some distance apart on holdings of land. Continuing into the village they will observe smallholdings that contain livestock, including horses and geese, and others that show evidence of different rural enterprises, or even of neglect. They are unlikely to realize that this scene evolved out of a unique experiment in planned settlement in the 1930s

The interest of this landscape lies in the radical transformation that was achieved in a comparatively short period of years. Because the change was effected so recently, we can follow it in documents, in photographs, and on the ground. But similar, comparatively abrupt reorganizations of land and people are known or suspected in the past, either carried out by authoritarian landowners, or effected more gradually over a longer passage of time, when, for example, a gentleman moved into a village community where none had resided before, or economic misfortune removed a substantial number of its families from the scene.

Before the Land Settlement Era

In the early 1930s, Fen Drayton was a small, quiet, insular village with a population of about 200, and with only sixty dwellings (the number had been falling for some years). One seventeenth-century farmhouse was reputed to have been the home of Sir Cornelius Vermuyden, whose drainage schemes made such an impact on the fenlands. There were a few farms, houses, cottages, and barns, the parish church, a Methodist chapel, a primary school, a blacksmith's forge next to the Three Tuns public house, and a post office and general store. Farming was the main economic activity in the parish with some gravel working but few other sources of employment. At this time, a period of worldwide industrial depression, the numbers of unemployed were rising to an all time peak and in northern England and south Wales, it seemed unlikely that the steelworks, coal mines, or shipyards, would ever reabsorb more than a small proportion of them. As an alternative means of employment, at a time when the future of agriculture was also being much debated, it was suggested by some, including Lloyd George, former Prime Minister, that the problem might in part be solved if a proportion of the labour force returned to the land to make a living. Others, it must be said, thought that the call of 'back to the land' was as inappropriate as the call of 'back to the handloom'.

Fen Drayton was one of twenty-one places chosen to accept a considerable group of newcomers, earning their living on smallholdings of 3–6 acres by horticultural activities. The radical import of the plan deserves some explanation. The Quakers had for some time been experimenting with allotment schemes for the unemployed. Mr (later Sir) Percy Malcolm Stewart, chairman of the London Brick Company, offered to donate £25,000, if the government would give a similar amount, and the Quakers would administer a scheme to settle unemployed industrial workers on full-time landholdings away from their home areas. Impatient of the delays that followed, Mr Stewart himself bought an estate at Potton, Bedfordshire, in January 1934, and this precipitated the setting up of a Land Settlement Association (LSA), which had government approval and was announced to the House of Commons in July 1934. In November Mr Stewart was appointed Commissioner for the Special Areas (England and Wales), areas where unemployment was especially intractable, and the LSA was asked, with the help of government money, to transfer some 800 families from Northumberland, Cumberland, and Co. Durham to smallholdings in other parts of England. A Welsh Land Settlement Society was also created to administer similar schemes in Wales.

The LSA planned to establish 2,000 holdings in different parts of the country, on different kinds of land. Not surprisingly, it experienced difficulty in finding estates for sale that were on suitable soils for division into smallholdings, that could be immediately occupied, and that promised a decent living. Fen Drayton House and its lands, in an area where many parishes, including Fen Drayton, had a tradition of smallholdings and market gardening, and which was near to road and rail links to national wholesale fruit and vegetable markets, completely fulfilled these requirements. Advertised to be sold by auction on 5 July 1935, the estate comprised gravel beds, Daintree's Farm with 65 acres of land, Middleton's Farm with 72 acres, Fen Drayton House, fifteen cottages, fen pastureland, and orchards—in all about 350 acres. At the beginning of the auction, it was announced to prospective buyers, that all but the gravel beds were already sold. The *Cambridge Chronicle* reported that it was unusual for land to be sold before auction, 'but a semi-public body which was concerned with settling families on the land from depressed areas, deserved special consideration'.

In all, twenty-one estates were developed by the LSA which differed from each other in various respects (a full list appears at the end of this chapter). Those on better soil, including Potton and Fen Drayton, were divided into holdings of 3–6 acres producing some pigs and poultry but with a bias towards horticulture, whereas Fulney estate (Lincolnshire) on rich silt land, was completely devoted to horticulture, with some 2-acre holdings eventually concentrating on the forcing of spring flowering bulbs. On poorer land, as at Abington (Cambridgeshire), larger holdings of 8–10 acres were biased towards pig and poultry keeping whilst Stannington's (Northumberland) 12-acre holdings were totally devoted to livestock. Many of the new houses that were needed were designed by the architects Packington and Enthoven to be compatible with their locality but other factors could influence the design. The estate's Local Advisory Committee at Fox Ash (Essex), considered non-flushing lavatories in sheds appropriate for 'that class of house', whilst timber-framed, asbestos-lined, prefabricated houses were erected at Andover (Kent) as the result of an offer, made by the Hurlingham Bungalow Company, of one free bungalow in every four purchased. Generally placed on individual holdings, houses on some estates were grouped in hamlets in order to help foster community spirit.

A transformation of Fen Drayton's landscape resulted from the new settlement scheme, which incorporated certain principles at the outset. It was laid down first that smallholdings should be established in groups around Central Farms and organized as administrative units under estate managers; second, training and supervision should be provided for the settlers; third, cooperative methods should be used to purchase the men's agricultural requirements and market their produce; fourth, costly tools and capital equipment should be available for them to hire; and last, loans at favourable rates should be available to aid them to equip their holdings. These principles had never before been combined in any land settlement scheme.

The newcomers to Fen Drayton, who were to be carefully selected, had to be men of between 30 and 50 years old, long-term unemployed, and, ideally, should have cultivated an allotment. They had also to be prepared to work long hours and it was vital that their wives should agree to work alongside them. Initially twenty men were to be installed on the estate as this was considered the largest number which could be trained at one time.

Setting up the LSA

Wardens, later to be called managers, were assigned to all the new estates. With a number of technical assistants, they had to train the settlers in various types of agriculture, horticulture, and animal husbandry, and be responsible for all the buying and selling on their estates (a later Director of the LSA was to describe the Warden as the 'lynch pin' of the whole organization). Mr Arthur Piper, previously Horticultural Manager of the South Eastern Agricultural College in Kent (Wye College), arrived in Fen Drayton in October 1935, and was joined in November by the first contingent of prospective settlers from the Special Area of Co. Durham.

Settlers remained 'on the dole' for between fifteen and eighteen months, after which they were expected to be competent to run their holdings. The first three months of their training was a time of preparation that allowed them to build up their strength before undertaking twelve months' intensive instruction on the holdings. Fen Drayton's trainees constructed roads and service tracks, and divided the estate's fields, including another field purchased from a local farmer, into fifty-four rectangular plots of between 3 and 6 acres. The holdings lay alongside each other in groups as instructed by Mr Piper. This was in order that acres of crops, such as potatoes, brassicas, or carrots, could be cultivated in the same place on each holding, and the land ploughed and harrowed as one unit in the manner of large, mechanized market garden farms. Thirty acres of soft fruit were planted in similar blocks. The system, which caused some antagonism between settlers, was discontinued sometime after the war and holdings were planted and cultivated separately, with a subsequent change in appearance. Piggeries and poultry houses were erected on each holding with a glasshouse, a smaller propagating house, and 100 Dutch Lights (cloches).

The LSA sold some original cottages on the estate but used suitable properties. At first, Fen Drayton House, built in grey stone, 'regardless of expense', in the 1840s, was used as a billet for trainee settlers, and Daintree's Farm, where the stables were located, provided a home for the estate manager and storage space for capital equipment and tenants' supplies. Situated close to the centre of the village, Middleton's Farm and its cottages and barns became the estate's Central Farm and the hub of the whole estate. As the compulsory buying and selling centre for the tenants, it soon accommodated grading and packing sheds, piggeries, a hatchery, an office, and an agricultural supplies shop.

On 17 January 1936, the *Cambridge Weekly News* reported the arrival of the men from Durham when Fen Drayton Village School was crowded out in celebration. 'The miners entertained the villagers with items and choruses which were greatly enjoyed, and the intention was expressed that more such entertainments would follow.' Because the LSA was intent on fostering a sense of community when bringing so many strangers together, it provided land for a village hall, the Carnegie United Kingdom Trustees Charity paid for it, and it was soon the venue

Mrs Carter and daughters on their Springhill Road Holding

In 1939, far from the Sunderland shipyard where her husband once worked, Mrs Carter enjoys a frivolous moment with her daughters, Rose (on the pig) and Barbara, whilst taking a break from feeding the pigs and poultry. The smallholding is in Springhill Road. Later, Barbara married 'Pip' Gill the Estate Propagating Manager who became an LSA tenant specializing in chrysanthemum growing. Eventually their son also became a Grower with the LSA.

for a Christmas party for 150 children. Thirty-seven families had arrived on the 'settlement' by June 1937, from Auckland, Sunderland, and West Hartlepool, and more were expected later. The existing village school became so overcrowded that a 'temporary' infants' classroom was erected and remained in use until 1970.

Mr Thomas Collier ('Tot') was among the first group of settlers. Aged 46 years, a coal-mine deputy until unemployed, he came from Perkinsville, where he had grown vegetables on an allotment, selling some to the local Co-operative Store. With his wife, a son, and two daughters, he had lived in a 'run-down' and pest-ridden, two-roomed house. His son, John Thomas Collier (Tommy), had left school when aged 14 to do odd jobs, including ice-cream vending and tea-making on a building site, before attending Dole School. There he soon tired of sharpening chisels and left after producing only a teapot stand. For four years, whilst receiving 4s. 2d. 'parish relief' per week, he had had little to do besides cultivating an allotment so he had spent his time with friends, looking out for girls. He remembered enduring strike periods with soup kitchens, and a diet consisting mainly of bread and jam and lentil soup which in later years he 'couldn't abide'. To get coal from the waste tips, for the family's needs, he had had to push an old barrow with buckled wheels for 9 miles. He felt extremely bitter. In 1936 the family moved onto a 6-acre holding in Springhill Road, Fen Drayton. They kept pigs and poultry, and grew carrots, potatoes, tomatoes, lettuces, celery, and soft fruit. Life was hard with little monetary reward but 'Tot' Collier was happy to be working. During 1936, 'Tommy' was employed by the builder erecting the Estate's houses and then became an LSA employee for the next 43 years engaged in general farming tasks. He thought his new life was definitely an improvement on the old.

The LSA Settlement made a dramatic impact on Fen Drayton. Stretching away from the village into the south-west quarter of the parish, over fifty houses of red brick and in two main styles were built within two years, nearly doubling the existing properties which were in close proximity, were unplanned, and varied in age, building materials, and architectural styles: the modern conveniences of the new homes were envied by some villagers, who were themselves poor, using earth closets and fetching water from the village pump. Fields and orchards which had previously been 'in poor heart', producing a low yield of corn and fruit, and supporting only a few pigs, a cow, and six horses, were soon transformed into smallholdings. These contained intensively sown crops, poultry houses with over 100 hens, piggeries with up to forty pigs, various glasshouses, and some goats and horses. The farms had become hives of industry, making daily life busier for all the villagers.

The Land Settlement Scheme in the Second World War: From 'Settlers' to 'Tenants'

With the outbreak of the Second World War, industrial unemployment disappeared and the government's priorities changed dramatically. The rapid expansion of food production was a prime necessity, and in order that the LSA estates could be utilized to their utmost capacity, recruitment policy was changed to favour applicants who had agricultural experience and some capital of their own. One newcomer at this time was Mr Madder, previously a head gardener and chauffeur in Harrow-Weald, who took over an Oak Tree Road holding in 1942 from a miner who had returned to Durham.

LSA tenants were encouraged to specialize where their greatest abilities lay, and any spare land on the estates was brought into production and cultivated by the Central Farms. The war situation created new markets for the settlers' produce. Fen Drayton's estate supplied huge amounts to numerous RAF stations, including those at Bassingbourne, Caxton, and Wyton. Mr Ward, Fen Drayton's deputy warden, explained that it was a very enjoyable time in many ways although he worked very long hours. Once, after arriving home late after a hard day's work, he was instructed to obtain 20 tons of potatoes immediately for a contingent of American airmen who had just arrived in the area. As animal feed became scarce, livestock numbers had to be reduced on the estates but Fen Drayton's pigs were fed on kitchen waste from the RAF bases. In 1943, Mr Madder successfully applied for his daughter to help on his holding as a Land Girl (later she married 'Tommy' Collier). He also had extra assistance from Italian, Austrian, and German prisoners of war. As the years passed, stability increased, a community spirit developed, and the original settlers and new tenants became less distinguishable either in the volume of crops they produced or in their incomes. Towards the end of the war, many from the Special Areas of unemployment in the 1930s had repaid enough of their loans (for equipment), whilst paying rent, to have full control over their finances. Some tenants chose to leave the LSA to take larger holdings with greater freedom and it was recognized that the estates could provide the first rung of an 'agricultural ladder'. By the end of the war, responsibility for the estates had passed from the Commissioner for Special Areas to the government.

The Post-War Period up to 1983: The End of the Three-Legged Stool

With the ending of the war came a revaluation of the role of smallholdings. Government proposals, ratified in the Agricultural Act of 1947, aimed to promote a stable and efficient agriculture. The Minister of Agriculture and Fisheries became the new owner of the estates and the Association acted as his agent, helping to achieve the objects of the government's agricultural policy by attracting men to the land, and by helping those with agricultural experience to become farmers on their own account. Policies were formulated with much greater government involvement and with an increasing emphasis on commercial criteria, although traditional welfare concepts continued to play a part. K. J. McCready wrote as follows in *The Land Settlement Association: Its History and Present Form*: 'Thus the period of the Second World War marked the development of the Association from its pre-war role as a social experiment, to its post-war form as a collection of agricultural co-operative communities, with co-ordinated supply production and marketing, combined to form a considerable force on the English agricultural scene.'

From 1951, smallholdings were let under annual tenancy agreements, which effectively gave the tenants security of tenure. Restrictions on the tenants' purchase of requisites were eased, but the Association kept control over the cropping of the smallholdings and the marketing of produce.

In 1967 a marked change in official government attitudes deeply affected Fen Drayton smallholders. A Committee under the Chairmanship of Professor M. J. Wise, reporting to the Minister of Agriculture and Fisheries, considered that the concept of the LSA Scheme as a 'gateway' into horticulture was no longer appro-

Family picking tomatoes

The Cash family in the fifties picking tomatoes on their Mill Road holding. Mothers, besides caring for the home and family, worked extremely long hours on the holdings, some with baby in the pram beside them. Children were expected to help on the land from an early age but enjoyed life in the country.

priate but that the Ministry had an obligation to those already in it. The Minister disagreed with the Committee's conclusion that the Scheme was no longer needed, but accepted that the compulsory use of the centralized services should be eased. He also agreed that the concentration on commercial horticulture be intensified with greater specialization in salad production. Estates which could not adapt should be withdrawn. This marked the end of the LSA's policy 'that diversity on the holdings provided security', the 'three legged stool' of horticulture, pigs, and poultry. In fact, estates were already changing. The keeping of poultry and pigs at Fen Drayton was being phased out, and four of the northern livestock estates of the LSA had closed. Soon, only eleven estates remained of the eighteen functioning after the war.

By 1973, the LSA had far surpassed its original aim of aiding settlers to secure the equivalent of an agricultural worker's wage. A minimum farmworker's wage was £974; an average farmworker's wage was £1,374; and an LSA tenant's average net income was £3,534. Visitors, including some from India, Korea, and Algeria visited Fen Drayton to observe a successful LSA estate in operation. The tenants were referred to as 'Growers' from 1972 because of their increasing specialization in horticulture and their dislike of paternalism. Preferred recruits were young men in their late twenties with at least five years' agricultural or horticultural experience, which could include three years at college. Growers, whether perceiving their holdings as providing a permanent occupation, or as a step on an agricultural ladder, made use of the Horticultural Improvements Grants Scheme and the LSA's credit facilities, as an aid to expansion. Homes were extended to provide extra bedrooms and larger kitchens and bathrooms, and new glasshouses were installed.

At the end of the 1950s the tenants' expansion was mostly in cold glass in the form of Dutch lights on metal or wooden structures. In the 1960s, the Association experimented with mobile greenhouses which were built as two or three stations with a span of 40 ft × 100 ft, and set on a central and two side rails. Weighing about 7 tons, they were designed to be towed over two or three plots by tractors, but Mr Ian Ruggles, a former tenant, said they were more easily moved manually with the help of neighbours. He explained that they were relocated with the changing seasons to advance the growth of a wide range of crops. Lettuces might be given frost protection by a mobile early in the year until it was moved to new ground for tomatoes to be planted beneath. Then after the lettuces had been cut, chrysanthemums could be planted in the cleared ground and the glasshouse returned to cover them once the tomato season was over. In the 1960s also, a wooden framed glasshouse called the 'Fenland Structure', measuring 18 ft 6 ins × 100 ft, was developed by Mr Scott, the estate manager with the firm Gabriel Wade and English of Ipswich. Several of these could be combined as a multi-bay to measure 100 ft square. Many were demolished in the 1970s, by which time the Growers were in the forefront of experiments with low-cost, intensive production techniques including the use of polythene tunnels. Then, large oil-heated 'Venlo' glasshouses were erected which covered either a tenth or a quarter of an acre. They had galvanized iron frames and aluminium roofing-bars and ridges. Light and airy, they provided a permanent, controlled environment, in which to grow cheap, clean, unblemished salads, all the year round, as required by the supermarkets. By the end of 1973 there were 34 acres of glass on Fen Drayton

'Jacktruck' on Mill Road holding

In 1968, Fen Drayton's growers began to stack their produce on pallets to facilitate collection from the holdings. Mr Jack Wilderspin, a local engineer, seeing their difficulty in handling the pallets, invented the 'Jacktruck' seen here on a Mill Road holding. Profits from sales of the unique pallet transporter, mainly to LSA estates, and engineering work for tenants, contributed to the success of his business and the subsequent building of larger premises on a new site in Fen Drayton.

Estate. Mr Hamlett, a former senior manager with the LSA, has lasting memories of installing more and more glass, which he explained, enabled the round lettuce to become 'the battery egg of the horticultural world'.

Various developments were undertaken by the LSA to meet its changing needs. In 1946, at Middleton's Farm, a new packhouse was built and in 1967 a propagating unit, 140 ft × 210 ft, was erected to produce quality plants unobtainable elsewhere in sufficient quantities. Cold storage facilities were provided for cut flowers and in the 1970s, a cottage was demolished to provide better access to the packhouse for delivery lorries of greatly increased size. At Fen Drayton House and in its vicinity various changes took place. In the 1940s an orchard was sold and replaced by a cul-de-sac of council houses; in the 1950s the main part of the house was turned into two flats for LSA staff and its left wing made into a separate dwelling for the estate manager; in 1969 land was sold for a village school and another small development of council houses; and in the 1970s, Park house was built for the estate manager in the area known as the Park, and Park Cottages built for LSA staff. Mixed ornamental woodland was removed at various times. Additional alterations were made to the landscape throughout these years. In 1947 long windbreaks of poplar trees were planted on the estate; then in 1956 two houses were built in Cootes Lane for LSA staff. During the 1960s free-standing corrugated-iron reservoirs were erected on the holdings but these proved insufficient for the needs of increased horticultural production, and in 1970 a butyl-lined reservoir was constructed to store 4 million gallons of water. Between 1973 and 1975 storage tanks for greenhouse heating fuel were installed on the holdings.

In the 1970s LSA Growers had mixed fortunes for reasons besides their varied growing abilities. These included weakness at the top of the organization (when strength was vital), increasing foreign competition, unusual weather patterns, and high fuel costs. Between 1973 and 1975, those who had, with the aid of government grants of 40 per cent, expanded their glass area and converted glasshouses to cheap oil heating to match those of the Common Market glasshouse industry, were faced by a 'dramatic' rise in the price of oil. They then experienced the 'long hot summer' of 1976 which produced a glut of tomatoes of which tons had to be dumped. Nevertheless, LSA Growers' average incomes rose until crop year 1979 to 1980 when they fell by 6.7 per cent, partly due to a large increase in the cost of non-returnable packaging, and partly to low profits made on produce for several reasons. In 1979, because of the high cost of heating, Growers delayed their planting programme which caused higher peaks in the production of lettuce and tomatoes. The high tomato yields, increased further by another uncharacteristically hot summer, had to compete during the critical months of July to September, with tomatoes imported from Holland. Once again tons of tomatoes were destroyed and then during that winter the price of oil rose seventeen times. An unusually warm spring followed which led to the first ever spring lettuce glut: too unprofitable to market they were ploughed into the ground. Returns from early glasshouse celery, a reliable crop for LSA Growers, were also badly affected because tomato growers in Guernsey, also faced by Dutch government-subsidized competition, began to produce celery too. The difficulties continued. For example, Ian Ruggles lost a chrysanthemum crop potentially worth £10,000 on the 1980 Christmas market. It was destroyed overnight when the oil supply to the glasshouse heating system froze. To compound the problems record rates of interest were being charged nationally. These were disastrous for the Fen Drayton tenants who had invested heavily in glass, and were, because of their concentration on mainly salad crops, vulnerable to the effect of the market. In 1981 five were forced out of business.

Many LSA tenants were increasingly unhappy. They lobbied Members of Parliament and accused the Association of mismanagement. They complained about having to compete with the government-subsidized Dutch glasshouse industry and demanded that the centralized marketing and packaging system should be returned to estate level. Ministry officials, the LSA's representatives, including Graham Ward the new Chief Executive, and advisers, were spending many hours working on solutions, when on 1 December 1982, Peter Walker, the Minister of Agriculture, Fisheries, and Food, dropped a 'bombshell'. He announced that the LSA estates would be sold and all services cease to exist on 31 March 1983.

The Sale of the LSA Estate

The announcement of the closure of the LSA scheme was relayed to one tenant by his daughter, who had heard it on the radio. Shocked growers, whose tenancy agreements stipulated they must give the Association twelve months' notice to vacate their holdings, considered that they should be allowed at least as much time to make alternative arrangements. Eventually the deadline for the removal of marketing services was extended to December 1983, to give people time to set up their own cooperatives. The Minister solved the problem of lifelong tenancy agreements

Aerial photo of Fen Drayton Estate, 1972

Part of Fen Drayton Estate in 1972. In the foreground in Cootes Lane, surrounding the village hall, is the Estate's Central Farm. It was one of the few with facilities for handling flowers. To its right is the large glass propagating unit. The glasshouses on the holdings along Middleton Way are prominent in the landscape. Even more glass was added to the holdings and fewer crops grown in the open ground.

by offering the tenants the chance to buy their holdings. They were pleased with this opportunity (some had long hoped for it), and it was extended even to those working their notice to vacate their holdings. However, it came too late for others who because of mounting debts or ill-health, had already severed their contracts.

The sale of the Estate had as marked an effect on Fen Drayton as did its creation. An outline planning application submitted by the Ministry of Agriculture, Fisheries, and Food (MAFF) in June 1983 resulted in various developments. All the Middleton's Farm central services buildings, and the adjacent village hall, were replaced by high-density housing which provided homes for some of Cambridgeshire's expanding population. A larger village hall with a car park, a bowling green, and two tennis courts was built in Cootes Lane, and in the garden of Fen Drayton House five large houses were built and more on the field opposite it. Parts of holdings 52 and 53 Springhill Road were amalgamated with a piece of grazing land, and part of a Mill Road holding became the site of a new pack-house. This, partly financed by MAFF and 'Food from Britain', was for the new consortium of former tenants, named 'Fen Drayton Growers'. The service trackways were incorporated within the holdings and sold, but the parish council successfully fought for some of them to be reinstated and designated as public rights of way.

Under LSA management, the smallholdings and houses had all looked similar. Since the sale the landscape's appearance, even that unaffected by new housing, has altered. Houses have assumed individual characteristics, a few are so disguised it is difficult to identify them as 'settlement' homes. Saleable glasshouses have been removed whilst others are derelict and submerged in weeds. Only a small proportion of holdings are used for commercial salad growing: the growers disbanded their cooperative after only five years. Several holdings have been amalgamated to form larger units, whilst others have been divided into smaller sections for letting. Some are used for the wholesale growing of container plants and trees whilst others contain horse paddocks. A motley collection of 'rescued' sheep and other animals identifies another group of holdings as the site of a farm animal 'sanctuary'. Planning restrictions limit the holdings to agricultural use only but the erection of a few temporary prefabricated homes has been allowed and attempts are being made to have the planning category changed.

Conclusion

Fen Drayton has undergone major changes in its landscape and in the livelihood of its inhabitants in a little over sixty years. But its story dramatizes the process of change which has transformed other landscapes in the past, though not usually as rapidly as here. People have removed themselves from many old settlements in the countryside, or have been removed at the will of landowners; their land has been put to a different use; and a fresh pattern has been imposed on the landscape. A present-day experience such as this is helpful in bringing the consequences for people and the land more vividly before our eyes. Frequently, some connecting threads are preserved that link the past with the present, and that is evident in Fen Drayton. The Land Settlement Association built on an existing disposition in this part of the country to cultivate land in small farms and to grow garden crops. The social principles that guided the LSA, on the other hand, were new, and while they invoked much sympathy in the mid-1930s, they lost their compelling relevance when the war created full employment between 1939 and 1945. For a while after 1945 the scheme was still valued for giving access to the first rung of an 'agricultural ladder'. Then it was promoted as a means of providing access to land and a reasonable livelihood for those with some capital and an educated background. But its aims by the later 1960s had completely changed from the original ones, and the reason for founding the LSA was no longer readily understood. Still, in a different world it had considerable successes. Even as late as August 1980 it was described in the *Grower* as 'this strange amalgam of public and private sector, which for all its shortcomings is a dominant factor in the fresh produce trade, as well as being a rare success story for state participation in private enterprise'.

The introduction of an LSA community invigorated Fen Drayton. It increased the population of about 200 to about 550, and provided a few manual jobs for local people (eventually reducing the number commuting to work). The village shops, the pub, and a new small engineering business profited from its presence, but essentially it was self-contained, drawing on resources from within its own organization. Inevitably, the villagers regarded those on the 'settlement', bound together by the organization, and by their shared purpose of extracting a living from the soil, as set apart from themselves. Yet, although the tenants worked for many hours isolated on their holdings, and drew on each other's support in times of difficulty, there was increasing interaction between the two groups. The more 'outgoing' 'men from the north' introduced football to the villagers, and the LSA community organized joint social events, the formation of a Women's Institute, and representation on the parish council (although this was not accepted at first) and the Parent–Teachers' Association. Some village and estate families worshipped together at the chapel, though most friendships were between the children.

The implanted settlement made the parish look and feel different from others around it. The regular land pattern and crops, the uniform housing, and diverse glasshouses, were distinctive in the surrounding farmland. Much traffic was generated, by horses, and then by tractors, between the smallholdings, the farms, and the packing shed, and by delivery vehicles and queuing, supermarket container lorries, carrying supplies and produce to and from the growers. All this created its own particular atmosphere.

The LSA's experiment was the source of many changes in Fen Drayton both in the landscape and in the community. In effect, it created two communities, with the Growers, under their LSA umbrella, retaining a separate identity, but the two grew closer as they shared in the upheaval resulting from the sale of the estate. Changes are still going on, as a consequence of the sale, but the landscape continues to provide a pictorial record of the LSA's policies over a period of forty-eight years, and evidence of a unique experiment in land settlement will remain for years to come.

The Full-time Smallholding Estates Developed by the LSA
*Abington
New House Farm, Great Abington, Cambridgeshire

+Andover
Little Park Farm, Andover, Hampshire

+Broadwath
Broadwath Farm, Heads Nook, nr Carlisle, Cumberland

*Chawston
Chawston Manor, Chawston, and Rookery Farm, Wyboston, Bedfordshire
(initially separate estates)

+Crofton
Crofton Hall, Thursby, nr Carlisle, Cumberland

+Dalston
Dalston Hall, Lingey Close Head, Dalston, Carlisle, Cumberland

Duxbury
Farnworth House Estate, Duxbury, Lancashire

+Elmesthorpe
Church Farm, Elmesthorpe, nr Earl Shilton, Leicestershire

*Fen Drayton
Middleton Farm, Fen Drayton, Cambridgeshire

*Foxash
Ardleigh, Essex

*Fulney
Dairy Farm, Lower Fulney, nr Spalding, Lincolnshire

+Harrowby
Harrowby Hall, nr Grantham, Lincolnshire

*Newbourne
Newbourn Hall, nr Woodbridge, East Suffolk

*Newent
The Scarr, Newent, Gloucestershire

+Oxcroft
Shuttlewood, nr Chesterfield, Derbyshire

*Potton
Home Farm, Potton, Bedfordshire

*Sidlesham
Keynor Farm (also Fletchers and Streetend Farms), Sidlesham, Chichester,
Sussex

*Snaith
West Bank Farm, Carlton, Snaith, nr Goole, Yorkshire

+Stannington
Moor Farm, Stannington, Morpeth, Northumberland

Yeldham
The Change, Great Yeldham, Essex

* Listed in LSA Annual Report 1979/80.
+ Estates recommended in the Wise Report in 1967 for early withdrawal from
the scheme.

Duxbury and **Yeldham** were in the scheme only until shortly after the war.

Further Reading

Introduction

ALLEN, D. E., 'Bramble Dating: A Promising Approach', in M. D. Hooper, *Hedges and Local History* (London, 1971).

BARTON, GRISELDA, and TONG, MICHAEL, *Underriver: Samuel Palmer's Golden Valley* (Westerham, Kent, 1995).

BONSER, K. J., *The Drovers: Who they were and how they went. An Epic of the English Countryside* (London, 1970).

GELLING, MARGARET, *Signposts to the Past* (London, 1978; 2nd edn., 1988).

—— 'Place-names', in David Hey (ed.), *The Oxford Companion to Local and Family History* (Oxford, 1996).

HENREY, BLANCHE, *No Ordinary Gardener: Thomas Knowlton, 1691–1781* (London, 1986).

HOOPER, M. D., *Hedges and Local History* (London, 1971).

HOPKINS, R. THURSTON, *Sheila Kaye-Smith and the Weald Country* (London, 1925).

MASSINGHAM, H. J., *The English Countryside: A Survey of its Chief Features* (London, 1939).

MITFORD, MARY RUSSELL, *Sketches of English Life and Character* (London, 1928 and other editions).

MOWL, TIM, and EARNSHAW, BRIAN, *Trumpet at a Distant Gate: The Lodge as Prelude to the Country House* (London, 1985).

NEAVE, DAVID, *Londesborough: History of an East Yorkshire Estate Village* (Londesborough, 1977; Driffield, 1994).

SACKVILLE-WEST, VITA, *Collected Poems*, vol. i (London, 1933).

THOMPSON, FLORA, *Lark Rise to Candleford* (London, 1939, and many later editions).

WOODWARD, H. B. (ed.), *Stanford's Geological Atlas of Great Britain and Ireland* (4th edn., London, 1913).

Chapter 1. The Downlands

ASTON, M., and LEWIS, C. (eds.), *The Medieval Landscape of Wessex* (Oxford, 1994).

BETTEY, J. H., *Wessex from AD 1000* (London, 1986).

—— *Rural Life in Wessex* (Thrupp, 1987).

—— *Estates and the English Countryside* (London, 1993).

BRANDON, P., *The Making of the English Landscape: Sussex* (London, 1974).

—— *The South Downs* (Chichester, 1998).

—— and SHORT, B., *The South-East from AD 1000* (London, 1990).

COBBETT, W., *Rural Rides*, 2 vols. (London, 1912; other editions).

CUNLIFFE, B., *Wessex to AD 1000* (London, 1993).

DARBY, H. C., and CAMPBELL, E. M., *The Domesday Geography of South-East England* (Cambridge, 1967).

DEFOE, D., *Tour through the Whole Island of Great Britain* (London, 1927; other editions).

DREWETT, P., RUDLING, D., and GARDINER, M., *The South East to AD 1000* (London, 1988).

EVERITT, A., *Landscape and Community in England* (London, 1985).

—— *Continuity and Colonisation: The Evolution of Kentish Settlement* (Leicester, 1986).

KERRIDGE, E., 'Agriculture 1500–1793', in *Victoria County History: Wiltshire*, ed. Elizabeth Crittall, iv. 43–64 (Oxford, 1959).

MASON, O., *South-East England* (Edinburgh, 1979).

MASSINGHAM, H. J., *English Downland* (London, 1936).

SAWYER, P. H. (ed.), *English Medieval Settlement* (London, 1979).

SHORT, B., 'South-East England', in J. Thirsk (ed.), *The Agrarian History of England and Wales*, v (i): *1640–1750* (Cambridge, 1985), 270–313.

TAYLOR, CHRISTOPHER, *The Making of the English Landscape: Dorset* (London, 1970).
—— *Village and Farmstead* (London, 1983).
TIMPERLEY, H. W., and BRILL, E., *The Ancient Trackways of Wessex* (London, 1965).
WHITE, GILBERT, *The Natural History of Selborne*, 1836 edn. (London, 1836; other editions).
WHITE, J. T., *South-East Down and Weald: Kent, Surrey & Sussex* (London, 1977).

Chapter 2. Wolds

The Wolds Before c.1500

ALDRED, D., and DYER, C., 'A Medieval Cotswold Village: Roel, Gloucestershire', *Trans. Bristol and Gloucestershire Archaeological Society*, 109 (1991).
ALLISON, K. J. (ed.), *Victoria County History: East Riding of Yorkshire*, esp. vols. ii (London, 1974) and iv (London, 1979).
BENNETT, S., and BENNETT, N., *An Historical Atlas of Lincolnshire* (Hull, 1993).
ELRINGTON, C. R. (ed.), *Victoria County History: Gloucestershire*, vol. vi (London, 1965).
EVERITT, A., 'River and Wold: Reflections on the Historical Origin of Regions and "Pays"', *Journal of Historical Geography*, 3 (1977).
FINBERG, H. P. R., *Gloucestershire: An Illustrated Essay in the History of the Landscape* (London, 1955).
FOX, H. S. A., 'The People of the Wolds in English Settlement History', in M. Aston, D. Austin, and C. Dyer (eds.), *The Rural Settlements of Medieval England: Studies Dedicated to Maurice Beresford and John Hurst* (Oxford, 1989).
HADFIELD, CHARLES, and ALICE, MARY, *The Cotswolds: A New Study* (Newton Abbot, 1973).
HARRIS, ALAN, *The Rural Landscape of the East Riding of Yorkshire, 1700–1850: A Study in Historical Geography* (Oxford, 1961).
HERBERT, N. M. (ed.), *Victoria County History: Gloucestershire*, vol. xi (London, 1976).
HOOKE, DELLA, 'Early Cotswold Woodland', *Journal of Historical Geography*, 4 (1978).
JONES, ANTHEA, *The Cotswolds* (Chichester, 1974).
NEAVE, S., and ELLIS, S., *An Historical Atlas of East Yorkshire* (Hull, 1966).
PLATTS, C., *Land and People in Medieval Lincolnshire* (Lincoln, 1987).
SLATER, T., 'More on the Wolds', *Journal of Historical Geography*, 5 (1979).

A Longer View of the Wolds

BERESFORD, M. W., *The Lost Villages of England* (Gloucester, 1983).
—— and HURST, J. G., *Deserted Medieval Villages* (Gloucester, 1989).
—— —— *Wharram Percy* (New Haven, 1991).
EMERY, FRANK, *The Oxfordshire Landscape* (London, 1974).
FINBERG, H. P. R. (ed.), *Gloucestershire Studies* (Leicester, 1957).
FINBERG, JOSCELINE, *The Cotswolds* (London, 1977).
HARRIS, ALAN, 'The Lost Villages and the Landscape of the Yorkshire Wolds', *Agricultural History Review*, 6 (1958), 97–100.
—— *The Rural Landscape of the East Riding of Yorkshire, 1700–1850* (Oxford, 1961).
—— ' "A Rage of Plowing": The Reclamation of the Yorkshire Wolds', *Yorkshire Archaeological Journal*, 68 (1996), 209–23.
MORIARTY, DENIS, *The Buildings of the Cotswolds* (London, 1989).
THIRSK, JOAN, *English Peasant Farming: The Agrarian History of Lincolnshire from Tudor to Recent Times* (London, 1957).

Chapter 3. Lowland Vales

BATES, H. E., 'The Hedge Chequerwork', in H. J. Massingham (ed.), *The English Countryside: A Survey of its Chief Features* (London, 1939).
BECKETT, J. V., *Laxton: England's Last Open-Field Village* (Oxford, 1989).
BERESFORD, MAURICE, *The Lost Villages of England* (rev. edn., Thrupp, 1998).
BILLINGTON, VIVIEN, 'Update on the Index of Woad-People', *Footprints: Journal of the Northamptonshire Family History Society* 17/1 (Aug. 1995). (Vivien Billington has deposited in the Northants Record Office a database of woad-growing parishes and families.)
DYER, C., *Warwickshire Farming*, Dugdale Society Occasional Paper, 27 (1981).
EVANS, HERBERT A., *Highways and Byways of Northamptonshire and Rutland* (London, 1924).
GOODACRE, JOHN, *The Transformation of a Peasant Economy: Townspeople and Villagers in the Lutterworth Area, 1500–1700*, Leicester Studies in English Local History (Aldershot, 1994).
GOODER, A., *Plague and Enclosure: A Warwickshire Village in the Seventeenth Century (Clifton upon Dunsmore)*, Coventry and North Warwickshire History Pamphlet, no. 2 (Coventry, 1965).
HOSKINS, W. G., *The Midland Peasant: The Economic and Social History of a Leicestershire Village* (London, 1957).
HOSKYNS, CHANDOS WREN, *Talpa: Or the Chronicles of a Clay Farm. An Agricultural Fragment* (London, 1852 and later editions).
HOWELL, C., *Land, Family and Inheritance in Transition: Kibworth Harcourt, 1280–1700* (Cambridge, 1983).
LEWIS, CARENZA, MITCHELL-FOX, PATRICK, and DYER, CHRISTOPHER, *Village, Hamlet and Field: Changing Medieval Settlements in Central England* (Manchester, 1997).
PHILLIPS, A. D. M., *The Underdraining of Farmland in England during the Nineteenth Century* (Cambridge, 1989).

Reed, Michael, *The Buckinghamshire Landscape* (London, 1979).

Steane, John, *The Northamptonshire Landscape: Northamptonshire and the Soke of Peterborough* (London, 1974).

Taylor, Christopher, *Village and Farmstead: A History of Rural Settlement in England* (London, 1983).

Thirsk, Joan, *English Peasant Farming: The Agrarian History of Lincolnshire from Tudor to Recent Times* (London, 1957), chs. 4, 8, and 13.

Thorpe, Harry, 'The Lord and the Landscape, Illustrated through the Changing Fortunes of an English Parish', *Transactions of the Birmingham Archaeological Society*, 18 (1962).

Wood, B. A., Watkins, Charles, and Wood, C. A. (eds.), *Life at Laxton, c.1880–1903: The Memories of Edith Hickson* (esp. pp. 10–21), Dept. of Adult Education, Nottingham University, 1983.

Chapter 4. Woodlands and Wood-Pasture in Western England

Bettey, J. H., *Wessex from AD 1000* (Harlow, 1994).

Dyer, C., *Hanbury: Settlement and Society in a Woodland Landscape*, Leicester University, Dept. of English Local History, Occasional Paper, 4th series, no. 4 (1991).

Gelling, M., *The West Midlands in the Early Middle Ages* (Leicester, 1992).

Palliser, D., *The Staffordshire Landscape* (London, 1976).

Rowley, T., *Landscape of the Welsh Marches* (London, 1986).

Taylor, C. C., *Dorset* (London, 1970).

Victoria County Histories, especially *Wiltshire* iv, *Staffordshire* vi, and *Shropshire* iv, and all volumes of *The Agrarian History of England and Wales*, ed. H. P. R. Finberg, later Joan Thirsk (Cambridge, 1967–).

Chapter 5. Forests and Wood-Pasture in Lowland England

Bazeley, M. L., 'The Extent of the English Forest in the Thirteenth Century', *Trans. Royal Hist. Soc.*, 4th Ser., 4 (1921).

Birrell, J., 'Deer and Deer Farming in Medieval England', *Agricultural History Review*, 40/2 (1992).

Bond, J., 'Forests, Chases, Warrens and Parks in Medieval Wessex', in M. Aston and C. Lewis (eds.), *The Medieval Landscape of Wessex* (Oxford, 1994).

Broad, J., and Hoyle, R. (eds.), *Bernwood: The Life and Afterlife of a Forest* (Preston, 1997).

Cantor, L., *The Changing English Countryside, 1400–1700* (London, 1987).

Cartmill, M., *A View to a Death in the Morning: Hunting and Nature through History* (Cambridge, Mass., 1993).

Cox, J. C., *The Royal Forests of England* (London, 1905).

Fisher, W. R., *The Forest of Essex* (London, 1887).

Grant, R., *The Royal Forests of England* (Thrupp, 1991).

Hart, C., *Royal Forest: A History of Dean's Woods as Producers of Timber* (Oxford, 1966).

—— *The Verderers and Forest Laws of Dean* (Newton Abbot, 1971).

James, N. D. G., *A History of English Forestry* (Oxford, 1981).

Jebb, L., *How Landlords can Create Small Holdings: Some Examples* (London, 1907).

Rackham, O., *Trees and Woodland in the British Landscape* (London, 1976).

—— *Ancient Woodland* (London, 1980).

—— *The Last Forest: The Story of Hatfield Forest* (London, 1989).

Schama, S., *Landscape and Memory* (London, 1996).

Short, B., *The Ashdown Forest Dispute, 1876–1882: Environmental Politics and Custom*, Sussex Record Society, 80 (Lewes, 1997).

Tubbs, C., *The New Forest: An Ecological History* (Newton Abbot, 1968).

Turbevile, G., *The Noble Arte of Venerie or Hunting* (1575, Tudor and Stuart Library Reprint, Oxford, 1908).

Vandervell, A., and Coles, C., *Game and the English Landscape: The Influence of the Chase on Sporting Art and Scenery* (New York, 1980).

Watkins, C. (ed.), *European Woods and Forests: Studies in Cultural History* (Wallingford, 1998).

Young, C. R., *The Royal Forests of Medieval England* (Leicester, 1979).

Chapter 6. Marshes

Bowler, E., 'For the Better Defence of Low and Marshy Grounds: A Survey of the Work of the Sewer Commissions of North and East Kent, 1531–1930', in A. Detsicas and N. Yates (eds.), *Studies in Modern Kentish History* (Maidstone, 1983).

Brandon, P., and Short, B., *The South-East from AD 1000* (London, 1990).

Dugdale, W., *The History of Imbanking and Draining* (London, 1662).

Dulley, A. J. F., 'The Level and Port of Pevensey in the Middle Ages', *Sussex Archaeological Collections*, 104 (1966).

Eddison, J. (ed.), *Romney Marsh: The Debatable Ground*, Oxford University Committee for Archaeology, Monograph 41 (Oxford, 1995).

—— and Green, C. (eds.), *Romney Marsh: Evolution, Occupation, Reclamation*, Oxford University Committee for Archaeology, Monograph 24 (Oxford, 1988).

——, Gardiner, M., and Long, A. (eds.), *Romney Marsh: Environmental Change and Human Occupation in a Coastal Lowland*, Oxford University Committee for Archaeology, Monograph 46 (Oxford, 1998).

Evans, J., 'Archaeological Horizons in the North Kent Marshes', *Archaeologia Cantiana*, 66 (1953).

Grieve, H., *The Great Tide* (Chelmsford, 1959).

PURSEGLOVE, J., *Taming the Flood: A History and Natural History of Rivers and Wetlands* (Oxford, 1988).

SILVESTER, R., 'The Fenland Project, Number 3: Marshland and the Nar Valley', *East Anglian Archaeology*, 45 (1988).

SMITH, J. R., *Foulness: A History of an Essex Island Parish* (Chelmsford, 1970).

WILLIAMSON, T., *The Norfolk Broads: A Landscape History* (Manchester, 1997).

Chapter 7. Fenlands

BLOOM, A., *Farm in the Fen* (London, 1944).

COWELL, R. W., and INNES, J. B., *The Wetlands of Merseyside* (Lancaster, 1995).

DARBY, H. C., *The Medieval Fenland* (Cambridge, 1940).
—— *The Draining of the Fens* (Cambridge, 1968).

HALL, D., and COLES, J., *Fenland Survey*, England Heritage Archaeological Reports, 1 (London, 1994).
——, WELLS, C., and HUCKERBY, E., *The Wetlands of Greater Manchester* (Lancaster, 1995).

HARRIS, L. E., *Vermuyden and the Fens* (London, 1953).

HILLS, R. L., *Machines, Mills and Uncountable Costly Necessities* (Norwich, 1967).

LAMBERT, J. M., JENNINGS, J. N., SMITH, C. T., GREEN, C., and HUTCHINSON, J. N., *The Making of The Broads*, Royal Geographical Society Research Monograph Series, 3 (London, 1960).

MIDDLETON, R., WELLS, C., and HUCKERBY, E., *The Wetlands of North Lancashire* (Lancaster, 1995).

SKELTON, R. A., and HARVEY, P. D. A. (eds.), *Local Maps and Plans from Medieval England* (Oxford, 1986).

THIRSK, JOAN, *Fenland Farming in the Sixteenth Century*, Leicester University, Department of English Local History, Occasional Paper 3 (Leicester, 1953).

THOMLINSON, J., *The Level of Hatfield Chase* (Doncaster, 1882).

WHEELER, W. H., *The History of the Fens of South Lincolnshire* (London, 1868).

WILLIAMS, M., *The Draining of the Somerset Levels* (Cambridge, 1970).

WILLIAMSON, T., *The Norfolk Broads* (Manchester, 1997).

Chapter 8. Moorlands

ATKINSON, J. C., *Forty Years in a Moorland Parish* (London, 1890).

FIELDHOUSE, ROGER, and JENNINGS, BERNARD, *A History of Richmond and Swaledale* (Chichester, 1978).

FLEMING, ANDREW, *The Dartmoor Reaves* (London, 1988).

FOX, H. S. A., 'Medieval Dartmoor as Seen through its Account Rolls', *The Archaeology of Dartmoor: Perspectives from the 1990s*, Devon Archaeological Society Proceedings, 52 (1994).

HARRISON, BARRY, and HUTTON, BARBARA, *Vernacular Houses in North Yorkshire and Cleveland* (Edinburgh, 1984).

HART, C. R., *The North Derbyshire Archaeological Survey to A.D. 1500* (Chesterfield, 1981).

HARTLEY, MARIE, and INGILBY, JOAN, *Life and Tradition in the Yorkshire Dales* (London, 1968).
—— —— *Life in the Moorlands of North-East Yorkshire* (London, 1972).

JENNINGS, BERNARD (ed.), *Pennine Valley* (Otley, 1992).

JOHNSON, NICHOLAS, and ROSE, PETER, *Bodmin Moor: An Archaeological Survey*, i: *The Human Landscape to c.1800* (Swindon, 1994).

RACKHAM, OLIVER, *The History of the Countryside* (London, 1986).

ROBINSON, BRIAN, *Walls across the Valley: The Building of the Howden and Derwent Dams* (Cromford, 1993).

VYNER, B. E. (ed.), *Medieval Rural Settlement in North-East England*, Architectural and Archaeological Society of Durham and Northumberland Research Report, No. 2 (Durham, 1990).

WINCHESTER, ANGUS, *Landscape and Society in Medieval Cumbria* (Edinburgh, 1987).

Chapter 9. Common Land

Note: Few scholarly works deal solely with common land. Books and articles bearing on specific aspects of its history are numerous, and a small selection of these is given here in Part I. Literary works of the nineteenth and early twentieth centuries are among the most important sources for the people and society of the commons; a brief selection is given in Part II. Much important work relating to local commons will be found in recent volumes of the *Victoria County History*. For place names, all local work must begin with the English Place-Name Society's county volumes, of which more than 60 have now been published, and with A. H. Smith, *English Place-Name Elements* (2 vols., Cambridge, 1956), also published by the Society.

Part I

BAKER A. R. H., and BUTLIN, R. A. (eds.), *Studies of Field Systems in the British Isles* (Cambridge, 1973).

BONSER, K. J., *The Drovers: Who they were and how they went. An Epic of the English Countryside* (London, 1970).

EDLIN, H. L., *Woodland Crafts in Britain: An Account of the Traditional Uses of Trees and Timbers in the British Countryside* (London, 1949).

FIELD, JOHN, *A History of English Field-Names* (Harlow, 1993).

HAVINDEN, MICHAEL, *The Somerset Landscape* (London, 1981).

NEESON, J. M., *Commoners: Common Right, Enclosure, and Social Change in England, 1700–1820* (Cambridge, 1993).

PALLISER, D. M., *The Staffordshire Landscape* (London, 1976).

RANSON, FLORENCE, *British Herbs* (Harmondsworth, 1949).

ROWLEY, TREVOR, *The Shropshire Landscape* (London, 1972).

SHORT, BRIAN (ed.), *The Ashdown Forest Dispute, 1876–1882: Environmental Politics and Custom*, Sussex Record Society, vol. 80 (Lewes, 1997).

STAMP, L. DUDLEY and HOSKINS W. G., *The Common Lands of England and Wales* (London, 1963).

TATE, W. E., *A Domesday of English Enclosure Acts and Awards*, ed. M. E. Turner (Reading, 1978).

TRINDER, BARRIE, *The Making of the Industrial Landscape* (London, 1982).

Part II

ATKINSON, J. C., *Forty Years in a Moorland Parish: Reminiscences and Researches in Danby in Cleveland* (London, 1891).

BORROW, GEORGE, *Lavengro* (London, 1851; later editions).

—— *Romany Rye* (London, 1857; later editions).

BUNTIN, TOM FLETCHER, *Life in Langdale: The Memoirs of a Lakeland Farmer* (Kendal, 1993).

COBBETT, WILLIAM, *Rural Rides* (London, 1830; later editions).

GREY, EDWIN, *Cottage Life in a Hertfordshire Village* (1934; repr., Harpenden, 1977).

HUDSON, W. H., *A Shepherd's Life* (1910; later editions).

JEFFERIES, RICHARD, *Nature Near London* (1883; later editions).

JENNINGS, LOUIS J., *Field Paths and Green Lanes: Being Country Walks, Chiefly in Surrey and Sussex* (London, 1877).

MITFORD, MARY RUSSELL, *Our Village* (5 vols., London, 1824–32; subsequent editions and selections).

STURT, GEORGE, *Change in the Village* (London, 1912; subsequent editions).

—— *William Smith, Potter and Farmer, 1790–1858* (London, 1919; subsequent editions).

THOMPSON, FLORA, *Lark Rise* (Oxford 1939; subsequent editions); as a trilogy entitled *Lark Rise to Candleford* (1945), with *Over to Candleford* (1941) and *Candleford Green* (1943).

—— *Heatherley* (Oxford; published posthumously in *A Country Calendar and other Writings*, ed. Margaret Lane, 1979).

THORNE, JAMES, *Handbook to the Environs of London* (1876; repr. Bath, 1970).

WILLIAMS, O. ALFRED, *Villages of the White Horse* (London, 1913).

Chapter 10. Frontier Valleys

BARROW, GEOFFREY, 'The Anglo-Scottish Border', in his *The Kingdom of the Scots* (London, 1973).

—— 'Frontier and Settlement: Which Influenced Which? England and Scotland, 1100–1300', in Robert Bartlett and Angus MacKay (eds.), *Medieval Frontier Societies* (Oxford, 1989).

BETTERTON, ALEX and DYMOND, DAVID, *Lavenham: Industrial Town* (Lavenham, 1989).

BRITNELL, RICHARD, *Growth and Decline in Colchester, 1300–1525* (Cambridge, 1986).

—— 'Boroughs, Markets and Trade in Northern England, 1000–1216', in Richard Britnell and John Hatcher (eds.), *Progress and Problems in Medieval England* (Cambridge, 1996).

CLACK, PETER A. G., 'The Origins and Growth of Darlington', in Philip Riden (ed.), *The Medieval Town in Britain*, Cardiff Papers in Local History, 1, University College, Cardiff (Cardiff, 1980).

DIXON, PHILIP, 'Towerhouses, Pelehouses and Border Society', *Archaeological Journal*, 136 (1979).

DYMOND, DAVID, and MARTIN, EDWARD (eds.), *An Historical Atlas of Suffolk* (Ipswich, 1988).

GELLING, MARGARET, 'A Chronology of Suffolk Place-Names', in Martin Carver (ed.), *The Age of Sutton Hoo* (Woodbridge, 1992).

LE GUILLON, M., *A History of the River Tees, 1000–1975* (Middlesbrough, 1978).

HEYWOOD, STEPHEN, 'The Round Towers of East Anglia', in John Blair (ed.), *Minsters and Parish Churches: The Local Church in Transition, 950–1200* (Oxford, 1988).

HOSKINS, W. G., 'The Origin and Rise of Market Harborough', in id., *Provincial England* (London, 1963).

MACKENZIE, W. MACKAY, 'The Debateable Land', *Scottish Historical Review*, 30 (1951).

McKINLEY, R. A., 'Market Harborough', in *Victoria County History: Leicester*, v (London, 1964).

McNEILL, PETER, and NICHOLSON, RONALD (eds.), *An Historical Atlas of Scotland, c.400–c.1600* (St. Andrews, 1975).

MOFFAT, ALISTAIR, *Kelsae: A History of Kelso from Earliest Times* (Edinburgh, 1985).

PHYTHIAN-ADAMS, CHARLES, 'The Emergence of Rutland and the Making of the Realm', *Rutland Record*, 1 (1980).

—— 'Introduction: An Agenda for English Local History', in id. (ed.), *Societies, Cultures and Kinship, 1580–1850: Cultural Provinces in English Local History* (Leicester, 1993).

—— *Land of the Cumbrians: A Study in British Provincial Origins, A.D. 400–1120* (Aldershot, 1996).

ROBSON, RALPH, *The Rise and Fall of the English Highland Clans: Tudor Responses to a Medieval Problem* (Edinburgh, 1989).

ROGERS, ALAN (ed.), *The Making of Stamford* (Leicester, 1965).

ROLLASON, DAVID, HARVEY, MARGARET, and PRESTWICH, MICHAEL (eds.), *Anglo-Norman Durham, 1093–1193* (Woodbridge, 1994).

PRYDE, G. S., 'The Burghs of Dumfriesshire and Galloway: Their Origin and Status', *Dumfriesshire and Galloway Natural History and Antiquarian Society Transactions*, 3rd ser., 29 (1950–1).

Thornton, G. A., *A History of Clare, Suffolk* (Cambridge, 1928).

Warner, Peter, *The Origins of Suffolk* (Manchester, 1996).

Watts, V. E., 'Place-Names', in *Durham County and City with Teesside* (British Association: Durham, 1970).

Chapter 11. Stonor

Baines, A. H. J., 'Turville, Radenore and the Chiltern Feld', *Records of Buckinghamshire*, 23 (1981).

Carpenter, C. (ed.), *Kingsford's Stonor Letters and Papers, 1290–1483* (Cambridge, 1996).

Hammond, M., 'The Anglo-Saxon Estate of *Readanora* and the Manor of Pyrton, Oxfordshire', *Oxoniensia*, 63 (1998).

Hepple, L. W., and Doggett, A. M., *The Chilterns* (Chichester, 1992; 2nd edn 1994).

Steane, J., 'Stonor: A Lost Park and a Garden Found', *Oxoniensia*, 59 (1994).

—— *Oxfordshire* (London, 1996).

Stonor, R. J., *Stonor: A Catholic Sanctuary in the Chilterns from the Fifth Century till To-day* (Newport, 1951).

Chapter 12. Hook Norton, Oxfordshire

Agricultural Economics Research Institute Oxford, *Country Planning* (Oxford, 1944).

Allen, R. C., *Enclosure and the Yeoman* (Oxford, 1992).

Blair, W. J., *Anglo-Saxon Oxfordshire* (Thrupp, 1994).

Dickens, M., *History of Hook Norton, 912–1928* (Banbury, 1928).

Tiller, K., *Hook Norton Village Trail* (Hook Norton, 1981).

Chapter 13. Eccleshall, Staffordshire

Brooks, Colin, 'Projecting, Political Arithmetic and the Act of 1695', *English Historical Review*, 97 (1982).

Holmes, G. S., 'Gregory King and the Social Structure of Pre-industrial England', *Transactions of the Royal Historical Society*, 5th ser., 27 (1977).

Palliser, D. M., *The Staffordshire Landscape*, Making of the English Landscape Series (London, 1976).

Pevsner, Nikolaus (ed.), *Staffordshire*, The Buildings of England Series, (Harmondsworth, 1974).

Spufford, Margaret, and Went, James, *Poverty Portrayed: Gregory King and the Parish of Eccleshall* (Keele, 1995).

Spufford, Peter, and Spufford, Margaret, *Eccleshall: The Story of a Staffordshire Market Town and its Dependent Villages* (Keele, 1964; repr. 1977 by the History of Eccleshall Fund, Sugnall Hall, Eccleshall).

Chapter 14. Staintondale, North Yorkshire

Harrison, B., 'The Alum Workers', *Cleveland Industrial Archaeologist*, 2 (1975).

—— and Hutton, B., *Vernacular Houses of North Yorkshire and Cleveland* (Edinburgh, 1984).

Hartley, M., and Ingilby, J., *Life in the Moorlands of North East Yorkshire* (London, 1972).

Lancaster, W. T. (ed.), *Cartulary of Bridlington Priory* (Leeds, 1912).

Marshall, G., 'The Ravenscar Alum Works: New Evidence from Documentary and Archaeological Sources', *Transactions Scarborough Archaeological and Historical Society*, 28 (1990).

Pickles, R. J., 'A Brief History of the Alum Industry in North Yorkshire', *Cleveland Industrial Archaeologist*, 2 (1975).

Rimington, F. C., 'Where and What was Fullwood?', *Transactions Scarborough & District Archaeological Society*, 3 (1974).

—— *The History of Ravenscar and Staintondale* (Scarborough, 1988).

Royal Commission on Historic Monuments (England), *Houses of the North York Moors* (London, 1984).

Spratt, D. (ed.), *Prehistoric and Roman Archaeology of North East Yorkshire*, British Archaeological Reports, British Series, 102 (Oxford, 1982).

—— and Harrison, B., *The North York Moors: Landscape Heritage* (Newton Abbot, 1989).

Turton, R. B., *The Honour and Forest of Pickering*, North Riding Record Society, New Series, vols. i–iv (London, 1894–7).

—— *The Alum Farm* (Whitby, 1938).

Chapter 15. Fen Drayton, Cambridgeshire

Clarke, Peter, 'Cooperative Working and the Land Settlement Scheme (LSA): A Historical Perspective', *The Yearbook of Agricultural Cooperation* (London, 1982).

Dearlove, Pamela, *The Effect of the Policies of the Land Settlement Association upon Fen Drayton*, Cambridgeshire, Local History Society Review, ns, 1–3 (Cambridge, 1992–4).

Fry, Joan Mary, *Friends Lend a Hand Alleviating Unemployment* (London, 1947).

Kitchen, Fred E. J., *Settlers in England* (London, 1947).

Land Settlement Association, *Illustrations of the Work of the Land Settlement Association* (London, 1937).

McCready, K. J., *The Land Settlement Association: Its History and Present Form* (London, 1974).

Ministry of Agriculture, Fisheries, and Food, *Departmental Committee of Inquiry into Statutory Smallholdings*, Chaired by Professor M. J. Wise, Final Report (London, 1967).

Picture Credits

The Publishers wish to thank the following who have kindly given permission to reproduce illustrations on the pages indicated.

Colour Plates
Plate 1 National Trust Photographic Library/Michael Walters.
Plate 2 (top) © Mick Sharp.
Plate 2 (bottom) © Norfolk County Council.
Plate 3 Norfolk Museums Service (Norwich Castle Museum).
Plate 4 © Colin Baxter.
Plate 5 C. Phythian-Adams.
Plate 6 National Gallery of Scotland.
Plate 7 © Lord Camoys.
Plate 8 Private collection.

In-text Illustrations
20 Courtesy of Tonbridge & Malling Borough Council.
20 From T.D.W. Dearn, *Sketches in Architecture* (London, 1807), Plate III. Print from the British Library.
22 Ian Johnston.
23 Ian Johnston.
24 (top) Ian Johnston.
24 (bottom) V&A Picture Library.
29 J.E. Hancock.
32 J.E. Hancock.
37 © Times Newspapers Limited.
40 Patrick Brown.
42 Patrick Brown.
45 Rural History Centre, University of Reading.
46 © Batsford (Photo: NMR).
49 © Salisbury & South Wiltshire Museum.
51 Copyright reserved Cambridge University Collection of Air Photographs.
58 The Bodleian Library, University of Oxford (J.J. Drayton d.35, pp.226-7).

60 The Bodleian Library, University of Oxford (Arch AA c.7, plate 16).
63 Bernard Jennings.
67 © Roy Nash.
72–73 Copyright reserved Cambridge University Collection of Air Photographs.
77 Bernard Jennings.
80 David Hall for the use of his map of Mears Ashby, Northants, reproduced from David Hall, *The Open Fields of Northamptonshire*, Northants Record Society, 38, 1995.
83 (both) Reproduced from C.S. and C.S. Orwin, *The Open Fields*, Plate 11 (OUP, 1938).
85 © Crown Copyright. National Monuments Record.
87 Reproduced from C.S. and C.S. Orwin, *The Open Fields*, Plate 11 (OUP, 1938). Photo © Aerofilms.
93 © Crown Copyright. National Monuments Record.
101 © Crown Copyright. National Monuments Record.
105 Christopher Dyer.
109 Shropshire Records and Research.
113 Copyright reserved Cambridge University Collection of Air Photographs.
116 The Illustrated London News Picture Library.
117 Rural History Centre, University of Reading.
121 Aerofilms.
124 Brian Short.
125 The J. Allan Cash Photolibrary.
126 From J.C. Cox, *The Royal Forests of England* (Methuen, 1905).
134 Private collection.
137 Victoria & Albert Museum, London/Bridgeman Art Library, London.
140 By permission of the Warden and Fellows of New College, Oxford.

143 The Bodleian Library, University of Oxford (Vet A4 e.2754).
151 Anne Reeves.
154 Hastings Museum and Art Gallery.
158 Anne Reeves.
162 Anne Reeves.
164 Anne Reeves.
166 Anne Reeves.
169 Copyright reserved Cambridge University Collection of Air Photographs.
172 Cambridgeshire Collection, Cambridge Central Library.
173 Cambridgeshire Collection, Cambridge Central Library.
177 Cambridgeshire Collection, Cambridge Central Library.
178 Norfolk County Council Library and Information Service.
183 Cambridgeshire Libraries.
187 Copyright reserved Cambridge University Collection of Air Photographs.
191 © Crown Copyright. National Monuments Record.
192 © Andrew Fleming.
195 © Crown Copyright. National Monuments Record.
199 David Hey.
201 The Bodleian Library, University of Oxford (Arch AA c.7, plate 27).
205 The Bodleian Library, University of Oxford (Arch AA c.7, plate 20).
208–209 Severn Trent Trent Water Limited.
213 © Barrie Trinder.
217 Copyright reserved Cambridge University Collection of Air Photographs.
219 Rural History Centre, University of Reading.
226 From *The Victoria County History of Essex*, *Volume IX* (University of London Institute of Historical Research and Oxford University Press, 1994).

229 From Louis J. Jennings, *Field Paths and Green Lanes* (John Murray, 1877).
232 Rural History Centre, University of Reading.
234 Rural History Centre, University of Reading.
239 C. Phythian-Adams from *Societies, Cultures and Kinship, 1580–1850: Cultural Provinces and English Local History* (Leicester UP, 1993.)
246 C. Phythian-Adams.
249 C. Phythian-Adams.
252–253 C. Phythian-Adams.
255 C. Phythian-Adams.
268 A. Doggett.
273 V&A Picture Library.
274 Rural History Centre, University of Reading.
277 Oxfordshire Photographic Archive, DLA, OCC.
278 Oxfordshire Photographic Archive, DLA, OCC.
280 Oxfordshire Photographic Archive, DLA, OCC.
281 Oxfordshire Photographic Archive, DLA, OCC.
291 top M. Spufford.
291 bottom M. Spufford.
292 M. Spufford.
304 M. Spufford.
308 B. Harrison.
314 B. Harrison.
317 B. Harrison.
318 B. Harrison.
326 Private collection.
329 From *Smallholdings with marketing and other services*.
330 Private collection.
332 Copyright reserved Cambridge University Collection of Air Photographs.

Maps pp. **243**, **244**, **258**
© C. Phythian-Adams, compiled and drawn by Ken Smith.
Maps pp. **55**, **168**, **266**, **282**, **284**, **295**, **301**, **309**, **311**, **320** drawn by Oxford University Press.

Index

Illustrations are in italics.

The abbreviations used for counties are:
Beds. = Bedfordshire
Berks. = Berkshire
Bucks. = Buckinghamshire
Cambs. = Cambridgeshire
Chesh. = Cheshire
Cornw. = Cornwall
Cumbd. = Cumberland
Dors. = Dorset
Derbys. = Derbyshire
Dumfries. = Dumfriesshire
Glos. = Gloucestershire
Hants. = Hampshire
Herefs. = Herefordshire
Herts. = Hertfordshire
Hunts. = Huntingdonshire
Lancs. = Lancashire
Ldn. = London
Leics. = Leicestershire
Lincs. = Lincolnshire
Mdx. = Middlesex
Mon. = Monmouthshire
Northts. = Northamptonshire
Nfk. = Norfolk
Northd. = Northumberland
Notts. = Nottinghamshire
Oxon. = Oxfordshire
Peebles. = Peeblesshire
Roxburgh. = Roxburghshire
Rutld. = Rutland
Salop. = Shropshire
Selkirk. = Selkirkshire
Sfk. = Suffolk
Som. = Somerset
Staffs. = Staffordshire
Surr. = Surrey
Suss. = Sussex
Warws. = Warwickshire
Westmd. = Westmorland
Wilts. = Wiltshire
Worcs. = Worcestershire
Yorks. = Yorkshire

Devon, Durham, Essex, and Kent are not abbreviated.

People and Places

Abbey Dore, Herefs., 107
Abbotsbury, Dors., 27, 39
Abington, Cambs., 325
Acle, Nfk., 156
Acton Scott, Salop., 120
Addington, Surr., 211
Aire River, 180
Albemarle, Duke of, 142
Alciston, Suss., 39
Aldbourne, Wilts., 37, 38, 233
Aldermaston, Berks., 34
Aldershot, Hants., 228
Aldred, David, 56
Aldsworth, Glos., 59
Alice Holt Forest, Hants., 123, 143, 144
Allendale, Northd., 203
Allen, David, 14
Allen, Robert, 287
Allerdale Forest, Cumbd., 139
Allertonshire, Yorks., 240, 248
Allesley, Warws., 118
Aln River, 259
Alnwick, Northd., 259
Alston, Cumb., 21, 253; Moor, 205
Althorp, Warws., 86
Alt River, Lancs., 171
Alvechurch, Worcs., 112
Amersham, Bucks., 233
Andover, Hants., 325
Andredsweald, 122, 132
Angmering, Suss., 34
Apes Hall, Cambs., 175
Apethorpe, Northts., 16, 147
Appleby, Westmd., 17
Appledore, Kent, 157
Archenfield, Herefs., 104
Arden Forest, Warws., 97, 98, 106, 107, 113, 118
Ardington, Berks., 47
Ardleigh, Essex, 335; Heath, 245
Argam, Yorks., 52, 68
Arras, Yorks., 68
Arundel, Earls of, 136–7
Ascot under Wychwood, Oxon., 134–5

Ashby Woulds, Leics., 223
Ashdown forest, Sussex, 12, 128, 130, 137, 142, 148, 224, 230
Ashdown House, Berks., 40, Aspley, Staffs., 290, *291*, 296, 299
Assendon, Lower, Oxon., 275; Spring, 267; Upper, 265; valley, 265, 267, 270, 276
Asterleigh, Oxon., 68
Astley, Worcs., 118
Aston Tirrold, Berks., 15
Athelney Is., Som., 173
Atkinson, Rev. J.C., 202, 204–5, 214
Aubrey, John, 28, 122
Auckland, Durham, 327
Audley End, Essex, 254
Austin family, 278, 286
Avebury, Wilts., 27, 30, 31
Avon River, 97
Axe River, Som., 176
Axholme, Lincs., 167, 170, 174, 177, 181, 183, 184, 187
Aycliffe, Durham, 247

Bagshot Heath, Surr., 210, 212, 225, 230
Bakewell, Robert, 96
Ballingdon, Essex, 250
Balliol family, 248, 250
Bamburgh, House of, 257
Banbury, Oxon., 18, 64, 280, 281, 289
Barham Downs, Kent, 228
Barnard Castle, Durham, 244, 249, 250, 251, 252
Barnes, William, 47, 118
Barnsdale, Yorks., 148
Barnsley, Yorks., 189
Bartindale, Yorks., 68
Barton Mills, Sfk., 186
Baslow, Derbys., 206
Bates, H. E.,13, 16, 78, 86
Battersea, Ldn., 215, 216
Battle Abbey, Sussex, 155
Battleburn, Yorks., 69
Beaconsfield, Bucks., 233

Beale, John, 118
Beckhampton, Wilts., 18
Bedford, Earl of, 180, 181
Bedford Level, 180, 181; Corporation, 182; River, 181, 185
Bedfordshire, 134, 220, 227
Beesley, Alfred, 283
Belbroughton, Worcs., 120
Bell, Adrian, 13
Belloc, Hilaire, 28
Belvedere, Kent, 17
Bere Forest, Hants., 129, 143
Beresford, Maurice W., 14, 61, 68
Berkhamsted, Herts., 212, 232, 233; Frith, 215
Berney, Thomas Trench, 165
Bernicia, 240, 256, 257
Bernwood Forest, Bucks., 123, 132, *140*, 141
Berwick, Northd., 259, 260
Berwick Down, Hants., 34
Bethnal Green, Ldn., 211
Betteshanger, Kent, 230
Bevill's Leam, 181
Bewcastle, Cumbd., 260
Bibury, Glos., 52
Bigmore, Peter, 95
Bignor, Suss., 33
Billingborough, Lincs., 178
Billingham, Durham, 246, 251
Bilsdale, Yorks., 196, 199
Binbrook, Lincs., 64, 74
Birchinlee, Derbys., 207
Birmingham, Warws., 213
Birrell, Jean, 230
Bishop Wilton Wold, Yorks., 69
Bisley, Surr., 66
Bix, Oxon., 270, 272, 275
Blackamoor, Yorks., 194
Blackburn Chase, Lancs., 136
Black Death, 11, 16, 179
Blackheath, Ldn., 211, 212, 228
Blackmoor Forest, Som., 98, 123; Vale of, 98, 115, 116, 118
Black Patch, Suss., 31

343

Blair, John, 237, 281, 283
Bleaklow, Derbys., 189
Blean Forest, Kent, 18, 230
Blenheim Park, Oxon., 136
Blewburton, Berks., 32
Blewbury, Berks., 15,
Blith, Walter, 118
Bloch, Marc, 52, 83
Bloxham, Oxon., 64, 69, 287
Bluntisham, Hunts., 170
Boarstall, Bucks., *140*, 141
Bodmin Moor, Cornw., 188,
190, *191*, 194, 203, 206
Bokerley Dyke, Dors., 34
Bolsterstone, Yorks., 204
Bonser, K.J., 14
Bordesley, Warws., 114;
abbey, 106, 107
Bordon Camp, Hants., 228
Borrow, George, 214, 235
Bostock, Chesh., 120
Boston, Lincs., 254
Bourne, Lincs., 174, 178
Bourton-on-the-Hill, Glos., 57
Bourton-on-the-Water, Glos.,
65
Bowden, Great, Leics., 245,
248, 249, 250; Little, 250
Bowland Chase, Yorks., 136;
Forest, 194
Box Hill Country Park, Surr.,
49
Box River, 242
Boxley, Kent, 39
Boxted Heath, Essex, 245
Boynton, Yorks., 67, 74
Boys, John, 230
Brabourne Lees, Kent, 228
Bradfield Heath, Essex, 245
Bradfield, Yorks., 200, 202, 204
Bradley, Leics., 225
Brailes, Warws., 106
Bramdean, Hants., 34
Brandon Creek, Cambs., 176
Brandon, Sfk., 232, 233
Brassington Moor, Derbys.,
189
Breckland, Nfk., 131, 132, 230
Brede River, 153
Breedon-on-the-Hill, Leics.,
79
Brett River, 242
Bridgwater, Som., 179
Bridlington Priory, Yorks.,
312, 313
Brill, Bucks., 132, 133
Broadland marshes, Nfk.,
160, 161, 162, 163, 164
Broadwath Farm, Cumbd.,
334
Broadway, Glos., 63–4, 77
Bromfield, Salop, 100, 104
Bromley, Kent, 211, 233
Bromley, Staffs., 293, 296, 303,
305; Hall, 296, 303–5, *304*

Bromsgrove, Worcs., 106
Bromswold, Hunts.-Beds., 50
Brontë, Emily, 207
Brookes, Colin, 305
Brook Green, Ldn., 211
Broomhead Moors, Derbys.,
206
Broomhill, Kent, 157
Broughton, Staffs., 295, 296;
Hall, 296
Brown, Capability, 149
Brundon, Sfk., 250
Buccleuch, Duke of, 261
Buckingham, 18
Bucklersbury market, Ldn.,
217
Buildwas, Salop, 107
Bulford, Wilts., 28, 48
Bulmer, Essex, 240
Bunyan, John, 95
Bristol, 213
Bure River, 151
Bures, St Mary, Sfk., 245, 250,
251; Hamlet, 250; Mount,
250
Burford, Oxon., 64
Burgh Castle, Kent, 165
Burghley House, Northts.,
254
Burlington, Earl of, 20
Burniston, Yorks., 310, 313
Burnt Fen, Cambs., *168–9*
Burrough Hill, Leics., 61
Burton Dassett, Warws., 60,
61
Burton Fleming, Yorks., 54,
67
Burwash, Suss., 130
Burwell, Cambs., 175
Bury St. Edmunds, Sfk., 240,
254
Butterwick, Yorks., 70
Buxton, Derbys., 189
Bykerdyke, Lincs., 180

Caird, Sir James, 71, 95
Caistor, Lincs., 64
Calder Valley, Yorks., 196, 200
Caludon, Warws., 107
Camberley, Surr., 228
Camberwell, Ldn., 211, 212
Cambridge, 178, 254
Cambridgeshire, 132, 242
Camden, William, 28, 189
Camoys, Lord, 275
Cam River, 179
Cannock Chase, Staffs., 12,
136, 138
Canterbury Cathedral Priory,
155
Car Dyke, Lincs., 172
Carew, Richard, 194
Carlbury, Durham, 245
Carlisle, Cumbd., 12, 259
Carlton, Yorks., 335

Cary River, Som., 176, 185
Castle Combe, Wilts., 65, 67,
77
Castle Donington, Leics., 232,
233
Castleton, Derbys., 200, 201
Catesby Priory, Northts., 60
Cary, Joyce, 16
Catherton Clee Hill, Salop.,
217
Catholme, Staffs., 107
Catmose, Vale of, Rutl., 248
Cavendish, Sfk., 251
Caythorpe, Yorks., 63, 68
Cerne Giant, Dors., 28, *29*
Chalton, Hants., 34
Chanctonbury Ring, Suss., 32
Chapel Ascote, Warws., 60, 61
Chapel-en-le-Frith, Derbys.,
201, 202
Chapman, John, 224
Charlbury, Oxon., 135
Charlcott, Salop, 120
Charminster, Dors., 41
Charnwood, Leics., 223, 230;
Forest, 123, 129, 130, 139
Chartham, Kent, 229
Chastleton, Oxon., 68
Chatham dockyard, 165
Chat Moss, Lancs., 171
Chatsworth, Derbys., 189
Chatteris, Cambs., 17, 181
Chawston, Beds., 335
Cheddar, Som., 15
Chelmsford, Essex, 18
Chesham, Bucks., 232, 233
Cheshire, 116, 202
Chesil beach, Dors., 160
Chester, 101
Chesterfield, Derbys., 189
Chetney marshes, Kent, *166*
Chilcomb Down, Hants., 32
Childs family, 318
Chilham, Kent, 18, 229
Chilterns, 27, *234*, 265, 275–6
Chingford, Essex, 129
Chippenham, Wilts., 116;
Forest, 123, 129, 132, 141
Chipping Campden, Glos., 77
Chipping Norton, Oxon., 277
Chipping Sodbury, Glos., 64
Chiseldon, Wilts., 48
Chisenbury, Wilts., 34
Chislehurst, Kent, 211
Chislet marshes, Kent, 163
Chitterne, Wilts., 39
Christmas Common, Oxon.,
267, 269, 276
Church Aston, Salop., 109
Chute Forest, Wilts., 141
Cinque Ports, Kent, 153
Cirencester, Glos., 64
Cissbury, Suss., 31, 32
Clapham, Ldn., 211
Clare, Sfk., 249, 251

Clare, John, 94
Clarendon, Wilts., 37; Forest,
133, 134, 136; Park, 142
Claridge, John, 44
Clarke, J.A., 71
Cleeve Hill, Glos., 62
Cleobury Mortimer, Salop.,
120
Cleveland Hills, Yorks., 253
Cleveland Way, 322
Cleveley, Oxon., 65
Cloughton, Yorks., 310, 312,
313, 315; Moor, 308;
Newlands, 316
Coalville, Leics., 214, 218
Cobbett, William, 45, 46, 47,
159, 214
Cockersand Abbey, Lancs.,
174
Colchester, Essex, 133, 226,
254
Cold Newton, Leics., 54
Coldstream, S.Berwick., 259
Coleshill, Herts., 233, 235
Collier, Thomas, 327
Colne River, 245; valley, 254
Combs Ditch, Dors., 34
Compton Verney, Warws., 85
Condicote, Glos., 52, 57
Conington fen, Cambs., 183
Coniscliffe, Durham, 245, 246
Coniston, Lancs., 205
Constable, Sir Marmaduke,
20
Cooke, Sir Brian, 317, 319; Sir
William Bryan, 321
Cook, Moses, *143*
Cornwall, 238
Cornwall, Dukes and Earls of,
136, 138
Cotes, Staffs., 290
Cotesbach, Leics., 89
Cotgrave, Notts., 223
Cotswolds, 16, 50, 52–4, 62,
75
Cottam, Yorks., 68, 69
Cottesbrook, Northts., 18
Cove Common, Hants., 233
Coveney, Cambs., 185
Coventry, Warws., 89, 254
Cowlam, Yorks., 69
Cowper, William, 10
Cox Heath, Kent, 225, 228
Cranborne Chase, Dors., 14,
27, 31, 34, 136, 138
Cranfield, Beds., 227
Craven, Yorks., 197
Crawford, O.G.S., 14
Crewe, Chesh., 12
Crimscote, Warws., 87
Croft, Yorks., 250
Crouch River, 158
Crowland, Lincs., 176, 244;
abbey, 173, 174
Croxton Kerrial, Leics., 59

Croxton, Staffs., 295, 302, 306
Croydon, Surr., 211, 212, 216, 232, 233
Cruikshank, George, 95
Cumberland, 198, 324; commons, 221
Curwen, C.E.,14

Dalston, Cumbd., 335
Danebury, Hants., 32
Danelaw, 281
Darby, Sir Clifford, 230
Darent valley, Kent, 233
Darlington, 244, 245, 247, 248, 251, 252, 254
Darnhall, Chesh., 112
Dartford Heath, Kent, 212
Dartmoor, Devon, 188, 190, 192, *192*, 194, 195, 215, 235; Forest, 123, 138
Davenant, Charles, 305
Daventry, Northts., 18
Davis, Thomas, 43
Dean Forest, Glos., 127, 128, 129, 131, 132, 133, 134, 139, 141, 142, 143, 144, 145, 146, 147, 218
Dearlove, Pamela, 10
Dearn, T.D.W., 19
Debateable lands, 261
De Brus family, 248
Dedham, Essex, 251; Heath, 245
Deeping Fen, Lincs., 182
Defoe, Daniel, 43, 44, 66, 115, 159, 189, 214
Deira, 240
Delamere Forest, Chesh., 203
De la Pole family, 285
De La Pryme, Abraham, 200
Dengie peninsula, Essex, 160, 164
Denver, Nfk., 181, 186
De Plesset family, 285, 286
Deptford, Ldn., 212; dock-yard, 165
Derby, 241
Derbyshire, commons, 220–1
Devizes, Wilts., 126, 129
Devon, 238; commons, 221
Dibden Purlieu, Hants., 128
Dickens, Charles, 166
Dinsdale, Low, Durham, 250
Dinsdale, Over, Yorks., 250
Ditchling Beacon, Suss., 49
Ditton Priors, Salop., *113*
Doddington, Cambs., 174, 175
D'Oilly family, 271, 285
Doncaster, Yorks., 202
Don Navigation, 202
Don River, 180–81
Dorchester-on-Thames, Oxon., 269
Douglas River, Lancs., 171
Downham, Cambs., 175

Downton, Wilts., 34, 38
Droitwich, Worcs., 101, 102
Dry Drayton, Cambs., 50, 54
Duffield Chase, Derbys., 136
Dugdale, William, 50, 183, 304
Dumfries, 258, 260, 261
Dungeness, Kent, 152
Dunnabridge, Devon, 195
Duntisbourne Abbots, Glos., 59
Dunwich diocese, Sfk., 240,
Duport, Dean, 170
Durham, 324, 326; commons, 221
Durham, County, 240; diocese, 240, 257; Palatinate, 247
Dutch River, 181
Dutt, W.A., 163
Duxbury, Lancs., 335
Dyer, Christopher, 56
Dymchurch, Kent, 163; seawall, 157

Earith, Hunts., 181
East Anglia, 241–2; fens, 167, 170, 173, 176, 177, 180, 181, 182, 183, 186, 187
East Bergholt, Sfk., 251
Eastbridge, Kent, 158, 163
Eastburn, Yorks., 68, 69
East Coulston, Wilts., 35
East Moors, Derbys., 190, 202
Eastnor, Herefs., 120
Eaton by Tarporley, Chesh., 102
Ebbsfleet, Kent, 165
Eccleshall, Staffs., 104, 290–306, *291*; Bishop's Wood, 292, 294, 297, 302; Great Wood, 294, 295, 297; maps, 309, 311, 320. See also Subject Index, under Eccleshall
Edale, Derbys., 197, 202
Egglescliffe, Durham, 246
Egypt Bay, Kent, 153
Eliot, George, 300
Elmesthorpe, Leics., 335
Elmham diocese, 240
Elmley marshes, Kent, *164*
Elsdon, Northd., 260
Elstow, Beds., 95
Ely, Cambs., 174, 179; Cathedral, 175
Emery, Frank, 70
Enderby, Lincs., 52
Enfield Chase, Mdx., 136, 144
English, Barbara, 71
Epping Forest, Essex, *125*, 134, 142, 143, 144, 145, 230
Epworth, Lincs., 181
Erbury, Sfk., 245
Erith, Kent, 17
Erringden Park, Yorks., 198
Eskdale, Cumbd., 197

Esk River, 258; valley, 260, 261
Essex, 131, 134, 137; marshes, 151, 156, 158, 160, 161, 163, 164; royal forest, 245
Evelyn, John, 142, 143
Everingham Hall, Yorks., 20
Everitt, Alan, 10, 16–17, 35
Evesham, Vale of, Worcs., 97
Exmoor, Devon, 123, 140, 188, 189, 203
Eyam Moor, Derbys., 189, 201
Eycot, Glos., 52

Fairfield, Derbys., 154, 196, 201
Fair Mile, 265, 267
Falconer, Matthias, 165
Falsgrave, Yorks., 310
Fane, John, Earl of Westmorland, 19
Farnborough, Hants., 228
Farndale, Yorks., 196
Fastolfe, Sir John, 65
Fawley, Bucks., 270
Feckenham Forest, Worcs., 98, 107, 118, 139, 141
Fen Drayton, Cambs., 10, 323–35. See also Subject Index under Fen Drayton *and* Land Settlement Association
Ffrancis, Wilson, 184
Fiennes, Celia, 183, 214
Figheldean, Wilts., 48
Finberg, Herbert P.R., 52, 66
Finberg, Josceline, 64, 72
Finchley common, Mdx., 223
Fitz Nigel, Richard, 133
Fitzwarin, Fulk, 148
Flambard, Bishop Ranulf of Durham, 259
Flamborough Head, Yorks., 63
Fleet, Hants., 228
Fleet River, 160
Fletton, Beds., 92
Foley family, 120
Fordington, Dors., 36
Forester, Lord, 120
Forest Hill, Kent, 211
Forty Foot Drain, Cambs., 181
Foulness island, Essex, 159, 162
Fountains Abbey, Yorks., 197
Fovant, Wilts., 28
Fowlholme, Nfk., 157, 159
Fox Ash, Essex, 325
Fox, Harold, 10
Foxley, Herefs., 120
Frampton, Dors., 34
France-Hayhurst family, 120
Frome, Som., 212
Frome Whitfield, Dors., 41
Fry, Charles, 13
Fuller, Thomas, 142
Fulney, Lincs., 325, 335

Gainford, Durham, 248, 250; -shire, 240
Gainsborough, Thomas, 149
Galashiels, Selkirk., 260
Galloway, 258–9
Galtres Forest, Yorks., 141
Gardham, Yorks., 68
Gardiner, Richard, 118
Garrigill, Cumb., *22, 23*
Gelling, Margaret, 13, 281
Gembling, Yorks., 53
Geoffrey of Langley, Chief Justice, 135
George, Lloyd, 324
Gerard family, 296, 300, 304; Thomas, 28, 41
Gillingham Forest, Dors., 118, 123, 136, 141
Gilpin, William, 122, 124, 147
Giraldus, 50
Glastonbury, Som., 171, 174, 179; abbey, 175, 176
Glemsford, Sfk., 251
Gloucester, Earls of, 138
Goathland, Yorks., 196
Goathurst Common, Kent, 228
Godmersham, Kent, 229
Godney Moor, Som., 176
Golden Valley, Herefs., 110, 118
Gonalston, Notts., 83
Goole, Yorks., 181
Gover Hill, Kent, 15
Gravesend, Kent, 166
Gray, H.L., 69
Great Casterton, Rutld., 245
Great Coxwell, Berks., 39
Great Witley, Worcs., 120
Great Yarmouth, Nfk., 152, 159, 160, 165
Guyhirn, Cambs., 179, 185
Great Yeldham, Essex, 335
Greenstead, Essex, map, 226
Greenwich, Lnd., 212; Hospital, 205
Greetham, Lincs., 57
Gretna Green, Dumfries, 262
Grey, Edwin, 235
Grigson, Geoffrey, 22
Grim's Hill, Glos., 51
Grim, the Collier of Croydon, 212
Grindslow Knoll, Derbys., 202
Groby, Leics., 129
Grovely, Wilts., 134
Guildford, Surr., 137
Guisborough, Yorks., 250
Gumley, Leics., 245
Gussage All Saints, Dors., 32, 34
Guthrum, Danish king, 245
Gwash River, 245, 251

Hackney Downs, Ldn., 211
Haddenham, Cambs., 184
Haddon, Derbys., 206
Hadleigh, Sfk., 245, 248, 251
Hadlow, Kent, 15, 19
Hainault Forest, Essex, 144
Haldane, A.R.B., 14
Halesowen, Worcs., 112, 120
Halifax, Yorks., 197
Hallaton, Leics., 251
Halvergate Marsh, Nfk., 150,
   151, 154, 155, 156, 158, 159,
   160, 161, 162, 162, 163, 165
Hambledon Hill, Dors., 30, 31
Hambleton, Rutld., 245, 248
Hammond, W.H., 321
Hampshire, 302; commons,
   221, 222
Hampstead Heath, Ldn., 211
Hampstead Marshall, Berks.,
   37, 127
Hanbury, Worcs., 106
Harbottle, Northd., 260
Hardy, Thomas, 28, 98
Harris, Alan, 52, 69
Harris, John, 279
Harrowby Hall, Lincs., 335
Hart, 246, 248
Harterness, Durham, 248
Hartlepool, Durham, 246, 251
Hartlib, Samuel, 141
Harwich, Essex, 251, 254
Hasted, Edward, 163
Hatfield Broadoak, Essex, 127
Hatfield Chase, 141; Forest,
   128; Park, 127
Hathersage, Derbys., 206
Haughmond, Salop, 107
Haverhill, Essex, 242, 251
Hawick, Roxburgh., 260
Hawkes, Jacquetta, 14
Hawling, Glos., 56
Hayes, Kent, 211
Hayfield, Derbys., 201
Heathfield, Suss., 130
Helperthorpe, Yorks., 69
Helpston, Northts., 94
Henley on Thames, Oxon.,
   237, 265, 272, 276
Henwood, Cornw., 205
Heptonstall, Yorks., 200
Herefordshire, 100, 117, 118
Hereward the Wake, 50, 148
Heritage Commission, 11
Hermitage, Dors., 38
Hertford, 18
Hertfordshire, commons,
   221, 224
Hewett, Samuel, 165
Hexham, Northd., 203, 258;
   diocese, 240
Hey, David, 10, 16
Heywood, Stephen, 242
High Halden, Kent, 231
High Halstow, Kent, 164

High Wycombe, Bucks., 234
Hindhead, Surr., 229, 230
Hindolveston, Nfk., 127
Hinton St Mary, Dors., 34
Hodskinson, Joseph, 220
Hog's Back, Surr., 28
Holcroft family, 119
Holdenby, Northts., 254
Holland, Earl of, 141
Holland, Lincs., 244
Holmes, Geoffrey, 305
Holmesfield, Derbys., 206
Honister, Cumbd., 205
Hood, Robin, 147–8
Hooke, Della, 52
Hook Norton, Oxon., 81,
   277–89, 277, 278–9, 280;
   maps, 282, 284. See also
   Subject Index, under Hook
   Norton
Hoo peninsula, Kent, 153, 154
Hope, Derbys., 201
Hope All Saints, Kent, 158
Horkesley, Great, 250; Little,
   250
Horncastle, Lincs., 52
Horninghold, Leics., 50
Horsey, Nfk., 159, 160, 162;
   common, 228
Horsey, Sir George, 160
Horsley, Staffs., 294
Hoskins, W.G., 11, 14, 235,
   283
Hoskyns, Chandos Wren, 95
Hound Tor, Devon, 195, 197
Hounslow Heath, Mx., 223
Howden reservoir, Yorks.,
   208–9
Hudson, W.H., 28, 214
Hugh, Bishop of Le Puiset,
   247
Hugh, Bishop of Lincoln, 139
Huntingdonshire, commons,
   221
Huntley Wood, Glos., 129
Hurstbourne Tarrant, Hants.,
   42
Hurst, J.G., 14, 61
Hyde Park, Ldn., 137, 211
Hythe, Kent, 153, 155

Idle River, 180
Ilsley, Berks., 15
Imber, Wilts., 48
Inclesmoor, Yorks., 188
Ine's Laws, 132
Ingarsby, Leics., 61
Ingatestone, Essex, 254
Inglewood Forest, Cumbd.,
   194
Ironbridge, Salop., 121
Islandshire, Northd., 259
Isle of Wight, 37, 38, 49
Iwade, Kent, 159, 164
Iwerne Courtnay, Dors., 41

Jebb, Louisa, 147
Jedburgh, Roxburgh., 257,
   259, 260
Jefferies, Richard, 28, 214
Jefferys, Thomas, 220
Jennings, Bernard, 10
Jennings, Louis, 214
Johnston, Ian and family, 21–2
Jordans, Bucks., 235
Juniper Hill, Oxon., 93

Kelso, Scotland, 237, 259
Kenchester, Herefs., 101
Kenilworth, Warws., 107
Kent, commons, 221, 222,
   227–8, 230
Kent, Nathaniel, 210
Kersey, Sfk., 251
Kesteven, Lincs., 241
Keston, Kent, 211
Key family, 299–300
Keythorpe, Leics., 225
Kilham, Yorks., 59, 63, 64, 67,
   70, 74
Kilnwick Percy, Yorks., 68
Kimcote Heath, Leics., 224
Kinder estate, Derbys., 207
King Charles I, 180, 181
King Edmund, 245
King Edward the Confessor,
   248
King Edward I, 134, 259
King Edward III, 260, 314
King Ethelred, 247
King Henry I, 134
King Henry II, 134, 138, 139,
   227
King Henry III, 138, 313
King Henry IV, 138
King John, 313
King Richard I, 134, 313
King William I, 134, 165, 310
King William Rufus, 134
King, Gregory, 222, 225, 303–6
King's Bromley, Staffs., 111
King's Cliffe, Northts., 233
King's Lynn, Nfk., 176, 186,
   254
King's Norton, Worcs., 112
King's Ripton, Hunts., 84
King's Somborne, Hants., 37
King's Stanley, Glos., 64
King's Sutton, Northts., 79
Kingswood Forest, Glos., 147
Kipling, Rudyard, 28, 124
Kirby Hall, Northts., 256
Kirkham, Lancs., 179
Kirk Leavington church,
   Yorks., 255
Kirkstead Abbey, Lincs., 57
Knap Hill, Wilts., 30
Knaptoft, Leics., 89
Knaresborough Forest, Yorks.,
   130, 141
Knarsdale, Cumbd., 203

Knight, John, 203; Frederic
   Winn, 203
Knights Hospitaller, 307–22
Knole Park, Kent, 125
Knowlton, Thomas, 20
Kropotkin, Prince, 11

Lake District, 190, 196, 207,
   210, 229
Lammermuir Hills, 257
Lancaster, Dukes and Earls of,
   136
Lancaster Forest, 135
Langdale, Westmd., 190, 229
Langham Moor, Essex, 245
Langholm, Dumfries., 261
Langland, William, 110, 148
Langport, Kent, 163
Langport, Som., 179
Langstrothdale, Yorks., 201
Langtoft, Yorks., 70
Lark River, 186
Laughton, Suss., 130
Launde, Leics., 225
Lavenham, Sfk., 251, 252–3
Laxton, Notts., 83, 91, 96
Leatham, Isaac, 69
Leavening, Yorks., 53
Leeds, Yorks., 213
Lee, Joseph, 89
Leicester, 213, 241, 281;
   Chase, 136, 138; Forest,
   139, 141, 213
Leicestershire, 132, 134;
   commons, 223–4
Leighton Buzzard, Beds., 18
Leland, John, 61, 79, 97, 98,
   105, 265, 271, 292
Leominster, Herefs., 106
Lewis, Samuel, 222
Lichfield, Staffs., 102, 103
Liddel valley, 238, 260, 261
Liddlesdale, Argyll., 258, 261
Lilleshall, Salop., 107, 120
Lincoln, 178, 240, 241
Lincoln, Earl of, 177
Lincolnshire, 132, 240; fens,
   158; wolds, 62
Lindisfarne, Northd., 257
Lindsey, kingdom of, 240
Lingfield, Surr., 212
Littleport, Cambs., 168–9,
   175, 176
Little Siblyback, Cornw., 191
Little Woodbury, Wilts., 32
Lochmaben castle, Dumfries.,
   260, 261
Lockinge, Berks., 47
Loder, Robert, 43
Londesborough, Yorks., 20
London diocese, 240
London Lead Company, 22,
   205, 253
Longdendale valley, Chesh.,
   202

Long Melford, Sfk., 245, 251
Long Mynd, Salop., 98
Longtown, Cumbd., 262
Lothian, 256–7
Louth, Lincs., 65
Lower Halstow, Kent, 153
Lower Hope Point, Kent, 166
Lowesby, Leics., *51*
Loyn, Henry, 54
Lubberland Common, Salop., *217*
Lubenham, Leics., 89, 250
Ludford Magna, Lincs., 68
Lullington, Suss., 39
Lutterworth, Leics., 89
Lydd, Kent, 163
Lyddington, Rutld., 247
Lydlynch, Dors., 112
Lympne, castle, Kent, 165
Lyne River, 258

Macclesfield Forest, Chesh., 194, 201
Madeley, Salop, 121
Maiden Castle, Dors., 30, *32*, 33
Maiden Newton, Dors., 34
Maidenwell, Lincs., 53
Maldon, Essex, 163
Malham, Yorks., 197
Malmesbury, Wilts., 105
Malvern Chase, Worcs., *101*
Manchester, Lancs., 207, 213
Manchester, Earl of, 182
Manningtree, Essex, 251, 254
Manwood, Sir William, 127, 136
Manxey, Suss., *151*
Marefield, Leics., 59
Margate, Kent, 159
Market Deeping, Lincs., 251
Market Harborough, Leics., 236, 245, 249, 250, 251, 252, 254
Markyate Street, Herts., 18
Marlborough, Wilts., 27, 33
Marner, Silas, 65
Marshall, William, 52, 59, 60, 62–3, 65, 71, 77, 149
Martin, Rev. John, 92
Massingham, H.J., 13, 28, 30, 274, 275, 276
Mauduit family, 249
Maxwell, Robert, Earl of Nithsdale, 261
McCready, K.J., 328
Meare, Som., 174
Mears Ashby, Northts., 80, *80*
Medbourne, Leics., 91, 245
Melchet Forest, Hants., 36
Melksham Forest, Wilts., 129, 141
Mendip, Som., 15; Forest, 133
Mercia, 237, 240, 281, 293

Merevale, Warws., 107
Mereworth, Kent, 19, *19*, *20*
Merrington Green, Salop., 17
Mersey River, 237
Meynell, Hugo, 94
Mickleover, Derbys., 84
Middle Ditchford, Glos., 61
Middle Level Main Drain, Cambs., 185
Middlesbrough, Yorks., 253
Middleton in Teesdale, Durham, 251, 253
Middleton St. George, Durham, 250
Middlewich, Chesh., 101, 102
Midhope, Yorks., 204
Midley, Kent, 158
Mile End Common, Ldn., 216
Millmeece, Staffs., 290, 294
Milne, A.A., 148
Milton Abbey, Dors., 37
Milton-under-Wychwood, Oxon., 64
Minchinhampton, Glos., 64, 65, 67
Minion, Cornw., 205
Minnis Bay, 161
Minster, Kent, 36
Mitford, Mary Russell, 214
Moira, Leics., 231
Moore, John, 89
More, Sir Thomas, 11, 87
Moreton-in-Marsh, Oxon., 64
Morfe Forest, Salop, 98, 141
Moriarty, Denis, 87
Morpeth, Northd., 259
Morridge, Staffs., 201
Mortimer Common, Berks., 16
Morton, John, 214
Morton's Leam, 179, 181
Mousehold Heath, Nfk., 223
Mowthorpe, Yorks., 61
Much Wenlock, Salop., 105
Muker, Yorks., 16
Munden, Great and Little, Herts., 230
Murston, Kent, 163
Mushroom Green, Worcs.-Staffs., *213*

Nantwich, Chesh., 101
Naseby, Northts., 92, *93*, 245, 251
National Trust, 319, 322
Naunton, Glos., 54, 58, *67*, 68
Nayland, Sfk., 242, 249, 250, 251
Needwood Chase, Staffs., 136, 144; Forest, 118
Neeson, J.M., 222, 224
Nene River, 179, 185, 244
Neroche Forest, Som., 129, 132, 140, 141, 144
Netherlands, 180

Nettlebed, Oxon., 265, 271, 275
Nettleton, Lincs., 64
Newbourn, Sfk., 335
Newbridge, Suss., 130
Newby, Yorks., 310, 313
Newcastle, Northd., 259
New Castleton, Roxburgh., 262
Newent, Glos., 335
New Forest, Hants., 14, 123, 124, 128, 136, 142, 143, 145, 147, 221, 230
Newminster Abbey, Northd., 197
Newton Bromswold, Northts., 52
Newton in Stokesay, Salop., 107
Newtown Linford, Leics., 129
Nisbet, Robert, 11
Norden, John, 126, 158
Norfolk, 238; Broads, 150, 167, 176, 187; marshlands, 153, 155, 156, 163
Norham, Northd., 259
Norhamshire, 259
Normanton Heath, Leics., 223
Northallerton, Yorks., 256
Northampton, 18, 281
Northamptonshire, 89–90; commons, 222, 224
North Cerney, Glos., 59
North Downs, 12, 27, 28, 35–6, 44, 47, 49
North Kent marshes, 151, 154, 155, 156, 160, 161, 164, 165
Northumberland, 324; commons, 221; Forest, 139
Northumberland, Duke of, 206; Earl of, 198
Northumbria, Earl of, 247
North York moors, 188, 190, 193, 194, 196, 197, 202, 204, 244
Norton, Durham, *246*, 247, 248
Norwich Cathedral Priory, 155
Norwich diocese, 240
Norwood, Surr., 211, 215
Nottingham, 213, 241
Nuthurst, Warws., 104
Nutley, Suss., 130

Oakham, Rutland, 248, 251
Octon, Yorks., 68
Oddington, Glos., 52
Offham, Suss., 30
Offley, Staffs., 293, 295
Old Melrose, Roxburgh., 257
Old Swinford, Worcs., 213
Old Wives Lees, Kent, 18, 228
Ongar, Essex, 18, 132
Oseney abbey, 280, 283, 285, 286

Ousefleet, Yorks., 177
Ouse River, 175, 179, 180, 181, 184; Little Ouse, 186
Overton, Hants., 33, 36
Oxleas Wood, 211

Packington and Enthoven, architects, 325
Padworth, Berks., 16–17
Paine, Tom, 149
Painswick, Glos., 64, 74, 75
Palmer, Samuel, 22, 23, *24*
Pamber Forest, Hants., 36
Paris, Matthew, 136
Parrett River, 182, 186
Paulerspury, Northts., 19
Peak District, Derbys., 189, 193, 199, *199*, 200, 205
Peak Forest, Derbys., 123, 128, 136, 141
Peak House, Staintondale, Yorks., 316, 317, 318, 321; estate, 318
Peak Hill, Yorks., 319. See also Raven Hall
Peakirk, Lincs., 176; Drain, 181
Peckham Rye, Ldn., 211
Peebles, Peebles., 260
Pegalotti, Francesco, 57
Pendock, Worcs., *101*
Pennenden Heath, Kent, 228
Penge, Surr., 211, 216
Penistone, Yorks., 204
Penkridge, Staffs., 104
Pennine dales, 244
Pennines, 188
Penselwood, Som., 231
Pepys, Samuel, 162
Perkinsville, Durham, 327
Pertwood Down, Wilts., 33
Peterborough, Northts., 174, 178, 179, 185
Petersfield, Hants., 42, 49
Pepys, Samuel, 43
Pett Level, Suss., 151
Pevensey Level, Suss., 151, *151*, 155, 157, 158, 160, 163, 165
Pevsner, Nikolaus, 292, 296
Pewsey vale, Wilts., 45
Pickering Forest, Yorks., 123, 130, 138, 141, 194, 310
Piercebridge, Durham, 244, 245
Pilling Moss, Lancs., 174, 175
Piper, Arthur, 325
Piper, John, 278–9
Pirbright, Surr., 228
Pishill, Oxon., 265, 267, 268, 269, 270, 272; Venables, 268, 275. See also Stonor
Plot, Robert, 201, 214, 267, 283
Plowman, Piers, 110, 148

Plumstead, Ldn., 212
Pocklington, Yorks., 17
Pockthorpe, Yorks., 68
Pope, Alexander, 274
Popham, Sir John, 180
Popham's Eau, Cambs., 180, 181
Postles, David, 58
Potton, Beds., 324
Poundbury fair, Dorchester, 46
Powte's Complaint, 170
Prees Heath, Salop., 17
Price, Uvedale, 120
Pusey, Philip, 95
Pyrford, Berks., 139
Pyrton, Oxon., 267, 268, 269, 270, 272; Hundred, 267

Quantockwood, Som., 132
Quarnford, Staffs., 196
Queenborough, Kent, 165
Quorn Hunt, Leics., 94

Rackham, Oliver, 124, 132
Raisthorpe, Yorks., 61
Ramsey, Hunts., 174; abbey, 175
Ramsgate, Kent, 159
Rankin, M.M., 216
Ransome, Arthur, 170
Raven Hall, Yorks., 321; Country House Hotel, 321–2. See also Peak House
Ravenscar, Yorks., 322; Estate Company, 321–2
Rawcliffe Moss, Lancs., 184
Reach, Cambs., 179
Read, Sewell, 273
Reculver, Kent, 161, 162, 165
Redbourn, Herts., 18, 90
Redesdale, Northd., 260
Reedham, Nfk., 165
Reeves, Anne, 10
Renishaw, Derbys., 17
Rhee wall, Kent, 157
Ribble River, Lancs., 237, 238
Richborough Castle, Kent, 165
Rickard, Gillian, 18
Risbridge, Sfk., 254
Rixton Moss, Lancs., 20–21
Roberts, Cecil, 275, 276
Rockall, Samuel, *274*, 274–5, 276
Rockbourne, Hants., 34
Rockingham, Northts., 16, 251; Forest, 123, 125, 128, 129, 131, 142, 143, 144, 147, 245
Rocque, John, 220
Rollestone, Wilts., 48
Romney, Kent, 153, 155, 157; New, 157; Old, 159; Marsh, 10, 151, 152, 153, 155, 156, 157, 158, *158*, 161, 163, 165, 230

Rossendale Forest, Lancs., 196
Rothbury, Northd., 205
Rotherfield, Suss., 130, 272
Rotherham, Yorks., 202
Rother Levels, Kent, 151; river, 153
Rous, John, 61
Roxburgh, 257, 258, 259, 260
Royal Commission on Common Land (1955–7), 210, 220, 235
Royal Commission on the Historic Monuments of England, (RCHM), 317
Rudston, Yorks., 51, 54, 64
Rushbourne Wall, Kent, 162
Rushock, Herefs., 115
Rushock, Worcs., 119
Russell's Water, Oxon., 275
Rutland, 240, 241, 248; Forest, 134, 139
Rutland, Duke of, 206
Rye, Kent, 153, 163

Sadberge wapentake, 241, 247, 248
Sackville, Lord, 125
Sackville-West, Vita, 23
St Benedict, 11
St Benet-at-Holme Abbey, Nfk., 155, 156, 176
St Benet's, Nfk., 174
St Briavels, Glos., 128
St Cuthbert, 247
St Chad, 293
St Guthlac, 173
St Ives, Hunts., 178
St James's Park, Ldn., 137
St Leonard's Forest, Suss., 130
St Margaret, 60
St Margaret of Scotland, 256
St Olaf, 50
St Paul's Cathedral, Ldn., 156
Salcey Forest, Northts., 123, 126, 128, 141, 143, 144, 147
Salmonby, Lincs., 54
Salisbury Plain, Wilts., 14, 27, 28, 48, 49, *49*, 134
Saltersbrook, Chesh., 202, 203
Sambourne, Warws., 114, 120
Sandal's Cut, Cambs., 181
Sandhurst, Berks., 228
Sark Foot, Dumfries., 262
Sark River, 262
Sarre, Kent, 153
Savernake Forest, Wilts., 27, 129, 133, 139
Saxton, Christopher, 126
Sayers, Dorothy L., 170
Scalby, Yorks., 307, 310, 313, 319; Soke of, 310, 313; Forest of, 310
Scarborough and Whitby Railway, 321, 322
Scots Dyke, 261

Seal Chart, Kent, 229
Seasalter, Kent, 151
Sedgemoor, Som., 180, 182, 185, 188
Sedgley, Staffs., 114
Seer Green, Berks., 233, 235
Selborne, Hants., 214
Selkirk, Selkirkshire, 259, 260
Selwood Forest, Som., 98, 118, 123, 132, 141, 216, 227
Sevenhampton, Glos., 51
Sevenoaks, Kent, 228
Severn River, 100
Shaftesbury, Dors., 105
Sharpe, Hilda D., 12
Shearsby, Leics., 88
Sheerness dockyard, Kent, 165
Sheffield, Yorks., 189, 213
Shenstone, William, 120
Shepard, E.H., 148
Sheppey, Isle of, Kent, 154, 164, *164*, 165
Sherborne, Dors., 103, 104
Sherwood Forest, Notts., 123, 124, 125, 129, 130, 135, 143, 148, 213
Shirburn, Oxon., 271
Shirlet Forest, Salop., 98
Shirley, Surr., 211, 212
Shooter's Hill, Ldn., 211
Shornmead, Kent, 166
Shotover Forest, Oxon., 123, 142
Shrewsbury, Salop., 118
Shrewton, Wilts., 39
Shropshire, 99, 100
Shrubland, Sfk., 215
Shugborough, Staffs., 120
Shuttlewood, Derbys., 335
Sidlesham, Suss., 335
Sidnall, Salop., *113*
Silbury Hill, Wilts., 31
Sileby, Leics., 91
Singleton, Suss., 49
Sitwell, Sir George, 17
Sixteen Foot Drain, 181
Skeetholme, Nfk., 157, 159
Skelton Moors, Yorks., 196
Skerne River, 249
Skipp, Victor, 29
Skirlaw, Bishop of Durham, 250
Slater, Terry, 52
Sledmere, Yorks., 64, 70, *72–3*, 74
Slindon, Staffs., 290, *291*
Smithfield, Ldn., 18
Snave, Kent, 163
Soar River, 248
Sockburn, Durham, 245, 247, 250
Soham Mere, Cambs., 171, 174, 182
Solihull, Warws., 109

Solway Firth, 12, 258, 260, 262
Somersby, Lincs., 53
Somerset, 132; coastal marshes, 160, 164; commons, 221; Levels, 131, 151, 153, 158, 164, 167, 170, 171, 173, 175, 177, 179, 180, 182, 184 186, 187
Sonning, Berks., 127
Sound Common, Chesh., 18
South Downs, 12, 28, 29, 37, *37*, 45, 49
Southlake Moor, Som., 175, 176
South Malling, Kent, 37, 147
Southrop, Oxon., 278, 279, 285
South Walsham, Nfk., 156, 176
Southwick, Suss., 34
Sowerbyshire Forest, 129, 194, 196, 198
Sowy Is., Som., 175
Spalding, Lincs., 244, 251
Speed, John, 28
Spencer, John, 86
Spetisbury, Dors., 33
Springfield, Dumfries., 262
Staffordshire, 137, 230, 232, 279, 281; moorlands, 196
Stainmoor, Yorks., *205*
Staindrop, Durham, 251
Stainfield Priory, Lincs., 57
Stainthorp, Durham, 247
Staintondale, Yorks., 307–22; maps, 309, 311, 320; *308*, *314*, *317*, *318*. See also Subject Index under Staintondale
Stalham Staithe, Nfk., *178–9*
Stalmine Moss, Lancs., 184
Stamford, Lincs., 240, 241, 245, 248, *249*, 250, 251, 254, 256
Stamp, L.D., 235
Stanage Moor, Yorks., 206
Stanford on Avon, Northts., 85
Stanley, Wilts., 107
Stannington, Northd., 325
Stanton Drew, Som., 15
Stanton Moor, Derbys., 190, 193
Stanton on the Wolds, Notts., 69
Staplegordon, Dumfries., 261
Stead, Ian, 54
Steane, John, 271
Steeple, Essex, 164
Stelling Minnis, Kent, 28
Stewart, Sir Percy Malcolm, 324
Stockton on Tees, Durham, *246*, 247, 249, 250, 251, 253, 254

Stoke-by-Clare, Sfk., 247, 249
Stoke-by-Nayland, Sfk., 247, 248
Stoke Dry, Rutld., 247
Stockbridge, Hants., 18
Stokesley, Yorks., 250
Stone, Staffs., 107
Stonea Is., Cambs., 171
Stonehenge, Wilts., 27, 31
Stoneleigh, Warws., 107
Stonor, Bucks., 265–76, *268*; map, 266; Park, 265, 267, 270, 271, 273, 276; valley, *268*. See also Subject Index, under Stonor
Stonor family, 269, 271–2, 276; Sir John, 269, 271
Stourbridge Fair, Cambs., 179
Stourbridge, Staffs., 231
Stourhead, Wilts., 149
Stour Level, 151, 156, 159, 165
Stour River, Sfk., 238, 240, 242, 245, 249, 250, 251; Navigation, 254; valley, map, *243*
Stowell, Glos., 74
Stow-on-the-Wold, Glos., 64
Strathclyde, 261
Street Heath, Som., *187*
Strickland, Henry., 69, 74
Stroudwater, Glos., 65–6, 68
Stuntney, Cambs., 174
Sturt, George, 214, 233, 235
Sudbury, Sfk., 245, 248, 250, 251
Suffolk, 238; commons, 222
Suffolk, Earl of, 279, 285, 286
Suffolk Sandlings, 151, 156, 164, 230
Sugnal, Staffs., 295
Sumner, Heywood, 14
Sunderland, Durham, 326, 327
Surbiton, Surr., 217
Surrey, commons, 221, 224
Sussex, *124*, 233
Sutton, Lincs., 183
Sutton Veny, Wilts., 39
Swaledale, Yorks., 16, 193, 198, 254
Swaythorpe, Yorks., 68
Sweet, Roy, 15
Sweet Track, Som., 15, 171
Swere River, 277
Swincombe, Oxon., 272
Swinefleet, Yorks., 177
Sydenham, Ldn., 211
Sydling St Nicholas, Dors., 33
Sykes, Sir Christopher, 70–1

Tadley, Hants., 17, 36
Tadmarton camp, Oxon., 281, 283
Talbot family, 119
Tamar River, 238

Tankersley, 189
Tanworth in Arden, Warws., 106, 112
Tarrant Gunville, Dors., 37
Tarrant Hinton, Dors., 34
Tees river, 238, 240, 241, 242; valley, 244, 245, 248, 250, 252–3, 254, 256; map, 244
Tempsford, Beds., 95
Teviotdale, Roxburgh., 257
Thames River, 237
Thanet, Kent, 153, 162
Theddingworth, Leics., 250
Thetford diocese, 240
Thingoe Hundreds, Sfk., 248
Thirkelby, Yorks., 68
Thomas, Sir Keith, 148
Thompson, Flora, 93, 96, 214
Thoralby, Yorks., 61
Thoresby, Lincs., 57
Thorne, James, 214
Thornton Abbey, Lincs., 174, 177
Thorpe, Harry, 85
Thortergill, Cumbd., 21, *24*
Throckmorton, Arthur, 19
Thunreslau, Essex, 240
Thwing, Yorks., 57
Thynne family, 120
Tidworth, Wilts., 48
Tilbury fort, Essex, 166
Tildesley family, 296, 299–300
Tillingham River, 153
Tiptree Heath, Essex, 215
Tisbury, Wilts., 39
Tonbridge, Kent, 211; Castle, 19
Tostig, Earl of Northumbria, 310
Totley, Yorks., 206
Toulson, Shirley, 14, 15
Towthorpe, Yorks., 61
Trent River, 180, 184
Trollope, Anthony, 95
Trowbridge, Wilts., 107
Trundle, Suss., 30, 32
Tugby, Leics., 225
Tunbridge Wells, Kent, 211
Turbevile, George, 146
Turville Heath, Bucks., 267
Tutbury, Staffs., 118
Tweed River, 237, 257, 259; valley, 238, 260
Tyneham, Dors., 48

Uffington, Berks., 28
Underriver, Kent, 22, *24*
Upavon, Wilts., 39, 48
Upchurch, Kent, 153
Upper Arley, Worcs., *105*, 105
Upper Derwent valley, 207
Uppingham, Rutld., 251
Upton Cross, Cornw., 205
Upwell, Cambs., 173

Vale Royal, Chesh., 119
Vancouver, Charles, 144
Vaughan, Rowland, 118
Vermuyden, Cornelius, 180–82, 323
Verney, Richard, 86
Vine, The, Hants., *45*

Wacton, Nfk., 164
Wadhurst, Kent, 147
Wakefield, Yorks., 189, 198
Walcher, Bishop, 247
Wales, South, 279, 281
Walkhampton Common Reave, *192*
Wall, Staffs., 103
Walland marsh, Kent, *154*, 155, 157, 161
Wallasea island, Essex, 160
Walter of Henley, 56
Waltham Forest, Essex, 140, 141, 143
Wansdyke, 34
Wantsum, marshes, Kent, 156, 159; river, 151, 153, 165
Ward, Graham, 332
Wargrave, Berks., 38
Wark Castle, Northd., 259, 260
Warkworth, Northd., 259
Warmscombe, Oxon., 267, 268, 269, 270, 275, 276
Warren, C. Henry, 13
Warter, Yorks., 70, 76, *77*
Wash marshes, 155
Watling Street, 103, 154
Watlington, Oxon., 267, 268, 269, 271, 272
Waugh, Evelyn, 17
Waveney River, 151, 238; valley, 131
Weald, 122, *124*, 128, 131, 132, 133, 147, 158; iron industry, 130, 142; parish shapes, 129
Weald of Kent, 16, 23, 215
Weald Moors, Salop, 188
Weardale, Durham, 253, 254; Forest, 194
Weaverthorpe, Yorks., 69
Welland River, 176, 236, 237, 238, 240, 241, 242, *243*, 244, 248, 250, 251, 252; valley, 254; map, *243*
Wellingborough, Northts., 18
Wells, Som., 15
Welton le Wold, Lincs., 50
Wenlock Edge, Salop., 98
Wensleydale, Yorks., 198
Wentlooge Levels, Mon., 154
Wessex, 237, 241
West Blatchington, Suss., 34, 41
West Haddon, Northts., 224

West Hartlepool, Durham, 327
Westhay, Som., 15
West Hythe, Kent, 163
Westmorland, 189, 190; commons, 221
Weston, Som., 126
Weston-under-Penyard, Herefs., 101
West Peckham, Kent, 15
West Wickham, Kent, 211
West Wykeham, Lincs., 68
Weymouth, Dors., 28
Whalley, Major-General, 89
Wharram Percy, Yorks., 61, 68
White, Gilbert, 29, 37, 214
Whitehawk, Suss., 30
White Horse, Berks., 28
Whiteparish, Wilts., 38
White Peak, Derbys., 189, 193, 204
Whitesheet, Wilts., 30
Whitley Row, 228
Whitley Scrubs, Kent, 215
Whitstable, Kent, 218
Whittlesea, Cambs., 183; Mere, 170, 185
Whittlewood Forest, Northts., 123, 127, 143, 147
Whittlingham, Nfk., 165
Widecombe, Devon, 195
Widerigga people, 240
Wiggonholt, Suss., 34
Wigmore, Herefs., 107
Wigston Magna, Leics., 95, 223
Willesborough, Kent, 218
Williams, Alfred, 214
Willingham, Cambs., 183
Wilmington, Suss., 28
Wilson family, of Sheffield, 206
Wilton, Wilts., 36
Wiltshire, chalk, 97, 119; cheese, 97, 110, 116, 119
Wimbledon Common, Ldn., 211
Winchcombe, Glos., 64
Winchelsea, Suss., 153, 157
Winchester, Bishop of, 137
Windmill Hill, Wilts., 30
Windsor, Berks., 272; Forest, 123, 125, 129, 133, 136, 142, 143, 144, 148
Winmarleigh, Lord, 21
Winstanley, Gerrard, 215
Winterborne Farringdon, Dors., 41
Winter, Sir John, 141
Wirksworth, Derbys., 200, 201
Wisbech, Cambs., 176
Wise, Prof. M.J., 328
Wishford, Wilts., 134, *135*
Wissey River, 186

Witham Fen, Lincs., 184
Wittering, Suss., 240
Wixoe, Sfk., 245
Wodehouse, Sir William, 164
Wold Dalby, Notts., 50
Wold Drayton, Cambs., 56
Wold Newton, Lincs., 50, 54, 63, 64
Wolseley Park, Staffs., *126*
Wombourne, Staffs., 117
Woodward, Horace B., 12
Woodsdale, Kent, 229
Woodstock, Oxon., 68, 126, 129; Park, 136
Woodville, Derbys., 214, 231
Woodville, Elizabeth, 271
Woolhope, Herefs., 111
Woolley, Wiliam, 201, 214, 218
Woolmer Forest, Hants., 123, 143, 144
Woolwich dockyard, Kent, 165
Wootton, Beds., 227
Worcestershire, 99–100, 132; commons, 221
Wordsworth, William, 16
Worlidge, John, 42
Wormwood Scrubs, Ldn., 211
Wroxall, Warws., 95
Wroxeter, Salop., 101, 103
Wyboston, Beds., 335
Wyburbury, Chesh., 108
Wychwood Forest, Oxon., 123, 125–6, 129, 131, 134, 142, 143, 144, 145
Wye, Kent, 28
Wye River, 100, 237
Wymondham, Nfk., 232, 233
Yare River, 151, 165
Yarm, Yorks., 250, 254
Yarranton, Andrew, 118
Yetholmshire, Roxburgh., 257
York, 241; diocese, 240
Yorkshire, commons, 221; dales, 193, 197, 198, 204; wolds, 50, 52, *60*, 62, 77
Young, Arthur, 44, 66, 189, 230, 278
Young, Rev. Arthur, 43

Subjects
absentee owners, 158
Adventurers in fens, 181; lands, *168*, 181
agricultural depression, 48, 75, 87, 96, 159, 186, 275, 279, 289, 324
Agricultural Labourers' Union, 289
Agricultural Land Commission Enquiry (1948), 160
ague, 162, 163
alehouses, 302

Anabaptists, 287
Andrews and Dury maps, 220
Anglo-Saxon Chronicle, 122, 149, 250, 281
Anglo-Scottish border, 256–62, 258
archaeology, scientific aids, 14, 15

Baptists, 287–8; Particular-, 287
barrows, long, 31; round, 31, 100
bastles, 261
Black Death, 11, 16, 60, 68, 84, 112, 179, 197, 198
Board of Agriculture, reports, 222, 230
boundaries, 105; hundred markers, 15. See also under parish boundaries, and trees
Brymbo Ironstone Works, 280, *281*

Cambridge Chronicle, 324
Cambridge Weekly News, 326
Carnegie United Kingdom Trustees' Charity, 326
chair bodger's hut, *234*
champion country, 97, 130, 147
chapels, dissenting, 17, 74
charcoal burners, *219*
chases, 123, 133, 200
Chilterns Management Plan, 276
churches, 152, 154, 155, 158; minster churches, 103, 247
cider, 117
clay vales, see lowland vales
COMMON LAND, 16, 18, 210–35; buildings, 218, 228; hospitals, 228; workhouses, 228; crafts, 212, 227, 231–4; chairmaking, 234; hoop-making, 232, *233*; enclosure, 224, 227–8; Parliamentary, 224; extent, 210–11, 214–5, 220, 222, 223, 230–1; farming, 212; gypsies, 234; industries, 212, 217, 227, 231–4; chain making, 213; chairmaking, *274*; charcoal burning, 212; *219*; common rights, 216, 221, 228; disputes, 217; legal status, 215, 221, 235; landscape, 211, 225; location, 210, 214; maps, 211; military use, 228; natural resources, 216, 226, 228, 229, 230, 233; place names, 211, 213, 215, 216, 226, 228, 229, 230, 233; plants, 216; population

pressure, 218, 220; social structure, 225–6, 233, 235; squatter settlement, 218, 227, 228, 231, 233
common rights, 123, 128, 130, 134, 135, 140, 141, 144, 145, 147, 175, 202, 206
Compton Census (1676), 287
cony warrens, see rabbit warrens
copperas gathering, 165
coppicing, 102, 124–5, 128, 131, 133
corn barns, 39
Council for Preservation of Rural England, 207
Courts of Sewers, 161, 176, 184
'cursus' earthworks, 31

dairying, 115–16, 156, 158, 290, 298–300
'Danish' five boroughs, 241, 248, 250
deer, 134, 136, 200; leaps, 126, *126*; parks, 37, 123, 126, 127, 133, 136, 137
deserted villages, 14, 39, 52, 59, 61, 67–8, 85, 86, 113
disafforestation, 138 –9, 200, 201
DOWNLANDS, 14, 27–49; barrows, 31; building stone, 28; common fields, 30; corn barns, 39; deer parks, 37; depopulation, 47–8; farming, prehistoric, 30–33; Roman, 33–4; Anglo-Saxon, 34–6; medieval, 36, 38, 40–41; 18th and 19th c., 43–7; modern, 47–9; grasslands, 30, 49; henges, 31; hillforts, 31; lynchets, 38; military use, 48, 49; monasteries, 35; rabbit warrens, 37; settlement pattern, 27–30; sheep-corn farming, 30, 36, 38, 48; social structure, 28–9, 44–5; soils, 27; tree cover, 27, 30; water meadows, 42–3
Drayton's *Polyolbion,* 58
drove roads, 14, 15, 52
drovers, 15, 18, 158

ECCLESHALL, Staffs., 290–306; alehouses, 302; assarting, 294; buildings, 290, 292, 299; bishop's castle, 292; church, 291, 292–4; common fields, 296, 297; craftsmen, 302; farming, 297–300; dairy-, 290, 298–300; economic depres-

sion in 14th c., 296; enclosure, 296, 306; fieldnames, 294; household furnishings, 299, 303; hunting, 295, 297; industries, 294, 302; landscape, 290, 291, *292*, 292–3; maps, 295, 301; poor rate, 300–1; population, 294, 296, 300, 302; poverty, 297; settlement history, 293; social structure, 293, 296–7; bishop, 293, 295; gentry, 296–7, 303–4; yeoman, 296, 297; the poor, 297, 300, 302–5; squatters, 302; soils, 290; transport, 298; woodland, 292–4, 295, 297, 302
enclosure, 70–71, 88–9, 91, 94, 114, 117–8, 179, 184, 198, 200, 201, 203, 204, 224, 227–8, 272, 277, 284, 287, 288, 296, 306, 312, 316, 319–21
English Place Name Society, 13
entrance lodges, 19, *19, 20*
Environmentally Sensitive Areas, 160

FEN DRAYTON, Cambs., 323–35, *332*; House, 324, 326, 331; administration, 325, 326; agricultural cooperation 325, 326, 328, 333; agricultural ladder, 328, 329, 334; farm incomes, 329, 331; farming, 324, 327, 328; glasshouses, 326, 329, 330–1, *330*; horticulture, 324, 326, 329, 330; housing, 325, 329, 331, 333; land auction, 324; land layout, 325–6; landscape, 323–4, 327, 331, 333–4; pigs and poultry keeping, 325, 326, 327, 328, 329; policy in wartime, 327–8; post-war, 328, 332–3; school, 326, 327, 331; smallholders, 324, 325, *326*, 326, 327, 329, *329*; consortium (1983), 333
FENS, 132, 167–87; administration, 167, 176, 184; causeways, 171; common rights, 175; drainage, 174, 180–2, 184–5; by Dutch, 180; Drainage Commissions, 184, 185; embanking, 176, 180–1; enclosure, 179, 184; farm crops, 182, 183, 184, 186; farming, 169, 170, 174, 175, 179, 183, 186; landscape, 170–71, 187;

monastic owners, 174, 175–6, 177, 179, 180; map, 168; natural resources, 174; peat beds, 170, 171, 172, 176–7, *177*, 187, *187*; levels, 182, 185; settlement patterns, 167; prehistoric, 171; Roman, 171–3; Saxon, 173; medieval, 174; modern, 184; reedcutting, *172*; soils, 170–1; steam engines, 185–6; transport, river, 178–9; road, 186; terminology, 150, 167; warping, 184; windmills, 183, *183*
feldon, see under champion
fen slodgers, *173*
field drainage, 95
field walking, 153, 155, 157, 282
Fitzherbert's *Book of Husbandry*, 142
fishing, 163–4, 170, 174
flint mines, 31
Forestry Commission, 145–6, 188, 206, 275
FORESTS, in moorlands, 194; in lowland England, 122–49; administration, 135–6, 141, 142; assarts, 129; boundaries, 126–7; charters, 139; common rights, 123, 128, 130, 134, 135, 140, 141, 144, 145, 147; disafforestation, 138–9; drainage, 128–9; farming, 129, 138, 139, 147; hays, 127; launds, 128; law, 123, 127, 128, 133, 134, 142; lodges, 128, 129; place names, 132; private, 123, 127; rabbit warrens, 123, 128, 133, 137–8; royal 123, 226–7; settlement pattern, 129–30, 133, 141; prehistoric, 131; Roman, 131; Saxon, 132; Victorian, 130; soils, 123; standings, 128; stud farms, 128; surveys, 141, 142; terminology, 128; wardens, 128, 129, 135. See also under Woodland
foxhunting, 94
FRONTIER VALLEYS, 236–62; boundaries, 237, 238, 240, 254, 257, 259, 262; cultural corridors, 237, 256; cultural provinces, 238–40, *239*, 262; farming, 251; ford crossings, 251; geography, 242–4, 259–60; industries, 251, 252, 254; language, 237, 255; local societies (*pays*), 240; market grants, 248, 249, 250; poor law

unions, 254; ports, 253; railways, 253; rivers and drainage patterns, 237–8, *239*; settlement, Celtic, 240; Anglo-Saxon, 241; Danish, 241; monastic sites, 246; north-south divides, 245, 247, 248, 250, 251, 254, 259, 262; place names, 245, 247, 261; physical relief, 245
fruit-growing, 116–7

gatehouses, see entrance lodges
Green Man, 148
Grovely Oak Ceremony, *135*
gypsies, 17

Hearth Tax, 296, 297
heathland, 123, 131, 188, 210, 212, 213, 226, *229*, 245
hedge dating, 14
henge monuments, 31
hill forts, 31, 32, 100
hogback gravestones, 241
HOOK NORTON, Oxon., 277–89 ; *277*, *278–9*; boundaries of parish, 281; brewery, 277, 284, 288; buildings, *277*, 277–80, 285, 286–7;common fields, 285, 287; common rights, 288; dovecote, 285; enclosure, 277, 287, 288; Parliamentary, 284; crafts, 279, 288–9; farming, in Domesday, 285; early modern, 287; modern, 288, 289; fieldnames, 283; geology, 277–8; ironstone mining, 279–80, *280*, 284, 288; landscape, 277, 282; maps, 282, 284; market rights, 279, 286; park, 285; population, 279, 280, 287, 288, 289; rabbit warren, 285; railway, 278, 279, 284; religious dissent, 280, 287–8, 289; settlement, prehistoric, 282; Iron Age, 281; Anglo-Saxon, 281, 285; social structure, 280, 288; survey (1943), 289; Viking raids, 281, 285; water supply, 279
hops, 116, *117*
Horticultural Improvement Grants Scheme, 329
horticulture, 116, 118
hundred boundary markers, 15; meeting place, 241
hunting, 140, 146, 148; scene, *137*; stations, 128; symbolism, 146, 148
Hurlingham Bungalow Company, 325

icemaking, 165
industries, 251, 252, 254; alum working, 308, 317–19, *318*; brick making, 92, 165, 201, 205, 271, 275, 294, 302; charcoal burning, 112, 197, 294, 302; clothmaking, 64–6, 114, 120, 121, 147, 197–8; coalmining, 112, 120, 201; glassmaking, 120, 128, 142, 294, 302; ironstone mining, 279–80, *280*, 288, 294, 302; ironworking, 101, 114, 128, 130, 142, 145, 147; knitting, 90, 92, 198; lacemaking, 90, 147; lead, 21, 128, 197, 198, 201; limeburning, 201, 321; metalworking, 128; millstone quarrying, 197, 201; potteries, 102, 112; straw plaiting, 90; tilemaking, 102; tin mining, 197; woodworking, 138, 147
inheritance, partible, 200, 260
intercommoning, 15, 215–6, 310

Knights Templar, 65, 106

landscape, archaeology, 14; conservation, 11; geology, 11–12; in art, 13; literary accounts, 13; women writers on, 10; making of, 9, 10, 16; regional differences, 9, 11; periods of change, 10, 13; socially differentiated, 16–19
Land Settlement Association, 323–35; settled estates, 324–5, 329, sale of estates, 332. See also Fen Drayton
lead mining, 21, 128, 197, 198, 201
leets, 241
Lincolnshire rebellion (1536), 61
London, food, 212; fuel, 212
London Brick Company, 324
lookers' hut, *158*
LOWLAND VALES, 78–96; buildings, 92; emigration, 85; enclosure, 89, 91; Parliamentary, 91, 94; farming, medieval, 81–3, 86; early modern, 87, 90; modern, 91–2, 94, 95–6; settlement pattern, medieval, 79, 81; social structure, 81–3, 90; village revolution, 79–82

malaria, see ague
marlpits, 88, 118, 129
MARSHES, 10, 150–66; cattle,

158–9; customary law, 161; dairying, 156, 158; defences, 165–6; drainage, 152, 155, 156, 157–60; mills, *162*; embanking, 152, 155, 160; farming, 152, 155, 156, 157–60; sheep-, 156, 158; fields, 156; industries, brickmaking, 165; cement works, 165; landscape, 150–51, 164; reclamation, 152–3, 155; settlement pattern, 150, 152–3, 156–7; Roman, 153–4; Saxon, 154–5; medieval, 155; early modern, 157; soils, 150–1; terminology, 150; windmills, *162*
martello towers, 165
moated sites, 109–10
Midland Revolt (1607), 88
monasteries 16, 35, 155, 158
MOORLANDS, 188–209; assarting, 194, 196, 200; boundaries, 192, 197; bracken, 207; buildings, 195–6, 198, *199*, 199–200, 204–5; barrows, 193; cairns, 190, 192, 195; common rights, 202, 206; enclosure, 198, 200, 201, 203, 204; farming, 192, 193, 194, 195, 196, 197, 198, 201, 204; cattle-, 196, 198, 200, 201, 203–4; sheep-, 194, 197, 201, 204; game, *205*, 205–7; holloways, 202; industries, 201, 205; charcoal burning, 197; clothmaking, 197–8; coalmining, 201; knitting, 198; lead, 197, 198, 201; lime burning, 201; millstone quarrying, 197, 201; tinmining, 197; inheritance customs, 200; landscape, 188, 189–90; literary accounts, 190; military use, 207; monastic owners, 194, 197, 198, 200; parish shapes, 194; peat beds, 189, 193, 197; -carts, 201; -ways, 202; place names, 189, 194, 196, 197; rabbit warrens, 206; reservoirs, 206–7; roads, 203; turnpike-, 202–3; settlement patterns, 189; prehistoric, 190–3; Roman, 193; medieval, 194–6, 198; early modern, 198; modern, 203; social structure, 194, 197, 198, 200, 203; surnames, 196; terminology, 188, 189

mosses of Lancs. and Cumbria, 167, 179–80, 184
Museum of English Rural Life, Reading, Berks., 274

National Forest, 145, 149
National Parks, 207
National Trust, 207

Oak Apple Day, *135*
Oxfordshire County Sites and Monuments Record, 282

parish boundaries, 15; shapes, 35, 55, 66, 68, 88, 129 ; maps, 55; sizes, 106
parks, 110, 120
peat, 170, 171, 172, 176, 187; carting, *201*; stacking, *177*
peel houses, 261
place names, 53–4, 104, 108, 132, 189, 194, 196, 197, 211, 213, 215, 216, 226, 228, 229, 230, 233; study of, 13
poor rates, 300–1
population of England, 220

Quakers, 235, 289, 315, 324

rabbit warrens, 123, 128, 133, 137–8, 206
reed cutting in fens, *172*
river drainage patterns, map, 239
Royal Commission on the Historic Monuments of England, 190, 192

salt working, 101, 163
Scotland, 237, 238; East and West Marches, 237, 256
seasonal settlement, 52–3
Sites of Special Scientific Interest (SSSI's), 125, 145, 160
Skimmington riots, 142

STAINTONDALE, Yorks., 307–22, *308*, *314*, *317*, *318*; assarting, 313, 316; buildings, 316, *317*, 317; enclosure, 312, 316; Parliamentary, 319–21; farming, 312, 313, 314, 321; cattle-, 312; sheep-, 315; industries, alum working, 308, 317–19, *318*; limeburning, 321; landscape, 307, 316; land occupation, owners, 312–16, 320; tenant farmers, 315, 316, 320; maps, 309, 311, 320; open fields, 313, 314; Quakers, 315; settlement pattern, to Domesday, 307, 308, 310; medieval, 312, 313; modern, 321; social structure, at Domesday, 310; medieval, 310, 314; early modern, 314–16; 'freeholders' republic', 315, 322; soils, 307–8; transhumance, 310
stannaries, 197
Statute of Winchester (1285), 129
STONOR, Oxon., 265–76; boundaries of parish, 267, 268–9; brick making, 271, 275; buildings, *273*, 273; common rights, 272; commons, 270; deerpark, 270, 271, 275; dykes, 270; enclosure, 272; farming, 270, 272, 273, 276; dairy-, 275; sheep-, 272; farms, 271; landscape, medieval, 269–72; early modern, 272–4; modern, 275–6; map, 266; open fields, 270, 272; origins, Anglo-Saxon, 267, 268, 269; routeways, ancient, 267; soils, 265, 267; woodland and wood products, 265, 268, 269, 271–2, 273, 274, 275

Stonor, *Letters and Papers*, 271, 272
storm damage, 125–6, 157, 159, 160, 161–2
strip lynchets, 38

timber, felling, *143*; naval, 125, 142, 143, 144
transhumance, 310
tree cover, 123, 132–3, 190; prehistoric, 131
trees, 27, 30, 52, 62, 74, 102; as boundary marks, 15; alder, 124, 131, 171, 188, 216; ash, 78, 94, 124, 125, 131, 234, 269; aspen, 131, 272; beech, 62, 124, 125, 131, 234, 272, *275*; birch, 123, 124, 131, 188, 190; chequer, 15; elm, 78, 94, 131; chestnut, 125; disease, Dutch elm, 126; elm, 234; fir, 206, 274; Douglas fir, 124; hazel, 125, 131, 188; holly, 131, 200; hornbeam, 15, *125*, *125*, 131; lime, 124, 131; maple, 124, 131, 269, 272; oak, 123, 124, 125, 131, 133, 188, 190; pine, 131, 190; Scots pine, 18, 124; pollarded, *125*, 133; rowan, 188; turkey oak, 20; walnut, 20; whitebeam, 15, 272; willow, 78, 94, 171, 188, 272

village greens, 42–3, 220–1
villages, close, 18, 90, 92 ; open, 18, 74, 90, 92, 277, 280

Warburg Nature Reserve, 275
water-meadows, 118
wildfowl, 163, 170, 174; decoys, 164; wildfowlers, 173
woad, 89–90, 118; families, 89
Welsh Land Settlement Society, 324

WOLDS, 50–77; Anglo-Saxon, 52; building stone, 63, 76; buildings, 66, 76; cloth industry, 64–6; depopulation, 58, 60, 68; farming, medieval, 56–7, 59–61; early modern, 66, 69–71, 74–5; in depression, 75; field systems, 54, 56, 67, 69; landscape, 50–51, 56, 61, 62–3; medieval boroughs, 64; parish shapes, maps, 55; Parliamentary enclosure, 70–71; Scandinavian immigration, 54, 56; settlement pattern, 53–4, 56; social structure, 16, 59; waggoner, *60*
woodland, old landscapes now commons, 224–7; on wolds, 51–2
WOODLANDS, in lowland England, see under Forests; in western England, 97–121; assarting, 110–1; buildings, 114; enclosure, 114, 117–8; farming, medieval, 108–9, 111–5; modern, 115–8; industries, 102; charcoal burning, 112; cloth working, 114, 120, 121; coalmining, 112, 120; glassmaking, 120; ironworking, 101, 114, 120; potteries, 102, 112; tilemaking, 102; monasteries, 105, 107; population, 111; settlement pattern, 98–9; in prehistory, 100, 102; Roman, 100, 101, 102, 103; at Domesday, 106; medieval, 103, 104, 106–8 110, 115; modern, 102, 107, 111, 115, 119, 120; social structure, 102, 107, 111, 119, 120; soils, 99
woodpasture, 97–121; terminology, 97
Wye College, Kent, 325